T5-DIE-857

Advances in

OPTICAL *and* ELECTRON MICROSCOPY

Volume 6

Advances in

OPTICAL *and* ELECTRON MICROSCOPY

Volume 6

Edited by

R. BARER

Department of Human Biology and Anatomy, University of Sheffield, England

AND

V. E. COSSLETT

Department of Physics, Cavendish Laboratory, University of Cambridge, England

ACADEMIC PRESS · 1975
LONDON AND NEW YORK
A Subsidiary of Harcourt Brace Jovanovich, Publishers

ACADEMIC PRESS INC. (LONDON) LTD.
24/28 Oval Road,
London NW1

United States Edition published by
ACADEMIC PRESS INC.
111 Fifth Avenue
New York, New York 10003

Library of Congress Catalog Card Number: 65–25134
ISBN: 0–12–029906–2

PRINTED IN GREAT BRITAIN BY
THE ABERDEEN UNIVERSITY PRESS LIMITED
ABERDEEN, SCOTLAND

Contributors

D. J. GOLDSTEIN, *Department of Human Biology and Anatomy, University of Sheffield, England* (p. 135).

KURT F. J. HEINRICH, *Institute for Materials Research, National Bureau of Standards, Washington D.C. 20234, U.S.A.* (p. 275).

R. W. HORNE, *Department of Ultrastructural Studies, John Innes Institute, Norwich, England* (p. 227).

PETER R. LEWIS, *Physiological Laboratory, University of Cambridge, England* (p. 171).

D. S. MOORE, *Vickers Ltd., Vickers Instruments, Haxby Road, York, England* (p. 135).

M. PLUTA, *Central Optical Laboratory, Warsaw, Poland* (p. 49).

F. H. SMITH, *Vickers Ltd., Vickers Instruments, Haxby Road, York, England* (p. 135).

HAROLD WAYLAND, *Division of Engineering and Applied Science, California Institute of Technology, Pasadena, California 91125, U.S.A.* (p. 1).

Preface

The present volume covers a wider range of topics than some previous issues in this Series. In recent years optical microscopy has been largely dominated by the development of complex electronic systems in which the microscope plays a relatively minor role. One article describes such an instrument (Smith, Moore and Goldstein) but the emphasis is on the way a practical commercial instrument has been developed by collaboration between workers in industry and a University. The other articles on light microscopy are much nearer to the classical "string and sealing wax" approach of a generation ago, though alas, few amateurs would have the resources to follow suit. Wayland discusses his approach to the microscopy of living tissues. There is obviously a big difference between doing the job properly and catching an occasional glimpse. Pluta describes a bewildering variety of phase contrast systems; even his compendium is not exhaustive but it shows how many ways there are of doing much the same thing.

In respect of electron microscopy the range of topics covered is equally wide, although here they are concerned more with the use of the instrument than with its construction or operation. Lewis describes the rapidly developing methods for localizing enzymes by modification of standard histochemical staining procedures so as to produce reaction products that are electron dense. Improvements in the complementary techniques of negative staining are surveyed by Horne, who also discusses in some detail the interpretation of the resultant electron micrographs. The concluding article, by Heinrich, is concerned with problems of imaging in the context of scanning electron probe micro-analysis. Although accurate quantitation is not often possible, the visual display (in black and white or colour) of the X-ray signals adds greatly to the value of this method of localized elemental analysis.

As previously we list below the titles of articles commissioned for future volumes. Suggestions for and offers of articles on other subjects of current interest in microscopy will be gratefully received by the editors.

Use of lasers in microscopy
New techniques in optical lens design
Fluorescence microspectrometry
Extraterrestrial microscopy
Reflectance microscopy

Microkymography
Optical transform microscopy
Tracking microscopes
Molecule microscopes
High resolution electron microscopy
Phase contrast electron microscopy
Cryomicroscopy
Reconstruction of macromolecular and virus structures
Image intensifiers for electron microscopy
Coherence effects in electron microscopy
Optimization of conditions for very high resolution
The phase problem in electron microscopy

July, 1975

R. Barer
V. E. Cosslett

Contents

Intravital Microscopy

HAROLD WAYLAND

Non-standard Methods of Phase Contrast Microscopy

M. PLUTA

Development of the Vickers M 85 Integrating Micro-densitometer

F. H. SMITH, D. S. MOORE and D. J. GOLDSTEIN

Electron Microscopical Localization of Enzymes

PETER R. LEWIS

Recent Advances in the Application of Negative Staining Techniques to the Study of Virus Particles Examined in the Electron Microscope

R. W. HORNE

Scanning Electron Probe Microanalysis

KURT F. J. HEINRICH

Intravital Microscopy

HAROLD WAYLAND

Division of Engineering and Applied Science, California Institute of Technology, Pasadena, California 91125, USA

I. Introduction

THERE is no need to justify the importance of intravital microscopy in furthering our understanding of the relationship of microstructure, particularly of the blood and lymph circulations, to physiological function. The excellent contributions to vital microscopy of such leaders in the field as Bloch, Brånemark and the Willnows have recently been summarized in the Proceedings of the Symposium on Light Microscopy held in connection with the VIIth European Conference on Microcirculation held in Aberdeen, Scotland, in the summer of 1972. Although not specifically aimed at a study of intravital microscopy, the majority of the methods of microcirculatory study discussed at the 1972 Tucson Symposium on the Microcirculatory Approach to Peripheral Vascular Function were dependent on intravital microscopy.

The reader is referred to the proceedings of these two symposia for excellent background summaries, as well as comprehensive bibliographies.

In connection with an ongoing joint research program involving the Laboratory of Hemorheology and Microcirculation of the California Institute of Technology and the Cardiovascular Research Laboratory of the Los Angeles County Heart Association and the University of Southern California, Wallace G. Frasher, Jr., and I set out to design and build an intravital microscope system which we hoped would greatly increase our capability of quantitative measurement in microcirculatory studies. The instrument we designed and constructed has been described elsewhere in considerable detail (Wayland and Frasher, 1973). Since that article was written we have further adapted the instrument to new types of measurement, and new instruments have been built in neighbouring laboratories, taking advantage of the basic principles we utilized, but with special adaptation to the specific needs of the particular programs of those laboratories.

This article will discuss the basic principles on which we based our design, as well as the way in which we have adapted the original instrument to meet the following problems: (1) measurement of microvascular organization and macromolecular diffusion by means of fluorescent tracers; (2) improved light pipe and fibre optic illumination; (3) simplified design for low-budget adaptations for a limited scope of problems; and (4) servo-control of focus for studying the microcirculation in moving tissue, and particularly the beating heart.

II. Basic Design Criteria

(a) *The stage.* The microscope stage must be extremely stable, large enough to carry a dog or mini-pig along with auxiliary equipment required for life-support as well as any micromanipulators which need to move with the animal. This entire platform must be capable of precise horizontal movement in two orthogonal directions. There should be free access to all parts of the animal during observation, with minimal interference from either the observing or illuminating systems.

(b) *Imaging system.* Flexibility in use of a variety of objectives, both refracting and reflecting, must be provided. The system should be optically indifferent to the vertical distance of the objective above the surface of the animal table.

(c) *Illuminating system.* Fixed mounting stations should be provided for a variety of light sources (pulsed flash lamps, high pressure mercury

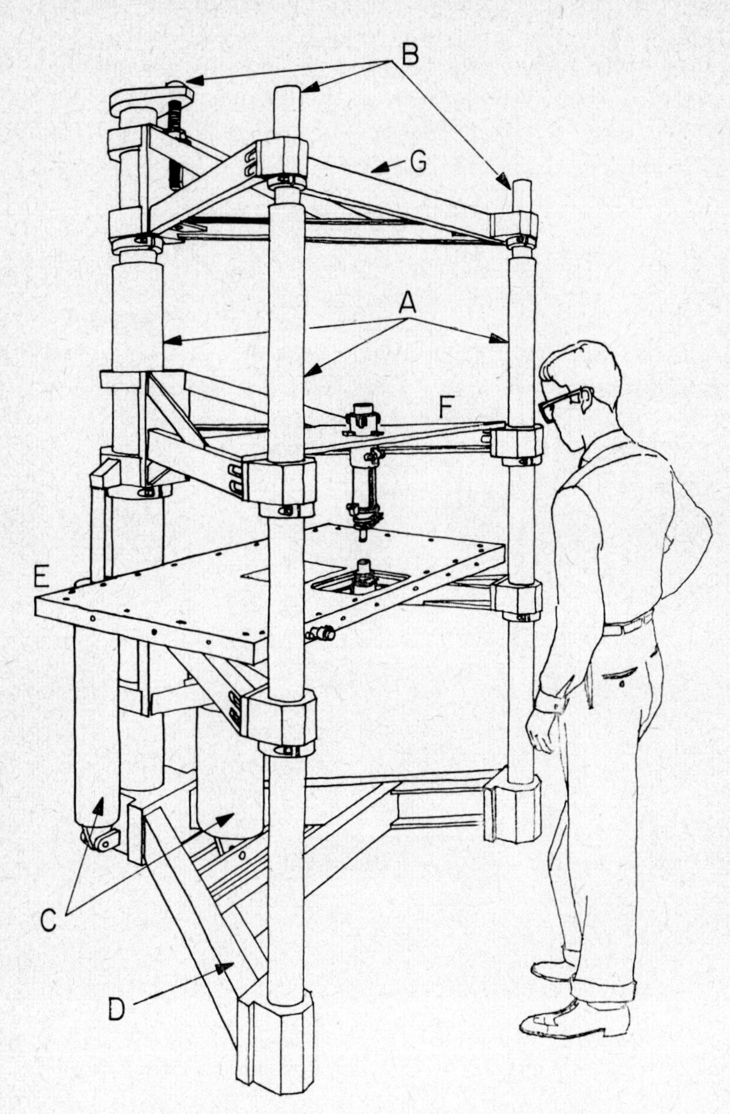

FIG. 1. Basic layout of Wayland–Frasher Intravital Microscope.

A. Main tripod support; B. Isolated tripod support for camera table; C. Pneumatic cylinders for adjusting height of animal table and optics table; D. Table for light sources; E. Animal table; F. Optics table; G. Camera table.

or xenon lamps, quartz-halogen incandescent lamps, lasers) and selection among them should be simply and rapidly accomplished. Adaptation between transillumination, epi-illumination and light pipe illumination should be easily and rapidly accomplished.

(d) *Recording equipment.* It must be possible to mount several different recording systems such as film cameras, TV cameras and photometric sensors with precision, and in such a way that rapid shift

Fig. 2. Photograph of completed intravital microscope system. (From "Modern Technology in Physiological Sciences", courtesy of Academic Press, London).

from one system to another is possible without the necessity of re-positioning or refocusing. Any mechanisms with moving parts, such as cine cameras, must be mounted so as to minimize the transmission of vibrations to the optical system.

III. Conceptual Solution

The basic mechanical design concept is illustrated in Fig. 1, and a photograph of the finished system in Fig. 2. The requirement for a large, stable animal table with free access to the animal for ease of manipulation is met by using a tripod support completely outboard of the principal working area. By using this same supporting structure to carry the basic optical train and illuminating systems, the optical elements can be rigidly interconnected to minimize relative motion

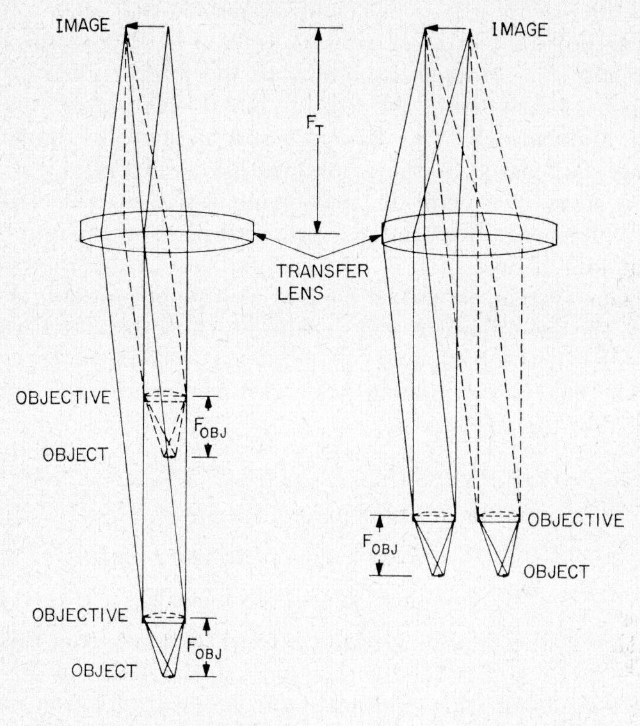

Fig. 3. Principle of telescopic optical system for microscopy. With the object at the front focal plane of the objective, the image will be formed at the back focal plane of the transfer lens with relative indifference to either vertical (a) or lateral (b) movement.

between condenser and objective and between light source and condenser. Observational devices which introduce no vibration, such as television cameras, are supported from the same basic tripod system. Film cameras, which can introduce vibration from their drive or shutter mechanisms, are carried on a separate tripod support system consisting of three tubular columns, each concentric with one of the main supports, and interconnected only at the floor level. Provision is made to permit rigid clamping of the tripod systems to each other, or to brace the camera system independently to the concrete slab on which the instrument is mounted, but so far this has not proved to be necessary.

The requirement for firm fixation of the recording system was met by the use of a telescopic optical system. Similar systems, the basic principles of which are shown in Fig. 3, have long been used in metallographic microscopes, but only with the appearance of the Microstar system of the American Optical Company has it been generally used in biological microscopes. It is especially surprising that such a telescopic system was not used in the elaborate Intravital Microscope made by Leitz. At last this principle is coming to the fore in some of the new microscope systems being offered by major microscope makers for biological applications. The basic idea is eminently simple: if the object plane is located at the front focal plane of the objective, light emanating from any point in that plane which is collected by the objective will emerge from the rear principal plane of the objective as a parallel bundle of rays (Fig. 3). If this bundle of rays is intercepted by a second lens system, it will be focused at a point (actually a circle of confusion, the size of which will depend on the characteristics of both lens systems) in the back focal plane of this second, or transfer, lens system. The final linear magnification will be given by the ratio of the focal length of the two lenses.

$$M = F/f \tag{1}$$

M = linear magnification

F = focal length of transfer lens

f = focal length of objective

The position of the final image is at a fixed distance from the transfer lens, hence if this lens is fixed with respect to the platform carrying the recording equipment, this equipment can be firmly fixed in place. The distance between the objective and transfer lens is immaterial over considerable limits, since the position of an image point in the final image plane is determined only by the direction of the parallel bundle of rays striking the lens, and not by the position on the lens at which it strikes (Fig. 3a). For a given size of transfer lens the only limitation on the distance between it and the objective is an eventual restriction on the size of observable field. As long as the full bundle of rays corresponding to a given image point is captured by the transfer lens there will be no loss in light gathering power or resolution except for those losses associated with poorer imaging qualities at the edge of the transfer lens. Magnification is, of course, not affected by the distance between the objective and transfer lens.

In order to have sufficient magnification in the final image to take advantage of the resolving power of the image recording medium (film, television camera, image intensifier) the transfer lens will generally be of relatively long focal length. We have used transfer lenses of focal length ranging between 300 and 1200 mm. The diameter of the transfer lens will be determined by the desired field and the maximum distance between the objective and transfer lens required. The diameter of each bundle of parallel rays from the objective will, in general, be quite small compared to the focal length of the transfer lens, so that the effective depth of focus in the final image plane is such as to permit the use of only a moderately corrected transfer lens. Since this lens is accepting parallel bundles of light, objectives for small refracting telescopes have proved ideal.

This relative indifference of the optical system to the distance between the objective and the transfer lens has proved to have many advantages. In our system the objective holder has a vertical motion of 200 mm, giving great freedom in the vertical location of the specimen on the animal table. Another advantage is that it permits the introduction of a vertical tracking servo system for following the motion of moving systems, such as the surface of the beating heart. With this optical system only the objective has to be moved, and the location and magnification of the final image remain constant (see Section IX). In fact, it is suitable for following both vertical (Fig. 3a) and horizontal (Fig. 3b) motion of the target, as long as these motions stay within design limits.

IV. The Animal Table

Since intravital microscopy by its very nature is concerned with microscopic studies in living systems, considerable thought was given to the animal table and auxiliary system for life support, mechanical support, accessibility for both macro and micromanipulation and, of course, accessibility of the field of interest for suitable illumination and microscopic observation. The size of the table (Figs. 1(E), 2, 4) was dictated by the desire to be able to handle animals as large as dogs, goats or mini-pigs. In this instrument it consists of a 75×120 cm sheet of stainless steel 25 mm thick. This is given additional stiffness by a frame of the same stainless steel 50 mm deep and 25 mm thick. This also serves to protect the traverse mechanism from fluids spilled on the table as well as carrying mounting holes for mechanical supports. The

FIG. 4. The animal table with a special animal holder for cat mesentery studies in place. The knobs at the lower left control manual horizontal adjustment of the table to ± 1 μm. The post to the right carries the joy stick controller for electrical drive of the table. The condenser is seen just below the plastic tray. The short knurled cylinder just above the objective provides micrometer motion for fine focusing, while coarse vertical positioning is done by means of a rack and pinion. The control knobs for vertical positioning are seen just below the upper platform. (From "Modern Techniques in Physiological Sciences", courtesy of Academic Press, London.)

table itself is supported through a ball bearing system on a base plate rigidly attached to the outer tripod system. The animal table is positioned horizontally by means of driving screws with a $\frac{1}{2}$ mm pitch, and can be moved either manually by turning the knobs at the lower left of Fig. 4 or electrically with a variable speed joy-stick control (on post at right of Fig. 4). The x–y coordinates of the table are indicated by graduated discs and revolution counters, so that its location can be recorded and retrieved within a few micrometers. The horizontal range is ± 50 mm in each direction.

The animal table is designed to handle a load of 120 kg, which permits it to carry considerable equipment in addition to even the larger animals for which it was designed.

Because of the basic indifference of the vertical position of the objective, no provision has been made for vertical motion of the animal table during an experiment. Pneumatic lifts (Fig. 1(C)) are provided to permit adjustment of the height of the animal table and optical table prior to any given procedure. The height of the animal table can be adjusted between 86 and 121 cm from the floor. Precise levelling of the table is important for satisfactory operation of the horizontal traverse mechanism. In practice, the clamps holding the support for the animal table are loosened, the basic level set with the pneumatic piston and the support clamped to the rear column. The table is then carefully levelled with the assistance of hydraulic jacks and the front clamps fixed. For many months we have found it completely satisfactory to keep the animal table in its lowest position, making coarse adjustments of the height of the animal preparation by blocking up the animal cradle.

For either light-pipe or epi-illumination a solid table would suffice. For transillumination optical access for the condenser system is supplied by two square openings 22 cm on a side, one located at the centre of the table and the other at the forward edge (Fig. 1). The openings are provided with cover plates and are sealed with O-rings when closed. In five years of operation we have not used the centre opening and would question the need of providing such access if another such instrument were to be built.

The particular stainless steel used for the animal table (304) was chosen for its ability to resist corrosion due to contact with various physiological solutions and blood. In this respect it has proved eminently satisfactory, and the provision of a set of threaded holes around the periphery, both on the surface of the table and in the apron, has given good flexibility in mounting various animal holders, life support systems, and light sources for light-pipe illumination. Greater flexibility would be provided, however, if the material of the table were magnetic so that magnetic clamps could be used to position equipment on the surface. On occasion we have clamped a thin steel plate to the surface to permit the use of magnetic clamps, particularly for positioning lenses for special illumination systems. Ordinarily an illumination system should be fixed with respect to the optical train and not move with the animal table. In the case of light-pipe or fibre optic illumination, however, it is desirable that the illumination system move with the animal, although it must be adjustable relative to the animal. The use of such a light pipe system for studying the microcirculation in the atrium of the cat heart is shown in Fig. 5.

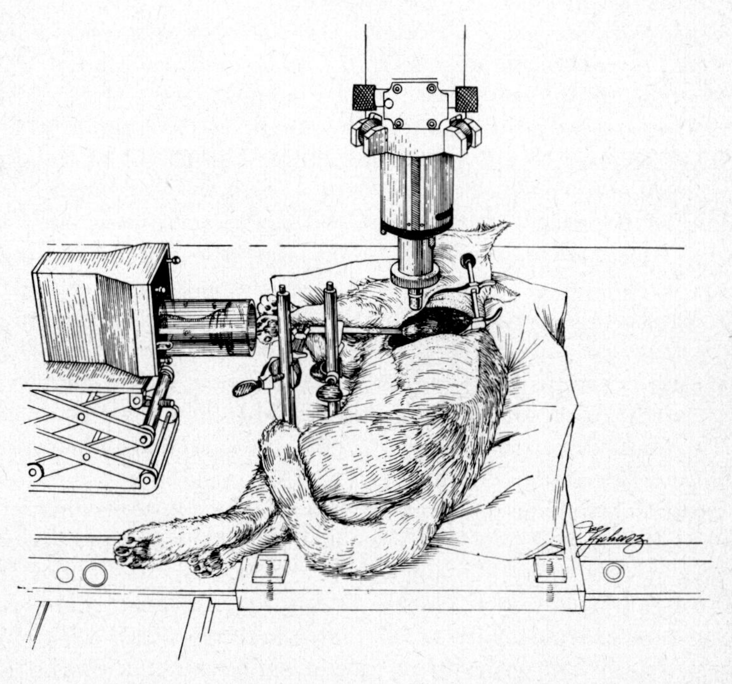

FIG. 5. Sketch of light pipe illumination of the wall of the atrial cavity of the cat heart.

We have found it necessary to have a specially designed animal holder for each type of preparation. With such specialized holders we find our large microscope convenient to use even for rodents. Three types of such holders are illustrated in Figs 6, 7, and 8. Figure 6 shows a system designed for the study of the microcirculation in the mesentery of the cat. This holds the anaesthetized animal firmly in place, with the entire small intestine floating in a temperature controlled bath of physiological solution with prescribed pH and crystalloid and colloid osmotic pressures. The fan of the mesentery to be studied is spread out over a transparent plastic sheet, which permits either condenser transillumination or epi-illumination or both. Details of this preparation and the holder are described by Frasher (1973).

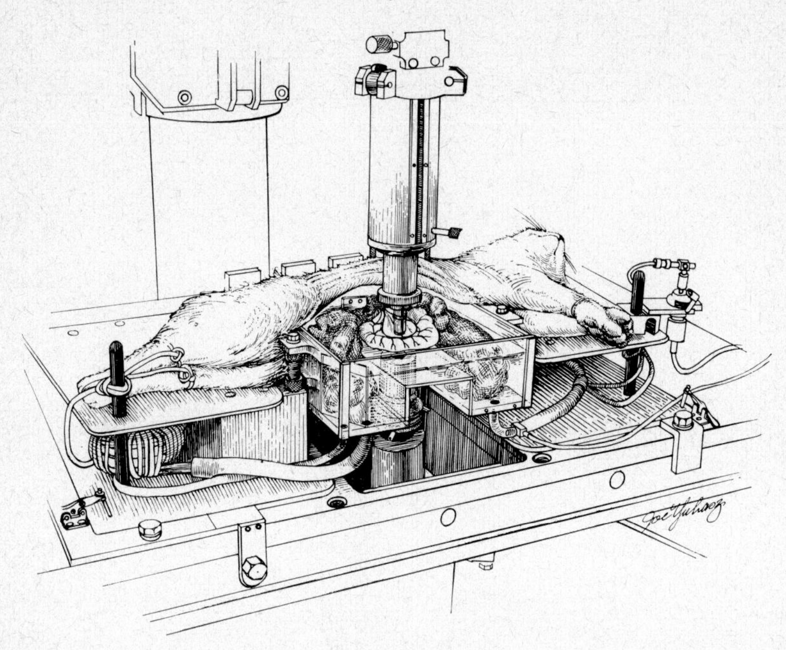

FIG. 6. Sketch of special cradle and plastic tray for studying the microcirculation in the cat mesentery. The entire small intestine is pulled out through a mid-line incision and floated in a temperature controlled, osmotically adjusted physiological solution. The mesenteric fan under observation is draped across a flat, transparent plastic sheet, permitting condenser trans-illumination. A special water-immersion objective is shown in this sketch.

A holder for observing the gingival microcirculation in the cat and small puppies is shown in Fig. 7. The main functions of this holder are to immobilize the head of the animal to eliminate respiratory movements, and to allow the surface of the buccal gingiva under observation to be positioned perpendicular to the optic axis of the objective.

A holder for studying the rat cremaster muscle is shown in Fig. 8. This is designed to use the split cremaster preparation of Baez (1973). In our application we are using fluorescent tracers stimulated by epi-illumination, so that condenser transillumination is not provided. The plastic block on which the cremaster rests is transparent so that the tissue can be adequately transilluminated for general exploration of

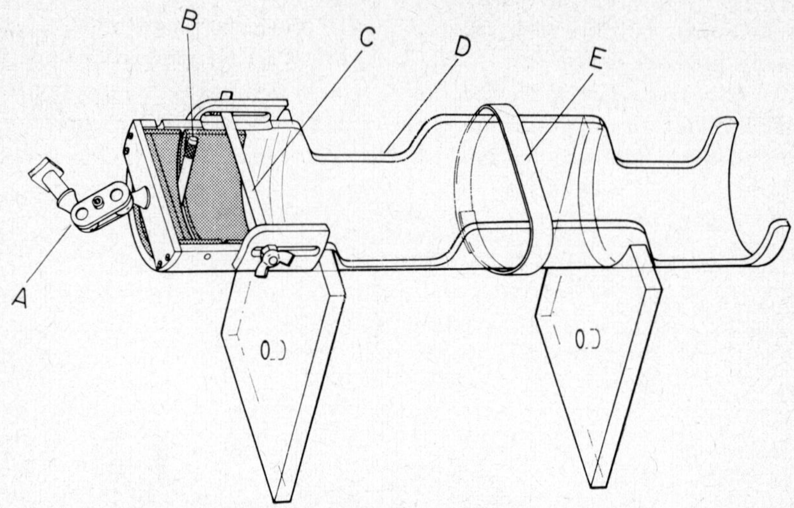

FIG. 7. Holder for observing the gingival microcirculation in kittens and puppies by incident light stimulation of fluorochromes injected into the blood stream.

A. Universal ball joint with snout clamp; B. Intra-oral pin to clamp against palate; C. Neck bar clamp; D. Plastic trough to hold animal; E. Velcro band to secure body.

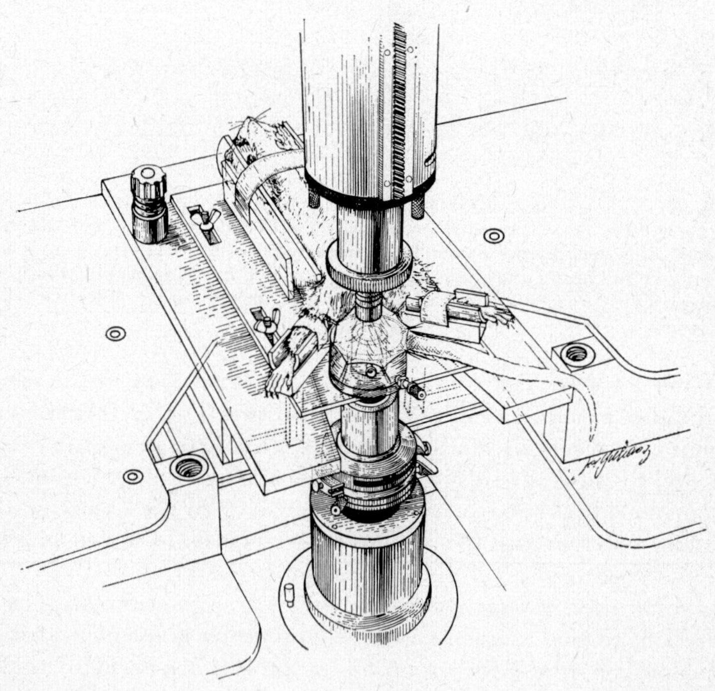

FIG. 8. Holder for studying the microcirculation in the opened cremaster muscle of the rat.

the vascular geometry. If higher resolution in transillumination is required, access would have to be provided for a condenser system, which could be provided by using a long working distance microscope objective for a condenser.

It has proved convenient to make most surgical preparations on a movable operating table, the height of which can be adjusted by means of a hydraulic lift. In many instances the animal is placed in the animal holder on the operating table and the completed preparation rolled up to the microscope system. The height of the table is then adjusted so that the animal holder with the animal in place can be slid onto the animal table of the microscope. For frequently used preparations, special guides and clamps are provided to permit rapid positioning and fixing of the specimen holder in proper relationship to the optical system. The positioning clamps for the mesentery holder are seen at the front edge of the table in Fig. 4.

V. The Imaging Optics

As discussed in Section III, the use of infinity corrected objectives and telescopic observation gives us great latitude in the vertical positioning of the object plane. Coarse positioning over a total range of 200 mm is done by a rack and pinion (Fig. 4) and fine adjustment by means of a $\frac{1}{4}$ mm pitch screw adjustment with a slotted guide so that the objective does not rotate. For some applications a multiple nosepiece would prove convenient, but we have preferred to avoid the added bulk, and possible mechanical interference with the specimen, of such devices. The necessity of screwing the objective into place is a real disadvantage, and we would recommend a quick-change mount, but one in which the objective is rigidly held rather than one with lateral spring loading. Provision has been made above the objective lens for the insertion of a K-mirror (Fig. 9) which permits rotation of the image through more than 180° without the chromatic aberrations which would be introduced with a dove prism. This is used whenever the precise orientation of a given vessel with respect to a sensor is necessary, such as for photometric determination of erythrocyte velocity (Wayland, 1973a, b) or the use of image-sweeping or television techniques for following diameter changes in a given microvessel (Johnson, 1973).

The "poor seeing" in living tissue due to light scattering by intervening layers of cells makes it undesirable to work at numerical apertures greater than 0·50 except in rather specialized situations. We have generally found it adequate to work at numerical apertures of

0·25 or 0·30 even in thin tissue such as the mesentery. The major gain with higher apertures is in light gathering power. Relatively little is usually gained in usable resolution, and the loss of depth of field is serious in most tissues, since the ideal flat field available to the histologist is difficult if not impossible to achieve in living tissue which is maintained in a reasonably viable metabolic state. In fact, the use of lenses corrected for extreme flatness of field is usually a waste of money for intravital studies.

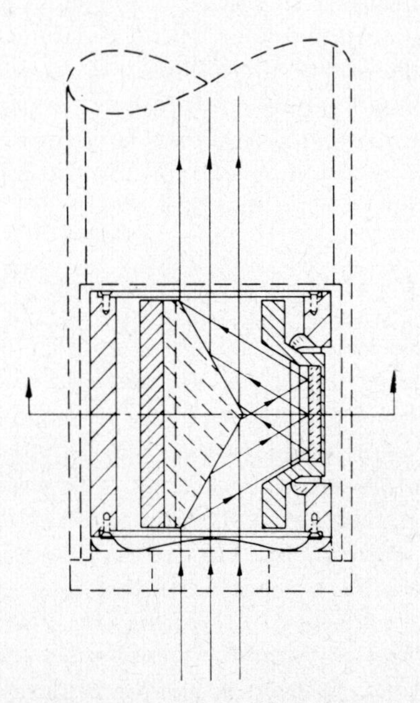

FIG. 9. K-mirror designed to fit just above the focusing tube shown in Fig. 4. A rotation of this mirror system by 90° rotates the image by 180°, permitting alignment of a given feature in the field with a measuring device in the image plane. (From "Modern Technology in Physiological Sciences", courtesy Academic Press, London.)

For many applications, long working distance high-dry objectives have proved most convenient since they permit great freedom of operation. When working through a layer of physiological bathing fluid care must be taken to avoid surface disturbances of the fluid layer. The aberrations due to the layer of material of different index of refraction can not be readily corrected with existing lens systems. If, however, one is observing blood vessels 15 μm below the surface of the mesentery, one is effectively working with a cover slip of that thickness when working high-dry.

Whenever it is technically feasible to have the lens in close proximity to the tissue, water immersion lenses are advantageous. Unfortunately, all commercially available water immersion lenses we have found have such large front ends that their use in intravital microscopy is severely limited.

For working with the cat mesentery, we have found it necessary to use a clamping system to hold the tissue quiet (Fig. 10). A large nose

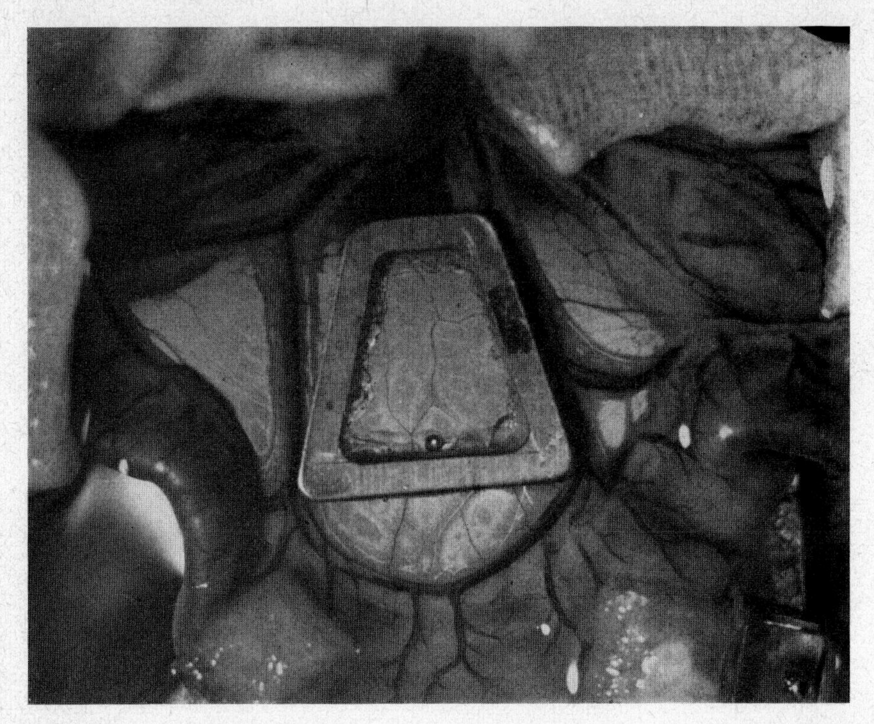

FIG. 10. Ring clamp to hold cat mesentery quiet. This is held in place by its weight. The surface in contact with the tissue is covered with Dacron felt. (From "Modern Technology in Physiological Sciences", courtesy Academic Press, London).

on the objective seriously limits the range of horizontal travel available for selecting a suitable field. Figure 11 shows two dipping objectives which we have made from Leitz UM and UMK objectives. These objectives are designed to be used in a universal stage in which the object is held at the centre of a glass sphere, permitting it to be rotated without changing the length of optical path or magnification. The hemispherical glass surface next to the objective is a basic part of the optical design of these lenses, and they do not reach their full aperture or magnification unless used with the appropriate hemisphere (e.g. a UM 32/0·30 attains its nominal magnification of 32× and numerical aperture of 0·30 only when used with the hemisphere: when used high-dry it has a magnification of 20× and a N.A. of 0·20). By cutting out a segment of

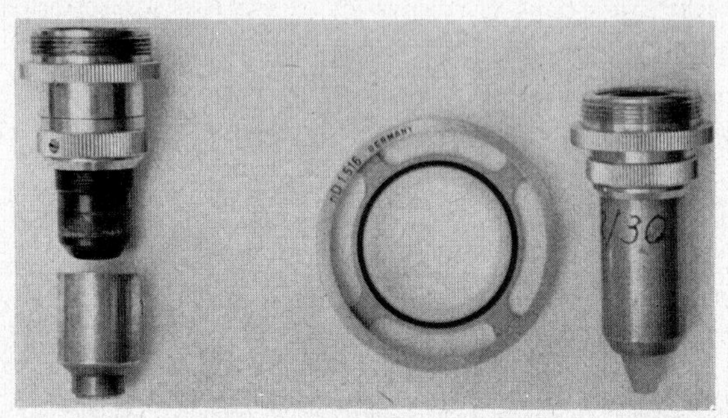

FIG. 11. Water immersion objectives made from Leitz UMK 32/0·50 objective on left, and UM 32/0·30 objective on right. A spherical segment for the UM series is shown in the center. The outside diameter of the metal ring is 41 mm.

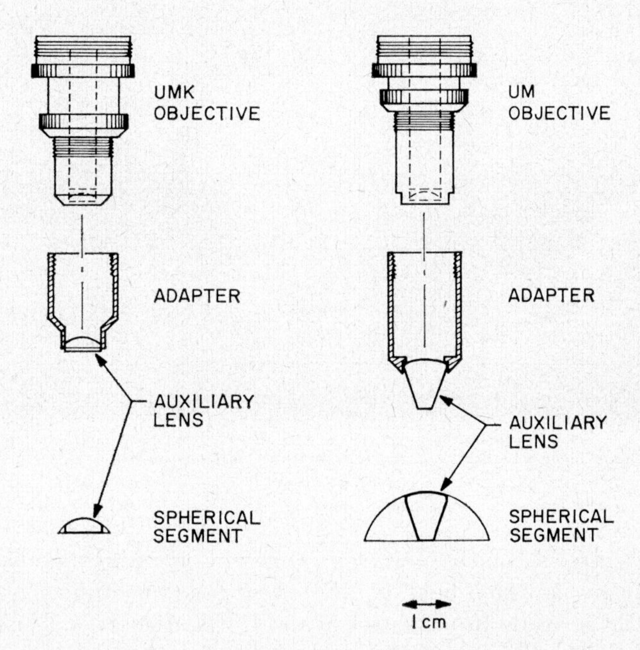

FIG. 12. Assembly of water immersion objectives shown in Fig. 11.

an appropriate hemisphere and mounting it on the objective so that the crown of the spherical surface is held at the appropriate distance (1·40 mm for the UM lenses), a useful dipping objective can be made with a small enough tip to give reasonable flexibility of movement

(Fig. 12). With the UM objectives, for which the radius of the hemisphere is 13·5 mm, the height of the spherical segment is only 10·0 mm (in its normal operation a slide is sandwiched between two segments), and it can be further cut down to permit micromanipulation between the lens and tissue. To take full advantage of the lens design, one should use immersion oil instead of water between the flat surface of the spherical segment and the tissue rather than a physiological bathing fluid, which will have a lower index of refraction. In practice, however, we have found the aberrations introduced by this mismatch are minor compared to those introduced by the tissue intervening between the tissue surface and the microvessels under observation. The segments are available in three different refractive indices and we have chosen the one with the lowest index, 1·516. Unfortunately, these lenses are corrected for a tube length of 170 mm. A negative lens can be used to give parallel bundles of rays, although for many purposes, and particularly for the lower apertures, we have used the lenses at their front focal plane without further correction.

Except when the preparation permits working with water immersion we have found long working distance objectives, a minimum of 5 mm, the most satisfactory. More and more infinity corrected objectives are reaching the market. The lower magnifications and apertures (up to 10× with N.A. 0·25) generally have adequate working distances. Objectives designed to be used with heating stages are available with apertures up to 0·60 and adequate free working distances.

True reflection objectives (in contradistinction to combination reflection-refraction systems) have the advantage of being completely achromatic. Unfortunately, the choice in commercially made reflection objectives is extremely limited. Because of their inherent bulk, they are useful for most intravital work only if they have unusually long working distances. The most useful commercial objective we have found is a 15×, N.A. 0·28 one made by Beck with a focal length of 13 mm, and a free working distance of 24 mm. If used with standard transillumination one must be sure that the aperture of the condenser matches that of the objective. The secondary mirror reduces the effective light gathering power of such objectives, but it can also be used to real advantage for special illumination requirements.

Darkfield transillumination is especially convenient with the highly parallel beam from a laser as shown in Fig. 13(a). The diameter of the secondary mirror is of necessity greater than the diameter of the field of observation, so that the back of this mirror can be used to vignette out the illuminating beam. Although the back of the secondary mirror can be made absorbing, it is better to mount a small mirror at 45° to the axis to deflect the beam out of the field, where it can be absorbed.

Such a mirror also permits this system to be used for incident light illumination, as shown in Fig. 13(b). If a laser beam is used for illumination, it can be brought down parallel to the axis of the optical train, so that vertical movement for focusing does not affect the area illuminated. Light which is reflected or scattered within the acceptance angle of this 45° mirror will be reflected away from the field. Light scattered

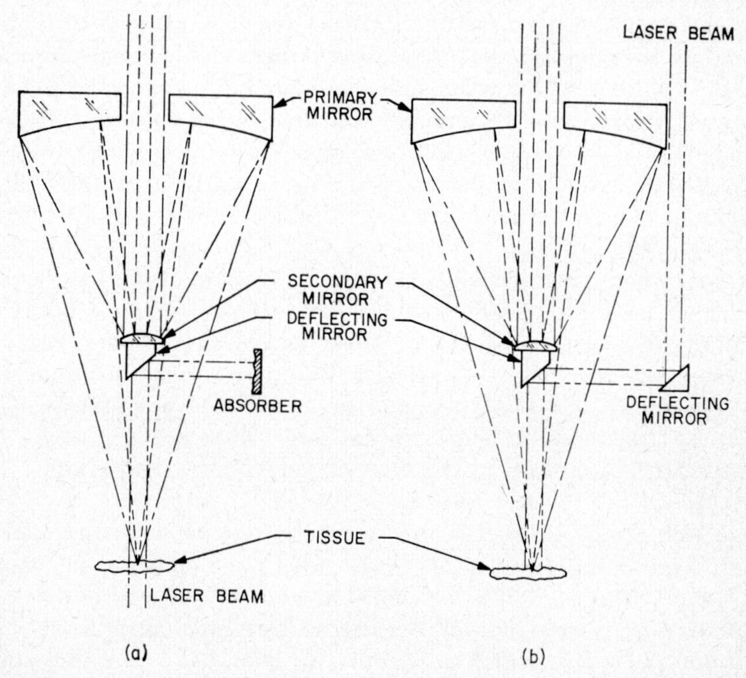

FIG. 13. Use of a mirror objective for darkfield illumination with laser light. (a) Trans illumination; (b) Epi-illumination.

outside the cone vignetted by the secondary mirror, and within the acceptance angle of the primary mirror, can be used to form a darkfield image for either trans- or epi-illumination. If an image due to fluorescent emission is desired, an appropriate barrier filter can be used to remove the exciting light.

In the case of fluorescence microscopy this particular system permits simultaneous imaging of both the darkfield image illuminated directly from the light source and the fluorescent image (Wayland, 1973a, b). In this case a dichroic mirror which reflects, say, the stimulated emission and transmits the exciting light is mounted above the transfer lens so that the fluorescent image is formed on one video camera and the darkfield image on a second. We have successfully superimposed these two images onto different phosphors of a colour television monitor so that the fluorescent image appears green against a purple-violet image of the field. In practice, however, we find it more convenient to present the two images side by side on separate black and white monitors. This gives better resolution for both pictures than is available with colour television systems.

In order to have a mirror objective with a long working distance and a relatively large useful field, we have constructed the mirror objective shown in Fig. 14. The mirrors are both spherical and the

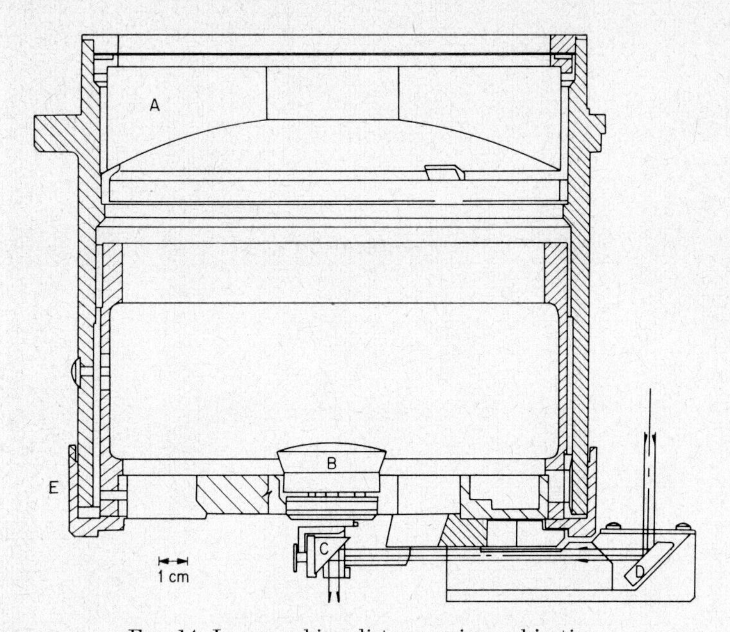

FIG. 14. Long working distance mirror objective.
A, Primary mirror; B, Secondary mirror; C, Deflecting mirror; D, Mirror to bring laser beam from above to mirror C; E, Adjusting screw to change inter-mirror distance.

ratio of the two radii of curvature was chosen to minimize transverse spherical aberrations. The specifications are as follows: Primary mirror, 185 mm radius of curvature, 164 mm diameter; secondary mirror 72·4 mm radius of curvature, 40 mm diameter. The focal length of system is 40 mm and its numerical aperture 0·33. The front focal surface of such a system is spherical, but in this case the radius of curvature is large enough that a 1 mm diameter field is flat to ± 1 μm which is well within the depth of field of the objective. A dipping cone has been

Fig. 15. Long working distance mirror objective mounted on intravital microscope.

provided to permit its use where a water surface must be penetrated. By proper choice of a concave dipping plate, it is theoretically possible to flatten the field out to a diameter of 3 mm.

The principal objection to this objective is its size and weight: it is about 200 mm in diameter, equally high and weighs 12 kg. This poses no serious problems on a microscope such as ours (Fig. 15), but it could hardly be adapted to a standard laboratory instrument!

VI. Illumination

A. *Transillumination*

1. *Employing a condenser*

(i) *Condenser system.* For specimens which are sufficiently transparent that light passing through them from a condenser is not highly scattered, the effective numerical aperture of a condenser-objective system is one-half the sum of the apertures of the objective and condenser or the aperture of the objective, whichever is smaller. For an object which scatters strongly, the effectiveness of the condenser is substantially reduced and in the extreme (which is also the case for illumination with parallel light) is reduced to zero, so that the effective resolving power of the system is determined by one-half the numerical aperture of the objective.

In spite of these considerations, the use of condenser illumination usually gives substantial gain in resolution whenever it is feasible. There is never any point in using a condenser at an aperture greater than that of the objective (except for fluorescence, darkfield or combined darkfield-brightfield illumination), and for intravital microscopy it is usually desirable to stop down the condenser below that aperture in order to reduce the amount of unwanted scattered light reaching the objective, which can seriously reduce contrast. For refracting objectives such stopping down of the condenser reduces the illumination only by the direct reduction of the condenser aperture. In the case of reflecting objectives, however, the relative proportion of the area vignetted by the secondary mirror to that of the primary mirror being filled with light can become too large for effective use of the objective—in the limit, in fact, the secondary mirror can vignette the entire primary beam from the condenser, so that only a darkfield image is formed by the primary mirror.

A long working distance condenser has proved to be essential in order to permit flexibility of design of the animal holders. Since we seldom use an objective with a numerical aperture greater than 0·60, we have found the Nikon condenser made for their inverted phase microscope particularly suitable. This has a maximum numerical aperture of 0·70 and a free working distance in air of 18 mm. If part of the path is in glass or plastic the total distance is, of course, increased. The condenser as it is used with our cat mesentery table is shown in Fig. 4. It is mounted in a centring mount, which contains an aperture diaphragm, which is carried on a tubular column with screw adjustments for aligning the axis of the condenser with that of the objective train (Fig. 16).

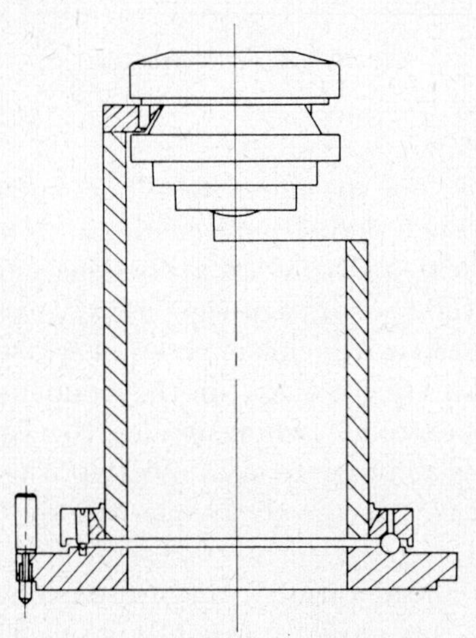

FIG. 16. Condenser mount. (From "Modern Techniques in Physiological Sciences", courtesy Academic Press, London.)

We have also made adapters to permit the use of long working distance objectives as condensers. For use in a restricted space, such as would be desirable with the rat cremaster mount, a tubular support of smaller diameter than the one shown in Fig. 16 would be desirable, and would be satisfactory when used with an objective as a condenser.

(ii) *Light sources for transillumination and their mounting.* A wide variety of useful light sources is now commercially available. For condenser transillumination we have found three types particularly convenient: the quartz-halogen incandescent lamp; the high pressure short arc with either mercury or xenon; and a pulsed xenon short arc. For normal observation alternating current excitation of the continuous sources is adequate, but for quantitative photometric work it is important to have an extremely well-filtered direct current source. For

Fig. 17. Source table. A pulsed xenon arc is mounted to the left, and a 200 W high pressure mercury arc to the right. The cylinder in the centre carries a front surface mirror to deflect the light vertically upward. (From "Modern Techniques in Physiological Sciences", courtesy Academic Press, London.)

cinematography we have found a pulsed light source with a pulse duration of 40 to 50 μsec. gives good images of microcirculatory flow. The use of pulsed sources is particularly desirable for photography, since the tissue is exposed to high light levels only during the actual recording of the image. We have found the pulsed light power supplies and xenon arcs produced by the Chadwick-Helmuth Company of Monrovia, California, particularly suitable, permitting us to make movies of blood flow in individual capillaries in the mesentery up to 500 frames per second.

Continuous sources have proved most useful for closed circuit television applications, although the blanking time between fields in a normal television scan (c. 1 ms.) is long enough to permit impression of the image on the image tube during this blanking period and subsequent scanning of an image which is not being continuously updated. This is not particularly useful with the standard Vidicon, but tubes with silicon targets can be used this way.

Although laser light sources are extremely monochromatic, it is essential to spoil their spatial coherence before they can be used with a condensing system for subsequent image formation. In principle this should be possible, generally with considerable loss of light, but we have no direct experience with laser illumination except for fluorescence studies. The use of the laser with a reflecting objective has been discussed in Section V above. Its use in incident light excitation with a refracting objective will be discussed below.

On our instrument the sources for condenser transillumination are carried on the lowest level of the microscope (Fig. 17). This source table is designed to carry up to five triangular optical benches each of which can carry a rigidly mounted source with lamp diaphragms, lens and filter holder. The sources are normally mounted on standard carriers for the optical benches, so they can easily be placed in any one of the five positions. In Fig. 17 a pulsed xenon arc source is shown on the optical bench to the left, and a 200 watt high pressure mercury arc on the bench to the right. The lower black cylinder in the centre contains a front surface mirror at 45° to the vertical to deflect the beam vertically. This is carried on a ball bearing assembly (shown in the lower part of Fig. 18) so that it can easily be rotated from one position to another. A series of ball detents (one of which is shown at the lower right of Fig. 18) permits rapid positioning of the mirror at any of the source stations. The ball carriers can be readily adjusted to give precise and repeatable alignment. (The upper mirror in Fig. 18 is part of an assembly used on the optical table and is not used on the source table).

The long optical path from the source to the condenser, plus the need

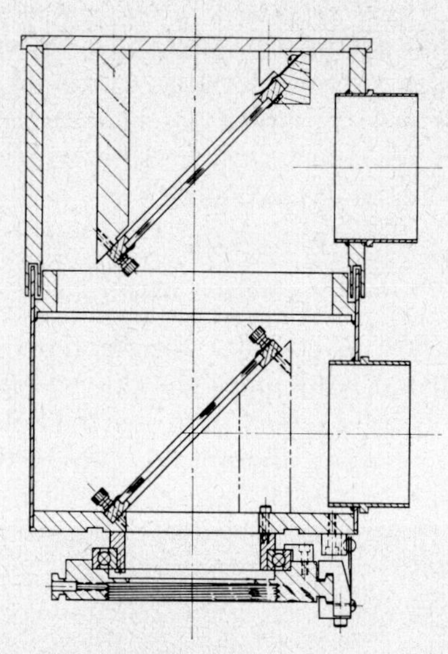

FIG. 18. Beam deflector assembly. The lower portion is carried on ball bearings and can be rotated about a vertical axis. Its position is determined by a series of detents, such as the one shown at the lower right, which can be fixed in prescribed positions. The upper mirror is fixed in position and reflects that light which is transmitted by the lower mirror into a television camera. (From "Modern Techniques in Physiological Sciences", courtesy Academic Press, London.)

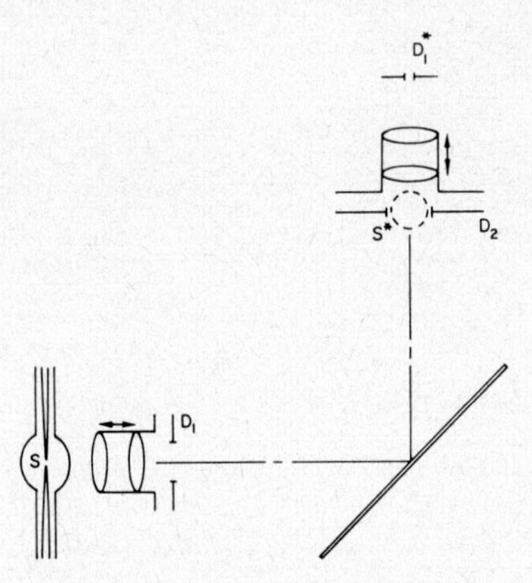

FIG. 19. Köhler illumination system used with the source table in Fig. 17. (From "Modern Techniques in Physiological Sciences", courtesy Academic Press, London).

to adjust the vertical position of the condenser to correspond to the level of the object would dictate a design employing a transfer lens system similar to that used in the imaging portion of the microscope. The fact that for each particular tissue which we transilluminate the level of the tissue is predetermined has made it possible for us to use conventional Köhler illumination (Fig. 19) with the necessary adjustments made prior to each experiment.

B. *Transillumination with Light Pipes*

For tissues which are inaccessible for standard condenser illumination, the use of clad rod light pipes has proved effective. Such a light pipe, with a 45° mirror on the end to divert the light vertically upward is shown in Fig. 20, with the illumination provided by a pulsed xenon

FIG. 20. Clad rod light pipe with 45° mirrored end mounted in tube for insertion into atrium of cat heart. (Courtesy Richard J. Bing and the Huntington Institute for Applied Medical Research.)

arc. In collaboration with Dr. Richard J. Bing, both in our laboratory and in that of the Huntington Institute of Applied Medical Research, such illumination has been used to study blood flow in the atrium of the cat heart using the arrangement shown in Fig. 5. With this illuminating system it has been possible to take colour movies up to 400 frames per second, permitting measurement of erythrocyte flow velocity in capillaries and small arterioles and venules. With this arrangement it is not

FIG. 21. Hypodermic needle light source, showing illuminated spot. (Courtesy Richard J. Bing and the Huntington Institute of Applied Medical Research.)

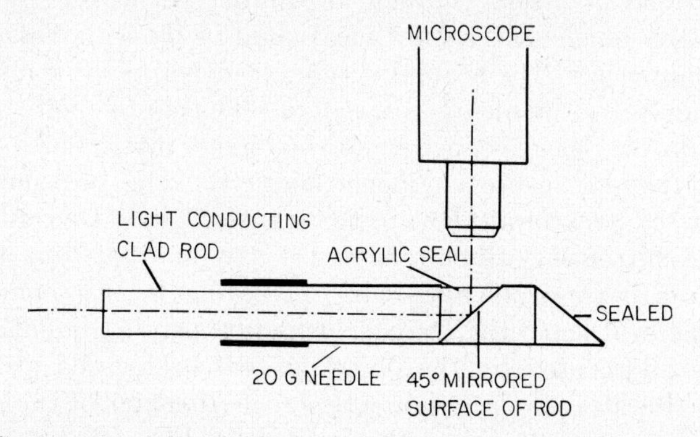

FIG. 22. Constructional details of hypodermic needle light source. (Courtesy Richard J. Bing and the Huntington Institute of Applied Medical Research.)

possible to keep the field in focus throughout an entire heart beat, so that bursts of film were exposed at varying levels of focus, so that data from each part of the beat could be recorded. This led to the design of the servo-controlled focusing system described in Section IX.

For tissues too thick for transillumination of the entire section of tissue, such as the ventricular muscle of the heart, Dr. Bing devised the hypodermic needle illuminator shown in Figs 21 and 22. The assembly is built into a 20 gauge needle, and the light conducting rod is 0·43 mm in diameter. The end of the rod in the needle is cut at 45° to its axis, polished and aluminized. The hole in the wall of the needle which transmits the light upward is filled with clear acrylic plastic to give a good light path. This seal, plus a similar seal at the end of the needle also prevents the entry of blood and tissue fluids into the light conducting system. For studies on ventricular microcirculation in the heart, the needle was inserted just underneath the superficial layer of the myocardium. Tissue motion is always a serious problem in studies on the beating heart. Our approaches to handling this problem will be discussed in Section IX.

C. *Epi-Illumination*

Epi-illumination with a mirror objective has been described in Section V. We have used epi-illumination exclusively for exciting fluorescence, and have adapted the system of Ploem (1967) to the use of a gas laser as the source of exciting radiation. A schematic diagram of the system we are using for stimulating fluorescein tagged molecules is shown in Fig. 23. An Argon–Krypton gas laser L′ furnishes a highly monochromatic beam of light at 488 nm. The parallel beam from this source is passed inversely through an infinity corrected objective O_2 to form a point image at the front focal point of the lens. This acts as a directed point source. The beam diverging from this point is intercepted by a diffusion disc D, which is rotated at 1800 rpm to spoil the spatial coherence of the laser beam. A narrow band interference filter F_2 peaked at 488 nm, insures monochromaticity. The diffusion disc is mounted at the back focal plane of the transfer lens. After reflection in the dichroic mirror M_2, which reflects the 488 nm light from the laser, but transmits beyond 500 nm, light emanating from a point on the diffusion disc will leave the transfer lens as a parallel bundle of rays. Those rays intercepted by the objective will be focused at its front focal plane. Because of the small acceptance diameter of the objective this is a very inefficient use of the laser light$_c$ but it has proved convenient and adequate. The amount of scattering by the diffusion disc was adjusted empirically by spraying it with a clear polymeric lacquer until maximum illumination was obtained in the front focal plane of the objective.

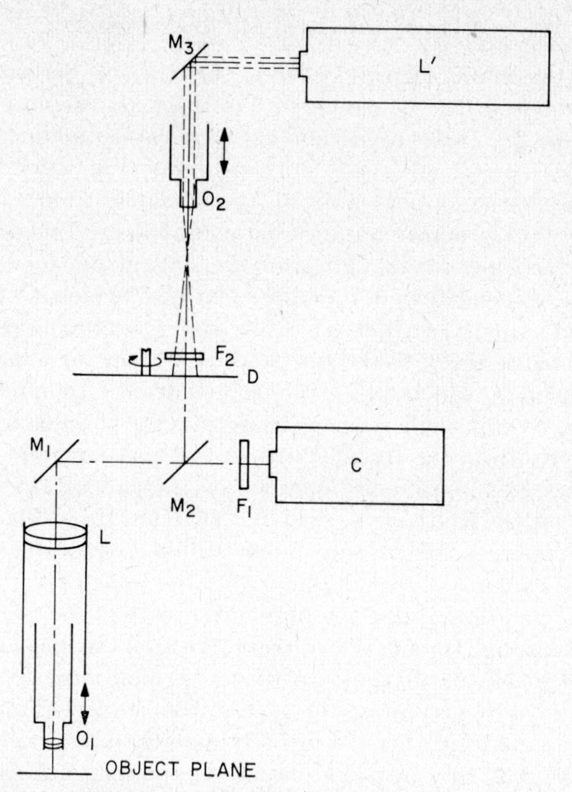

FIG. 23. Adaptation of Ploem incident light illumination system for laser light stimulation of fluorescent emission.
O_1, Infinity corrected objective; L, Transfer lens; M_1, Fully reflecting front surface mirror; M_2, Dichroic mirror. Reflects 488 nm, transmits 520 nm; F_1, Barrier filter. Transmits beyond 500 nm; D, Rotating diffusion disc; F_2, Narrow band interference filter peaked at 488 nm; O_2, Infinity corrected objective; M_3, Fully reflecting front surface mirror; L′, A-Kr gas laser tuned to 488 nm; C, TV camera.

The illuminated area on the disc is adjusted by axial motion of the lens O_2 so that the image of this area on the object corresponds to the desired field of view. The image formed in the stimulated emission by the objective O_1 is projected onto the face of the SIT in the TV camera C after being passed by the dichroic mirror M_2 and further filtered by the barrier filter F_1.

For use with fluorescein tags, an argon laser would be adequate, but the A–Kr mixed gas laser permits shifting of the wavelength to allow excitation of both rhodamine and Evans blue, when the dichroic mirror and barrier filters are suitably chosen.

VII. Image Recording Systems

With a telescopic viewing system the various observational and recording devices can be fixed with respect to the location of the transfer lens. In order to permit rapid selection among various recording systems, we use a 45° mirror mount (Fig. 18) which can be rotated about the optical axis of the imaging train, allowing the cameras, etc., to be spread out on a horizontal plane. Television cameras or viewing screens are mounted on the optical platform, while film cameras are suspended from the upper platform, which is carried on the inner tripod support, so that vibration from the moving parts will not be transmitted to the optical train, (Figs 24, 25 and 26). With a diversity of image formats, which may vary from that of a 16 mm cine camera to that of a 70 mm single frame camera, it is not feasible to project the image directly from the transfer lens to the final image plane in every case. In Figs 24 and 25, additional lenses are shown, which have been chosen so that the width of the field in the 70 mm Hasselblad is approximately the same as that in the 16 mm Milliken cine camera. With the 25 mm SIT camera in Fig. 26, however, the image from the transfer lens is projected directly on the fibre optic plate of the SIT tube.

When recording from a TV camera, we use the image on the TV monitor to check for sharpness of focus. A more objective indication can be obtained by presenting the output from a selected scan line on an oscilloscope, and adjusting the focus to give the most rapid rise of the signal as the scan line crosses a distinct boundary, such as the edge of a blood vessel. A more sophisticated version of this method is used in the servo-controlled focusing system described in Section IX.

The ability of the eye to accommodate makes it extremely difficult to determine the position of sharpest focus with camera viewfinders, so that we have found it more satisfactory to use a television monitoring system, which has been adjusted to be par-focal with the camera for determining sharpest focus. For this purpose, the rotating head (lower part of Fig. 18) carries a mirror which reflects 70% of the light horizontally and transmits 30% vertically. The vertical beam is deflected horizontally by a fully reflecting mirror (upper mirror, Fig. 18) into a vidicon camera (Fig. 25). Since a TV camera has no capability of optical accommodation, this difficulty has been eliminated. If a greater proportion of the light is needed for photography, a more sensitive TV tube could be used, such as a Silicon Vidicon or a SIT.

FIG. 24. Optical and camera platforms from front. A 16 mm pin registry cine camera is seen to the left and a 35 mm pin registry cine camera to the right. The telescope in right centre is used to check quality of focus when the image is projected onto a viewing screen. (From "Modern Techniques in Physiological Sciences", courtesy Academic Press, London.)

Fig. 25. Optical and camera platforms from left rear. The 16 mm cine camera is mounted at the right; a 70 mm Hasselblad camera just to the left of centre, and a TV monitor shows above and to the left of the Hasselblad. (From

Fig. 26. Optical and camera platforms from right rear. The gas laser is mounted at the upper left and a 25 mm SIT television camera is housed in the cylindrical container to the left of centre.

With modern fine grain films it is possible to record a great deal more information on film than can be seen by eye at the same magnification. It is much more efficient to record photographically at relatively low magnification and use further enlargement when viewing the film as long as the desired information has been captured, since doubling the linear magnification on the film cuts the light intensity by a factor of four. Over magnification can also be responsible for blurring due to red cell motion or even tissue motion. With the Chadwick–Helmuth flash system, which gives a pulse duration of between 40 to 50 μsec., we have had no problem with blurring of red cell images due to flow. A pin-registry cine camera with synchronized flash is strongly recommended for cine recording. Events associated with red blood cell flow in the microcirculation can be adequately recorded, even for frame by frame analysis of flow velocity, at framing rates of 400 frames per second or less as long as the individual pictures are sharp, which they should be with pulses of 50 μsec. or less at magnifications adequate to identify individual cells.

Television cameras have inherently poorer resolution than any but the grainiest of films. If the limitations of field size can be accepted, enough magnification can be used with such a television camera to utilize the full resolution of the optical system and still require less total illumination than a film camera. Commercially available videotape recorders, however, have a severe limitation on effective frame rate. In the U.S. the standard system gives a 525 line picture 30 times a second, but each picture is made up of a $262\frac{1}{2}$ line field each 1/60 of a second, with two such frames interlaced to form a full frame. If any sort of measurements are to be made from videotaped data, it is advisable to use a precise interlace rather than a random interlace. With custom built electronics, it should be possible at least to double the frame rate.

The extremely high sensitivity of the SIT system is especially valuable in fluorescence studies, since this system has essentially the same sensitivity as a photomultiplier, but permits the formation of an image of the entire field. The sensitivity is not uniform over the entire surface, so that precise quantitative work requires point by point correction. We have found this most easily done by converting the image information into digital form (from single-frame playback of videotaped data) so that subsequent operations can be performed on a digital computer.

Another advantage of the high sensitivity of such recording systems is the ability to work with monochromatic light at a variety of wavelengths to accentuate specific features of the field. The low light levels required permit the use of sources with an essentially continuous spectral distribution. For most of the visible spectrum and on into the near infrared, the quartz halogen lamp, suitably filtered or monochromatized, is suitable and convenient. As one moves to the violet end of the visible spectrum and on into the near ultraviolet, a high pressure xenon arc is probably the source of choice. Higher intensity for a given voltage can be obtained with the mercury arc at its emission lines. The 404·7 and 546·1 nm lines of the mercury arc are particularly suitable for giving high contrast to red cells against the tissue background, since these are close to specific absorption bands of hemoglobin.

Greater monochromaticity can be obtained with laser sources, and with the rapid development of dye lasers, it is now possible to obtain tuneable systems to cover virtually any wavelength region. Care must be taken to spoil the spatial coherence of any laser source when used merely as a light source for conventional imaging or there will be serious trouble from diffraction patterns associated with every speck of dust or minor anomaly in index of refraction in the system. The spatial coherence is, of course, essential for holography, but this is not the province of this paper. So far, the demonstrations of the application of holography to microcirculation studies have been far from convincing. It seems to this author that one is faced with the same fundamental limitation found in attempting to use phase contrast on thick tissue. There is so much information associated with index of refraction changes throughout the entire depth of the tissue that in the current state of the art, it is not yet possible to select out the desired information and reject that which is not wanted.

For epi-illumination the combination of being able to select an appropriate wavelength with suitable polarization of the incident light should make it possible to increase contrast. Stromberg and Shapiro (1973) have successfully used polarized light, plus an appropriately adjusted analyzer in front of the viewing system, to improve the seeing in studies of the pial circulation. The great sensitivity of the SIT should permit much greater flexibility in this type of system.

The image intensifier is another imaging device of potentially great usefulness at low light levels. Three stage intensifiers are available with optical gains of 100 000 or more. The final presentation is an image on a fluorescent screen. The basic resolution of a three stage image intensifier is about half that of the SIT system but it has the advantage that no additional electronics is required to present a field visible to the eye. Recording can be done either with a video camera focused on the screen of the image intensifier or by photographing it.

We made a careful analysis of the pros and cons of the image intensifier against the SIT when setting up our system for fluorescence studies, and concluded that the SIT system had comparable overall sensitivity with higher resolution. The rapid advances in the state of the art, however, would dictate a reanalysis for each new application.

VIII. Simplified Systems

The Wayland–Frasher Intravital Microscope described in the preceding sections was designed as a prototype instrument. We had little expectation that it would be copied directly, but hoped that our design concepts could be tested out in our own ongoing research programme, and that useful features could be adapted to specialized applications in a simplified form. One example of such spin-off is shown in Fig. 27.

This is an intravital microscope designed for the Huntington Institute of Applied Medical Research for their studies of the microcirculation of the heart. The basic requirements were established during a two-year period in which the programme was carried out at Caltech on the prototype instrument. The fact that only light pipe or epi-illumination was contemplated eliminated any need for a standard condenser system; hence a solid animal table was adequate. This permitted us to design the system around a machine tool base, in this case a surface grinder, which we were able to obtain second hand at a modest cost.

Three degrees of freedom are provided for the object table, although the height adjustment is seldom used. The fore and aft adjustment controlled by the right front hand wheel proved sufficiently fine. The lateral adjustment was too coarse, so that a worm drive was added, shown to the right of the left front hand wheel. A 1 cm thick steel plate was bolted to the table top to provide a smooth working surface which also permits the use of magnetic clamps, such as shown at the right front edge of the table, for positioning and holding auxiliary equipment such as light pipes and lenses.

Telescopic observation is used to permit fixed mounting of recording equipment. The objectives are carried in a rotating nosepiece mounted on a conventional microscope tube assembly. A coarse rack and pinion movement is used for major changes in height, while focus is accomplished with the combined coarse and fine adjustments of the microscope tube assembly. This mounting is not so stable mechanically as that on the prototype instrument, but has proved adequate for most purposes.

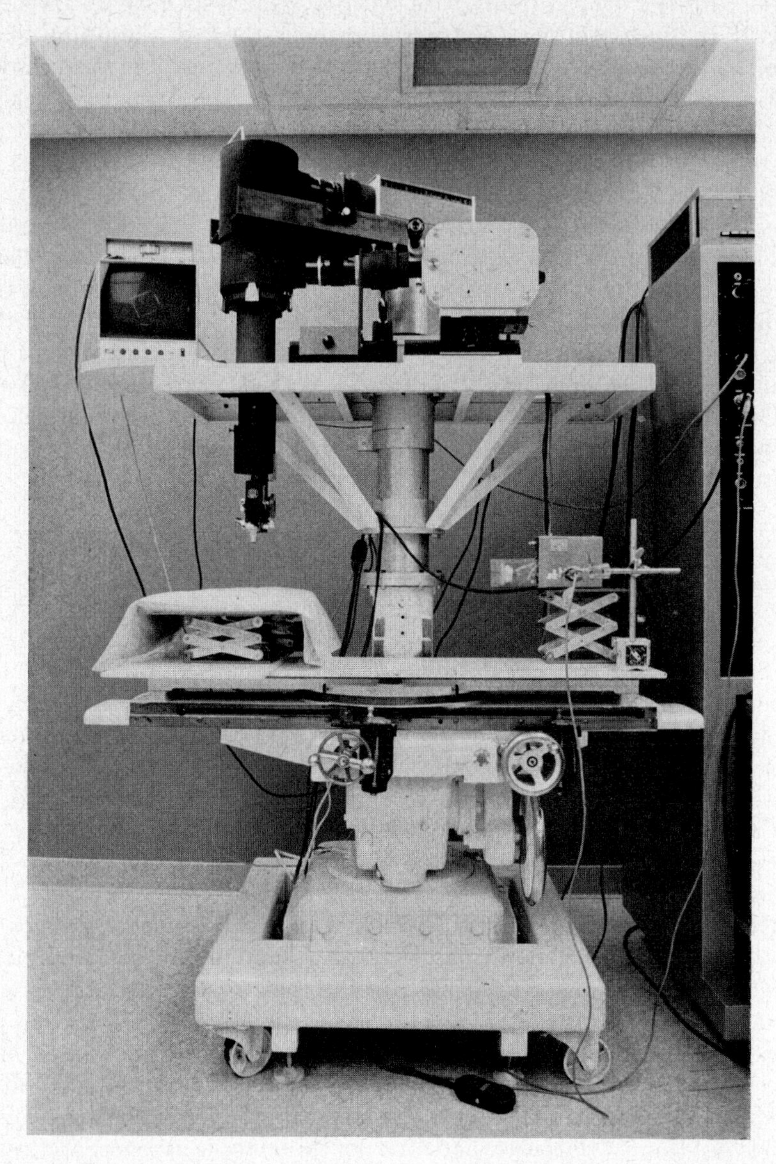

FIG. 27. Simplified intravital microscope using a surface grinder base and table. This was designed to be used with light-pipe illumination and epi-illumination.

The transfer lens, in this case with a focal length of 365 mm, is carried at the top of the tube which extends through the upper platform. This is capped with the same type of double mirror system shown in Fig. 18. This telescopic imaging system permits the cameras, such as the Milliken high speed pin registry cine camera shown at the right on the upper platform, and the television camera in the upper centre, to be firmly fixed in advance, after careful adjustment to be parfocal. Selection of field and determination of sharpness of focus are normally done by means of the television monitor shown in the upper left, although the camera field can be viewed through the lens system on the beam splitter just to the left of the cine camera. The mounting of the cine camera so that it is rigidly interconnected to the rest of the optical train is not ideal, since some vibration is transmitted to the objective.

This same system can readily be adapted to incident light excitation of fluorescence, using the same basic configuration as employed on the prototype microscope.

IX. Servo-controlled Focusing

A. *The Problem*

One of the major difficulties in intravital microscopy is that living tissue seldom stays in a precisely fixed position, due primarily to breathing, the beating of the heart, and muscle twitch. In many tissues this can be largely overcome by gentle clamping, and this has been quite successful with the mesentery, particularly of animals as large as the rabbit and cat or larger. Whenever clamps of any kind are used there is always question as to the effect of the clamps on the physiological state of the tissue.

The use of infinity corrected, long working distance objectives is ideally suited to automatic following of tissue motion, since only the objective has to move either vertically or laterally, and the image remains fixed over a wide range of objective positions (Fig. 3). Furthermore, the magnification, resolution, and light gathering power are not affected as long as the movement is kept within readily ascertained limits.

Any automated field-tracking system must perform three basic functions: it must (1) sense whether or not the image is in proper focus and location, (2) be a logic system capable of converting this information into an action signal, and (3) be a drive system which will bring the image back into proper focus and location.

In a collaborative programme between the Jet Propulsion Laboratory of the California Institute of Technology, the Huntington Institute of

Applied Medical Research, and the author's laboratory at the California Institute of Technology we have built a prototype system for vertical tracking of the surface of a beating heart, and have established the feasibility of a three dimensional tracking system with sufficient rapidity and accuracy of response to permit maintenance of adequate focus to measure red cell velocity by sequential frame analysis in capillaries of the unrestrained heart of the cat. This work has been done under the direction of Momtaz Mansour, of the Jet Propulsion Laboratory.

A power spectrum of the frequency distribution for vertical motion of a point on the surface of an unrestrained heart of a cat, with the pericardium removed, is shown in Fig. 28. Even though the heart rate in this case was 240 beats per minute, very little energy is found above 5 Hz, and no measurable amount above 20 Hz. On the basis of such experimental evidence, we set as design criteria a frequency response of 0–20 Hz and a displacement coverage (peak to peak) of 10 mm.

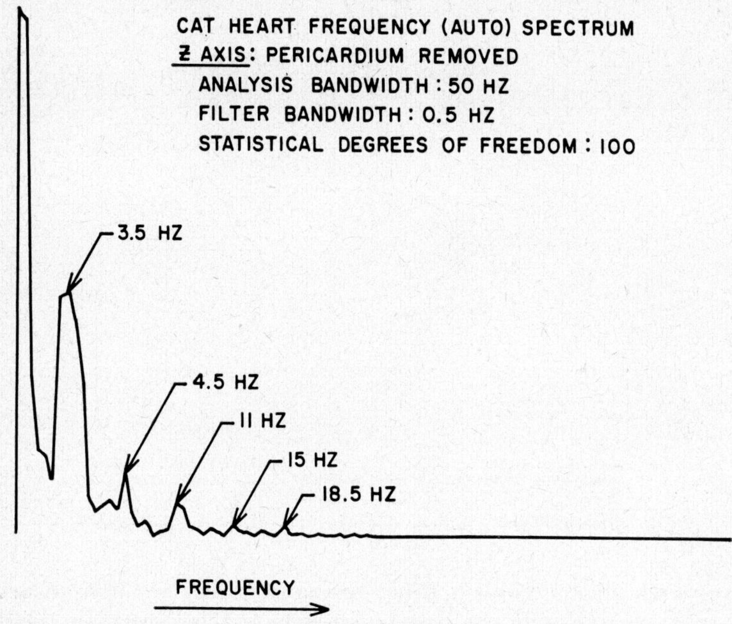

FIG. 28. Power spectrum for vertical motion of a cat heart with the pericardium removal.

B. *Focus Sensing*

Since we are basically interested in recording an optical image of a field being viewed through a microscope, we sought a property of the image itself to use for sensing the sharpness of focus. This is readily provided from the spatial high frequency content of the light distribution in the image plane of the microscope system. If the spatial high frequency content of the light intensity distribution is explored in a series of planes parallel to the plane of best focus, it will be found to possess a maximum at the plane of best focus. These higher frequencies fall off essentially symmetrically on either side of focus, so that additional information is required to direct the driver system to move the objective in the correct direction. We have chosen to do this with a pair of television cameras, one set in front of the plane of best focus and the other behind it. The physical arrangement is shown schematically in Fig. 29, and the actual prototype system in Fig. 30. Three television

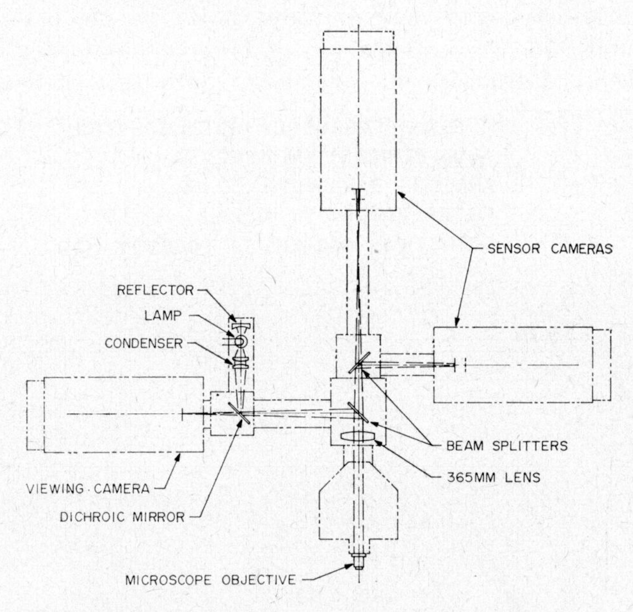

FIG. 29. Schematic diagram of Z-axis servo microscope system.

cameras with silicon vidicon tubes are used, with two non-absorbing beam splitters to divide the light among them. The camera to the left is the viewing camera, in which we wish to maintain a sharp image of the field, so its target is shown at the principal focus of the transfer lens. The upper camera has its target set behind the focal plane of the transfer lens (image plane), while the camera to the right has its target ahead of the focal plane. Any movement of the tissue toward the objective lens will bring the image formed on the upper camera more

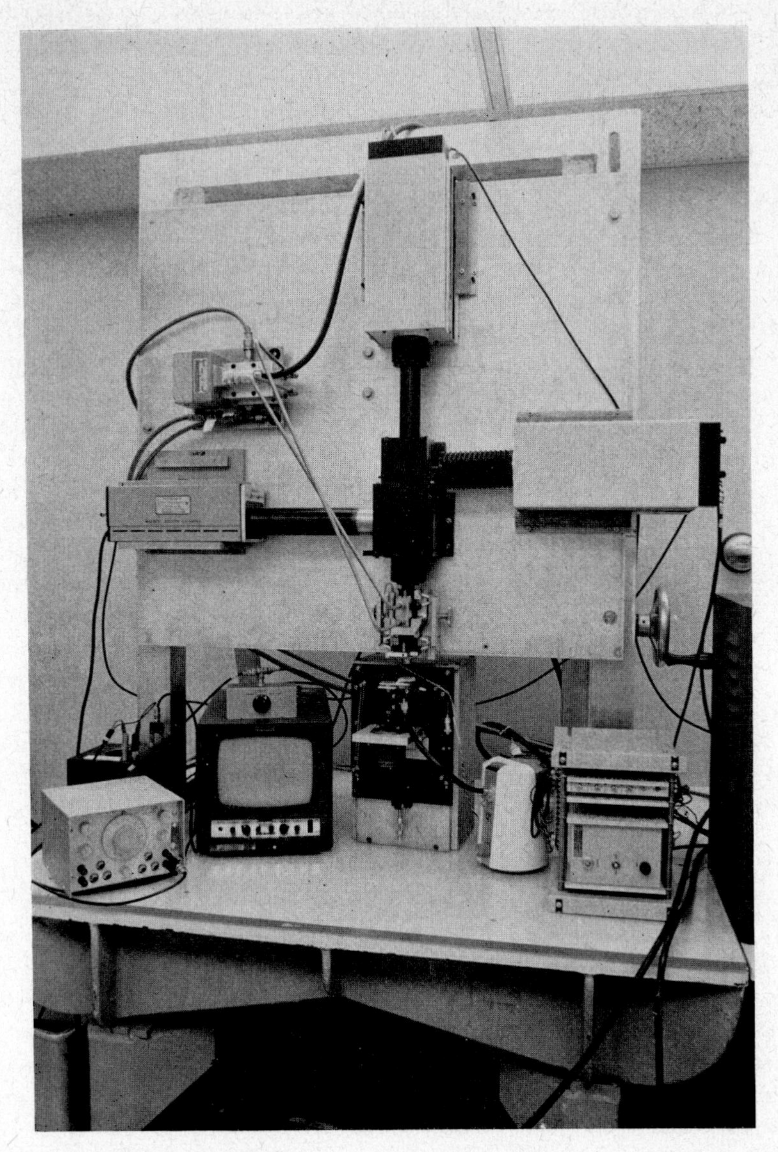

FIG. 30. Photo of prototype Z-axis servo microscope. The general layout is the same as that of Fig. 29.

nearly into focus and the camera to the right further out of focus. The video signals from these sensor cameras are then processed as described in the following Section C to provide a suitable focus error signal.

C. *Signal Processing*

The use of television monitoring provides a simple means of measuring the spatial frequency distribution of light intensity variation over the picture since the electronic scanning of the target in the TV camera converts the spatial frequencies into the time domain. In practice it has proved best to pass the signal from the TV camera through a band pass filter with a range from 100 to 300 KHz. This signal is then differentiated, the differentiated signal rectified, and the rectified signal integrated over a predetermined time interval and normalized against the integrated intensity signal over the same interval. Considering the frequency demands, a standard field ($262\frac{1}{2}$ lines in the US), which is scanned in 1/60 of a second, gives a good balance between sufficiently rapid acquisition of new data and utilization of the focus information in the entire picture.

The sharper the picture the larger this signal will be, while the lower limit of the band pass filter insures that temporal fluctuations of the light intensity will not influence the signal, and the normalization takes care of both slow drift or changes in reflectivity due to flooding with irrigating fluid, and since the signals from two different cameras must be compared, differences in sensitivity and gain of the two cameras will be largely eliminated. If $\bar{V}_1$ and $\bar{V}_2$ are the processed signals from the two cameras, these will both be positive. Thus the differential signal $\bar{E} = \bar{V}_1 - \bar{V}_2$, which can be made to vanish when best focus is achieved, will change sign as the image passes through the position of best focus. This then affords the basic information to drive the servo system. As with all servo systems, there is always the problem of balance between sensitivity, range, and stability. It may not be appropriate to use the same focal length lens system for both the imaging camera and the position sensing camera. The magnification on the imaging camera must be chosen to record the appropriate field for study, while the signals reaching the position sensing cameras must be chosen to give adequate sensitivity while maintaining stability.

The signal provided by the detection system will not have the correct phase for stable response of the tracking system. The addition of a displacement transducer between the objective holder and ground provides an additional signal, the derivative of which is used to provide the necessary damping of the system for stability.

D. *The Drive System*

In order to keep the lateral position of the image fixed in the recording plane it is imperative that the objective move strictly parallel to its optic axis. This can be accomplished by supporting the objective assembly by a set of reinforced flexure springs, shown diagrammatically in Fig. 31. In this system a rather large mass must be moved, but

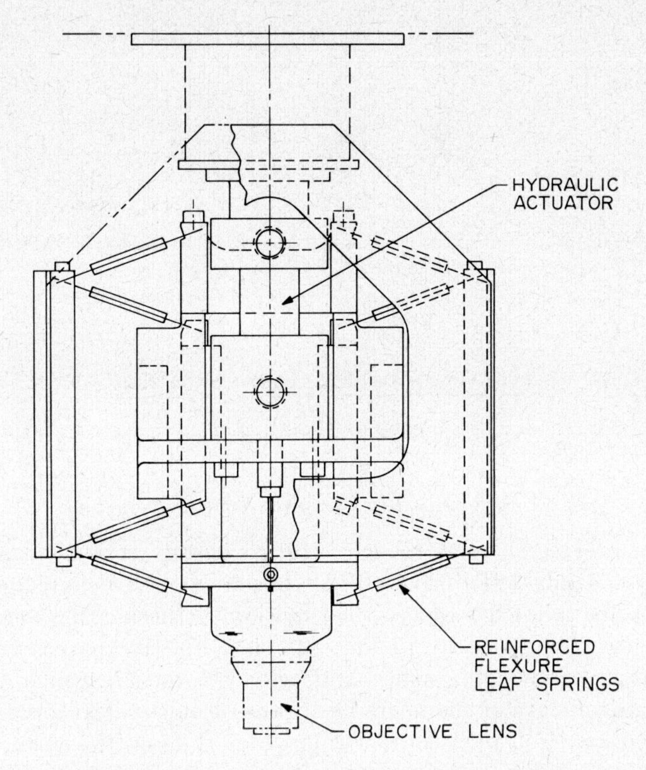

FIG. 31. Flexure system for precise Z-axis movement of objective.

since a hydraulic drive mechanism was chosen which can deliver a force of over 80 kg, this has posed no serious problem. If electromagnetic or piezoelectric drives were to be used a much lighter assembly would be desirable, so that little more than the objective itself would have to be moved. For this case, a cylindrical air bearing could be used to provide low friction and precision axial movement.

Our prototype was designed to use a commercially available pulse width modulation (PWM) electronic-hydraulic servo-control system made by Minneapolis–Honeywell, a schematic diagram of which is shown in Fig. 32. This particular system has a frequency response from 0–80 Hz, and the hydraulic system operates in a pressure range between 1000–3000 psi. This has proved to have both the speed and precision of response necessary for this application.

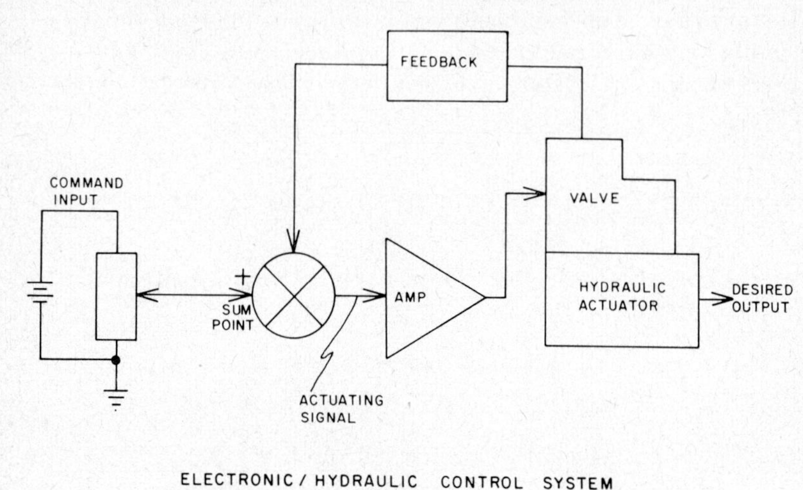

FIG. 32. Block diagram of electronic-hydraulic servo control system.

E. *A Mechanically Coupled Focusing System*

As an interim system, Richard Bing's group at the Huntington Institute of Applied Medical Research has developed a simple mechanical device for following the vertical motion of the beating heart. This is shown diagrammatically in Fig. 33. The objective moves vertically on a system of swinging arms, and is driven in synchrony with the vertical movements of the heart by means of mechanical force derived from the heart itself by a contact ring which rests on the surface of the heart. The lens support is counter balanced so that the pressure on the surface of the heart is kept to a minimum compatible with accurate maintenance of contact. The static pressure exerted on the heart surface is less than $3{\cdot}5\ \mathrm{gm/cm^2}$, which does not observably interfere with capillary flow although the inertial loading is considerably higher. Fine focus is adjusted by means of a screw in a sleeve driven by a chain (Fig. 33). The maintenance of focus is not so good as achievable with a good servo system, but this mechanical device is far less expensive to construct and maintain, and has proved adequate for much exploratory work, although requiring separate bursts of film at different focal positions to obtain data at all parts of the heart cycle.

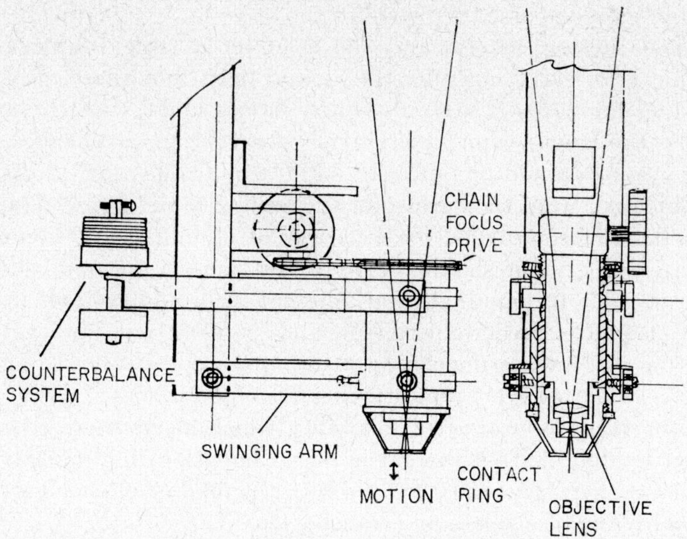

FIG. 33. Mechanical tracking system for heart studies. (Courtesy Richard J. Bing and the Huntington Institute for Applied Medical Research.)

F. *Three Dimensional Stabilization*

Any point on the surface of the heart will, in general, show both lateral and vertical movement, so that correction in three dimensions is desirable. In such a combined system, the vertical sensing and drive can be accomplished as described above. The frequency of heart movements is such that an image dissector tracking system can be used for following the horizontal motion and providing the appropriate command signals. Since image dissectors work best with a target with a sharp edge between dark and light, best results will be obtained if a small, high contrast target can be placed in the microscope field to give the lateral tracking signal.

The main constraint on the design of the motion mechanism is to uncouple the three degrees of freedom so that motion in any one of the three orthogonal directions does not perturb the motion in either of the others, and maintains the optic axis of the objective strictly parallel to its initial direction.

X. SUMMARY

This chapter presents a set of design criteria for a microscope system for use primarily in the study of blood flow in the microcirculation in living animals, although the same criteria are applicable to most microscopic studies of living systems.

The basic design criteria are: the need for a large, stable animal table which can carry not only the animal itself, but those measuring devices and life support systems which should move with the animal whenever the microscopic field under observation is changed. The imaging system should be optically indifferent to the vertical distance of the objective from the surface of the animal table. Fixed mounting stations should be provided for a variety of illuminating systems and selection among them should be easy and quickly accomplished. It must be possible to mount several different recording systems in fixed position prior to a given experiment, and, if more than one system is needed in a given experiment, rapid shift from one system to another must be possible without repositioning or refocusing.

A prototype system of great flexibility which meets these criteria is described in detail. This system is based on using infinity corrected objectives at their front focal planes, with the final image formed in the back focal plane of a telescopic transfer lens.

In addition to the prototype system, a simplified instrument employing the same basic design principles, as well as the application of this same telescopic observation system to a servo-controlled focusing system to maintain sharp focus on moving tissue, such as the beating heart, is described.

Acknowledgements

The author wishes to thank Dr. Wallace G. Frasher for his valuable contribution to the design of the prototype instrument; Dr. Richard J. Bing for both his technical input into the design of the simplified instrument, and for his permission to publish details of his light pipe illumination system and mechanical system for following the beating heart; and Dr. Momtaz Mansour of the Jet Propulsion Laboratory, the project engineer on the development of the servo microscope system, without whose ingenious design capabilities this instrument would never have succeeded. He is also grateful for financial support from the Alfred B. Sloan Foundation and from the National Institutes of Health, US Public Health Service, Grant No. HL 08977.

References

General References

The light microscope in microvascular research *In* "Symposium on Light Microscope in Microvascular Research" (1973). *Bibl. anat.* **11**, 185–243.

"Microcirculatory Research Methods" Proceedings of a Symposium on "The Microcirculatory Approach to Peripheral Vascular Function". *Microvascular Research* **5** (3), 229–435 (1973).

Detailed References

Baez, Silvio (1973). *Microvascular Research* **5,** 384–394.

Frasher, W. G. (1973). *Microvascular Research* **5**, 376–383.

Johnson, P. C. (1973). *Microvascular Research* **5,** 292–298.

Ploem, J. (1967). *Zeits. f. Wiss. Mikroskopie u. mikroskop. Tech.* **68,** 129–142.

Stromberg, D. D. and Shapiro, H. M. (1973). *Microvascular Research* **5,** 410–416.

Wayland, Harold (1973a). *Microvascular Research* **5,** 336–350.

Wayland, Harold (1973b). *Bibl. anat.* **11,** 19–24.

Wayland, Harold and Frasher, W. G. (1973) *In* "Modern Technology in Physiological Sciences" (J. Gross, R. L. Kaufmann and E. Wetterer, eds), pp. 125-153. Academic Press, London.

Non-standard Methods of Phase Contrast Microscopy

M. PLUTA

Central Optical Laboratory, Warsaw, Poland

I. Introduction

A. *Phase and Amplitude Objects*

Objects which change the intensity of transmitted or reflected light are called *amplitude objects*. They are directly observed in a conventional bright-field microscope, as a consequence of modulation in the amplitudes of light waves which produce the image (the intensity of a light wave is proportional to the square of its amplitude).

Objects which do not change the light intensity, but only shift the phase of the light wave are called *phase objects*. They are normally invisible because the human eye, as well as other light receptors, is insensitive to changes in the relative phase shifts of light waves. For transparent specimens the phase shifts are mainly due to different refractive indices, and/or varying thickness of the objects and their surrounding medium. In particular, living cells and organelles usually belong to this class of object.

Consider a parallel light beam with the plane wavefront $\sum$ (Fig. 1) passing through a transparent particle P of refractive index n_p and thickness t. This object is surrounded by a medium M having equal transparency but different refractive index n_m. If $n_m < n_p$ (Fig. 1(a))

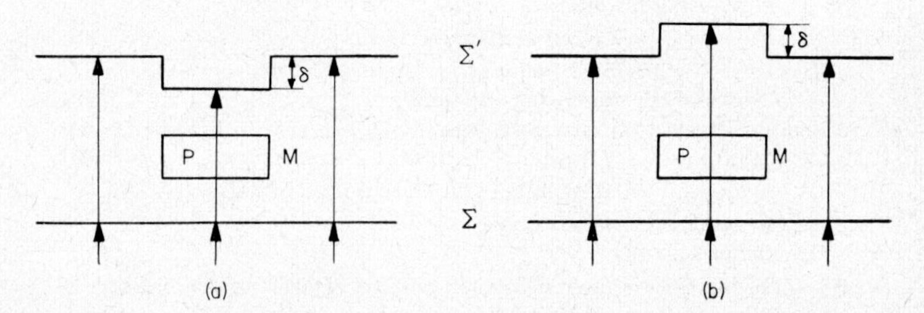

Fig. 1. Phase-retarding (a) and phase-advancing (b) objects.

the wavefront $\sum'$ immediately above the object P is retarded in phase by an amount

$$\phi = 2\pi\delta/\lambda, \text{ expressed in terms of radians} \tag{1a}$$

or

$$\phi = 360°\delta/\lambda, \text{ expressed in terms of degrees of arc,} \tag{1b}$$

where λ is the wavelength, and δ is the optical path difference (OPD) defined as

$$\delta = (n_m - n_p)t. \tag{2}$$

Conversely, when $n_m > n_p$ the wavefront is advanced in phase (Fig. 1(b)).

Consequently, objects with $n_p > n_m$ are called *phase-retarding objects*, and physically they give a negative phase shift ($\phi < 0$), whereas those with $n_p < n_m$ are called *phase-advancing objects*, and they produce a positive phase shift ($\phi > 0$).

In the case of light-reflecting specimens the phase shifts of the wavefront are, in particular, due to microsteps and other variations in specimen flatness. Depressions behave as phase-retarding objects, elevations as phase-advancing objects.

In reality, there are no ideal amplitude or phase objects. Real specimens show, to some degree, both amplitude and phase properties. However, it is convenient to categorize all microscopic specimens into one or the other group, according to whether amplitude or phase changes of light are predominant.

B. *Image Contrast*

Optical contrast is produced by differences in light absorption or reflection, spatial variations in refractive indices, light scattering, and other phenomena. If the intensity of light is reduced by neutral absorption *brightness contrast* occurs. The same kind of contrast is created by interference of monochromatic light and, in some circumstances, white light.

Brightness contrast (in brief, contrast) specifies the differences in light intensities between the object and its surround. Actually, this quantity is defined in various ways, but for the purpose of phase contrast microscopy the following definitions are the most convenient

$$C = \frac{I_b - I_p}{I_b} \tag{3}$$

$$C = \frac{I_b - I_p}{I_b + I_p} \tag{4}$$

where I_b is the light intensity (or brightness) of the background, and I_p is that of the object image. If the image is darker than the background ($I_p < I_b$) then the contrast is positive ($C > 0$), and conversely when the background is darker than the object image ($I_p > I_b$) the contrast is negative ($C < 0$).

If the light intensity is changed by selective spectral absorption or reflection, which modifies the colour of light, then *colour contrast* occurs. This kind of contrast is also produced by interference of white light. The definition and measurement of colour contrast are more complicated than those of brightness contrast. In relation to colour phase contrast methods this problem is discussed in detail by Beyer (1965).

C. *Zernike's Phase Contrast Method*

1. *General principle and typical realization*

Broadly speaking, the phase contrast method discovered by Zernike (1935) enables invisible optical path differences or phase shifts occurring in the specimen to be transformed into visible differences of light intensity in the image plane of the microscope.

A typical phase contrast system for transparent specimens is not far removed from a conventional bright-field microscope. Only an annular diaphragm (AD) (Fig. 2) and ring-shaped phase plate (Ph) are arranged

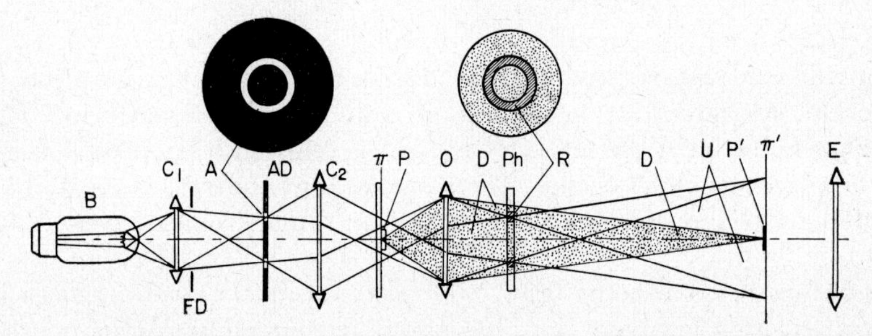

Fig. 2. Typical system for phase contrast microscopy.

in or near the focal planes of the condenser C_2 and objective O, respectively. The light issuing from the transparent annulus A of the condenser diaphram is diffracted by a transparent particle P under examination, placed in the object plane π of the objective. The diffracted light D does not pass through the phase ring R but through its surrounding areas in the exit pupil of the objective. The portion of this pupil occupied by the image of the condenser annulus is called the *conjugate area*, whereas the remaining portion is called the *complementary area*. The former is completely covered by the phase ring R which changes the phase of the direct undiffracted light U with respect to the diffracted light by $+90°$ or $-90°$. The phase contrast image P′ appears in the image plane π' as a result of interference between the direct and diffracted light beams, and it is observed through an eyepiece E. For most specimens the intensity of diffracted light is low, and there is a great difference in the amplitudes of these two interfering beams; in order to make their interference more effective the intensity of the direct light is reduced by adding an absorbing thin film to the conjugate area. In typical phase contrast devices this reduction is 75–90%. For phase contrast observation a Köhler illuminator consisting of a low-voltage bulb B, collector C_1, and field diaphragm FD is commonly used.

The phase plates are nowadays made by vacuum deposition of thin dielectric (DF) and metallic (MF) films onto a glass plate (Fig. 3), or

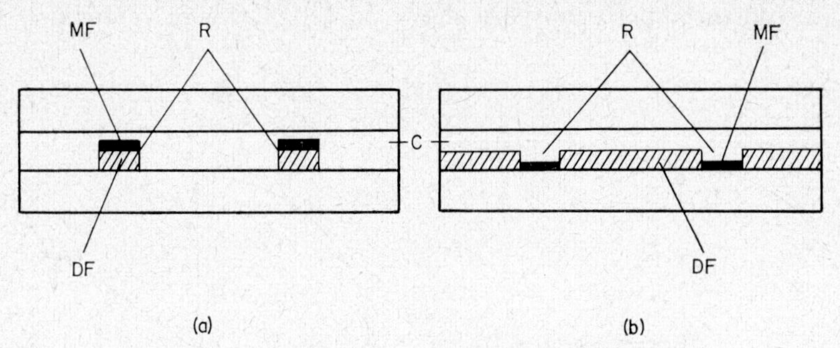

FIG. 3. Typical phase plates ((a) is more useful than (b)).

directly onto one of the lenses of the microscope objective. The dielectric film shifts the phase of the light, whereas the metallic film attenuates its intensity. The phase ring R is usually glued between two glass plates (or two lenses) by means of a suitable optical cement C. The thickness t and refractive index of the dielectric film, as well as the refractive index of the cement, are chosen so as to obtain a required phase shift ψ between the complementary and conjugate areas. This phase shift is defined as

$$\psi = 2\pi\frac{\delta}{\lambda} = \frac{2\pi}{\lambda}(n_c - n_r)t \tag{5}$$

where n_c and n_r are the refractive indices of the complementary and conjugate areas, respectively. The metallic film usually introduces a small additional phase shift which can be neglected or compensated by using a suitably thicker or thinner dielectric film than calculated from Eqn (5). The phase plate is positive ($\psi > 0$) if $n_c > n_r$. Then the direct light is advanced in phase compared to the diffracted one, and, vice versa, the phase plate is negative ($\psi < 0$) if $n_c < n_r$, when the direct light is retarded in phase relative to the diffracted light. Phase plates which change the phase of the direct light relative to the diffracted one by $+90°$ or $-90°$ will be called *quarter wavelength phase plates* because they produce an OPD equal to $\lambda/4$.

2. *Interpretation*

For a general insight into the physical principles and some practical aspects of phase contrast a vector representation is generally used. It has been formulated, in the most detailed manner, by Barer (1952a, e, 1953b, 1954, 1955). According to this representation a light wave is described by a vector whose length denotes the wave amplitude, and direction corresponds to the angular phase shift.

Vector diagrams illustrating the most typical situations occurring in phase contrast microscopy are shown in Fig. 4, where vector **p** = OQ

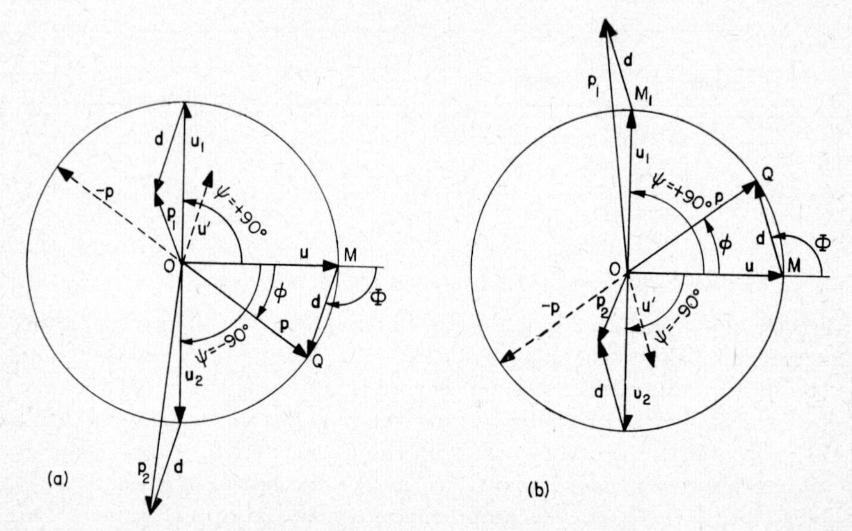

FIG. 4. Vector representation of the phase contrast method: (a) phase-retarding object; (b) phase-advancing object. Broken vectors apply to the theory of Nomarski's variable phase contrast system (see Section IV.B).

corresponds to the light passing through a phase object under examination, vector **u** = OM represents the light passing only through the surrounding medium, and vector **d** = MQ corresponds to the diffracted light.

If the refractive index of the object is greater than that of the surrounding medium then the object introduces a negative phase shift ϕ. This corresponds to the rotation of the vector **p** relative to **u** through the angle ϕ in the clockwise (negative) direction (Fig. 4(a)). When the object is less refractile than its surround the phase shift ϕ is positive, and the rotation of the vector **p** is counter-clockwise, i.e. positive (Fig. 4(b)).

Referring to the image plane of the microscope, the lengths of the vectors **u** and **p** define the amplitudes of light from the background and object image, respectively. If the transmittance of the object is the same as that of the surrounding medium, the vectors **u** and **p** have the same length, thus the intensity within the object image is equal to the background intensity, and the object is invisible. According to the diffraction theory of image formation in the microscope, the vector $\mathbf{p}$ can be treated as the sum of vectors $\mathbf{u}$ and $\mathbf{d}$. The latter is shifted in phase relative to $\mathbf{u}$ by an angle Φ equal to $90°+\phi/2$ (Fig. 4(b)) or $-90°+\phi/2$ (Fig. 4(a)). For phase objects producing very small phase shifts ϕ the angle $\Phi \approx \pm 90°$.

The phase contrast method enables the direct and diffracted light waves to be separated, so that one of these waves can be altered in phase and/or amplitude without affecting the other. This is done by means of a condenser diaphragm and phase plate. As mentioned, in a typical phase contrast microscope the direct light is shifted in phase by an angle $\psi = +90°$ or $-90°$. Then, for $\psi = +90°$ the vector $\mathbf{u}$ is transformed into $\mathbf{u}_1$. This latter and $\mathbf{d}$ give a new resultant vector $\mathbf{p}_1$, which is shorter than $\mathbf{u}_1$ for phase-retarding objects (Fig. 4(a)), and longer for phase-advancing objects (Fig. 4(b)). Consequently, these latter objects appear brighter, and the former darker than the background. If a $-90°$ phase plate is used the vector $\mathbf{u}$ transforms into $\mathbf{u}_2$. Now, the image of a phase-retarding object is brighter, whereas that of a phase advancing object is darker than the background.

Let the image contrast be defined by formula (1), then

$$C = \frac{|\mathbf{u}_1|^2 - |\mathbf{p}_1|^2}{|\mathbf{u}_1|^2} \quad \text{for } +\text{ve phase plate} \tag{6a}$$

and

$$C = \frac{|\mathbf{u}_2|^2 - |\mathbf{p}_2|^2}{|\mathbf{u}_2|^2} \quad \text{for } -\text{ve phase plate.} \tag{6b}$$

Calculating $|\mathbf{p}_1|^2$ and $|\mathbf{p}_2|^2$ by trigonometry, and assuming $|\phi| \ll 90°$, the following approximate values for image contrast can be obtained

$$C = -2\phi \quad \text{for } +90° \text{ phase plate} \tag{7a}$$

and

$$C = +2\phi \quad \text{for } -90° \text{ phase plate.} \tag{7b}$$

As mentioned previously, the image contrast of slightly dephasing objects can be magnified by weakening the intensity of the direct light. In the vectorial diagrams (Fig. 4) this corresponds to the diminution of the length of the vectors $\mathbf{u}_1$ and $\mathbf{u}_2$. Let N be the factor by which the intensity of the direct light is reduced. Then Eqns (7) become

$$C = -2\phi\sqrt{N} \quad \text{for } +90° \text{ phase plate,} \tag{8a}$$

$$C = +2\phi\sqrt{N} \quad \text{for } -90° \text{ phase plate.} \tag{8b}$$

Up to now it has been assumed $|\phi| \ll 90°$. Taking into account phase objects giving larger phase shifts ϕ, the relationship between the contrast C and ϕ is not linear, but resembles a sine function. For non-absorbing quarter wavelength phase plates the relation $I_p(\phi)$ extended over the whole periods of phase angle $\phi = 0$ to $\pm 360°$ is shown in Fig. 5. As seen, both positive and negative phase plates can give positive

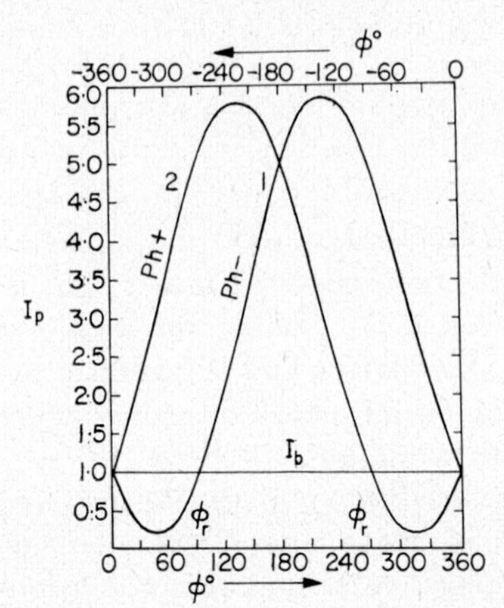

FIG. 5. Relationship between the phase shift ϕ produced by phase-advancing ($\phi > 0$), or phase-retarding ($\phi < 0$), objects and intensity I_p of their images observed by phase contrast with a non-absorbing quarter-wave-length phase plate. Curve 1 relates to the negative phase plate, and curve 2 to the positive phase plate (based on Barer, 1952a, 1952e, 1953b).

or negative image contrast depending on the value of ϕ. Following ϕ, from zero to $\pm 360°$, the difference $I_b - I_p$ between background and image intensities reaches a maximum, then diminishes, and for some value $\phi = \phi_r$ it becomes zero, and then its sign changes. The phase shift ϕ_r defines the range of unreversed contrast. For phase-retarding objects ($n_p > n_m$, $\phi < 0$), $\phi_r = -90°$ and $-270°$ for positive and negative quarter wavelength phase plates, respectively. In the case of phase-advancing objects ($n_p < n_m$, $\phi > 0$) the situation is contrary. Values of ϕ_r are $+270°$ (positive $\lambda/4$ phase plate) and $+90°$ (negative $\lambda/4$ phase plate).

If the phase plates absorb some direct light, the ranges of unreversed contrast diminish as N increases. This is illustrated by Fig. 6. for phase-retarding objects and positive $\lambda/4$ phase plate.

Vector diagrams (Fig. 4) show that for each OPD in the specimen there is an optimum value ψ_{op} of phase shift between the direct and diffracted light which gives the maximal or optimum image contrast. This value is that for which one obtains parallelism and antiparallelism between vectors $\mathbf{d}$ and $\mathbf{u}_1$ or $\mathbf{u}_2$. Such a situation satisfies, in fact, the conditions for the most effective light interference, and it occurs when

$$\psi = \psi_{op} = \pm 90° + \phi/2 \tag{9}$$

where signs "+" and "−" relate to the positive and negative phase plates, respectively.

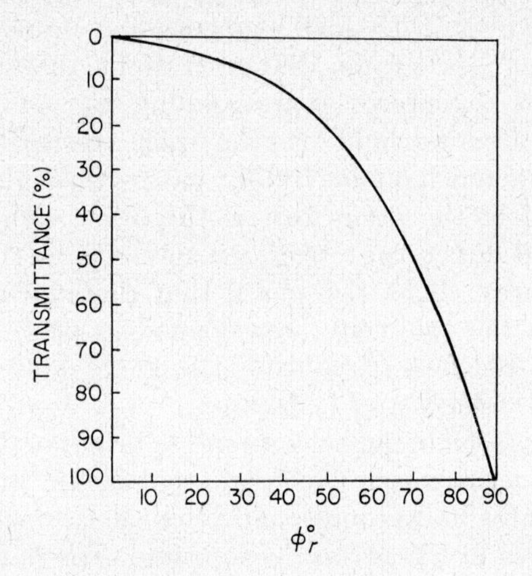

FIG. 6. Relationship between the transmittance of a positive quarter-wavelength phase plate and the range (ϕ_r) of unreversed contrast for phase-retarding objects (based on Barer, 1952a, 1952e).

Analysis of the vectorial diagrams also shows that for a given phase shift ϕ in the specimen there is an optimum value of transmittance T of the phase plate. This value is equal to

$$T = \frac{1}{N} = 4\sin^2 \frac{\phi}{2} \tag{10}$$

or

$$T = \frac{1}{N} = 4\cos^2 \psi_{op} \tag{11}$$

where ψ_{op} is the optimum phase shift of the phase plate (Eqn 9). When $N > 1$ the direct light is reduced, and if $N < 1$ the diffracted light should be weakened. Relations (9) to (11) are usually called Richter's condition (Richter, 1947; Barer, 1952e, 1953b, 1955).

3. *Imaging properties*

So far it has been assumed that there is complete separation of the direct and diffracted light. In principle this condition could be approached if both the clear area of the condenser diaphragm and its conjugate area on the phase plate were made extremely small. In practice, these elements must have a finite size and suitable shape which are determined by various factors such as adequacy of illumination, light coherence, image quality and resolution. The annular diaphragm and ring-shaped phase plate have evolved as a practical compromise. However, with such elements one cannot separate completely the direct light from diffracted light, because microscopic specimens include structures corresponding to various spatial frequencies which diffract the light over the whole aperture of the objective. Some of the diffracted light inevitably passes through the phase ring and is modulated in the same way as the direct light. On the other hand some of the direct light tends to spread into the region of the complementary area. Thus the phase ring should be wider (usually about 25%) than the conjugate area defined by the image of the condenser annulus. All these circumstances create optical artefacts, so called "*halo*" and "*shading-off*" effects.

The halo effect is seen in the background as bright or dark zones which surround the object images. The bright haloes are attached to dark images in the lighter background, and, vice versa, dark haloes accompany bright images in the darker background. The dark haloes are, in general, less disturbing that the bright ones.

The shading-off effect occurs in the object image and exhibits itself as brighter and darker central areas in dark and bright images, respectively. This effect causes the distribution of light intensity in the image not to correspond to the distribution of OPD in the object. This can best be illustrated by means of a thin phase object of uniform OPD, as in Fig. 7. If there were no shading-off and halo effects the image of the particle P appears uniformly dark or bright as shown in the top of Fig. 7(a) and (b), respectively.

Taking into account the halo and shading-off effects, one must distinguish four different light intensities, I_b, I_h, I_p and I_s, related to the far background, halo region near to the object image, object image near the edge and centre of the shading-off area, respectively. The visibility of the phase object is mainly defined by the edge contrast, that is by the difference in intensities I_h and I_p, whereas the difference in intensities I_p and I_s, as well as, to some degree, that between intensities I_b and I_h, characterizes *image faithfulness.* This feature can be quantitatively evaluated by means of a suitably constructed specimen containing thin strips of a transparent material, e.g. vacuum evaporated onto a glass slide.

The halo and shading-off effects are in general harmful, and together with the non-linearity between the image intensity and OPD in the specimen, make a correct interpretation of images of more complicated objects difficult. On the other hand, the halo often emphasizes the contrast between the object image and its background, and accentuates thin edges and border details of some objects, e.g. biological cells. This relates, in particular, to negative phase contrast, when a dark halo appears around the image detail.

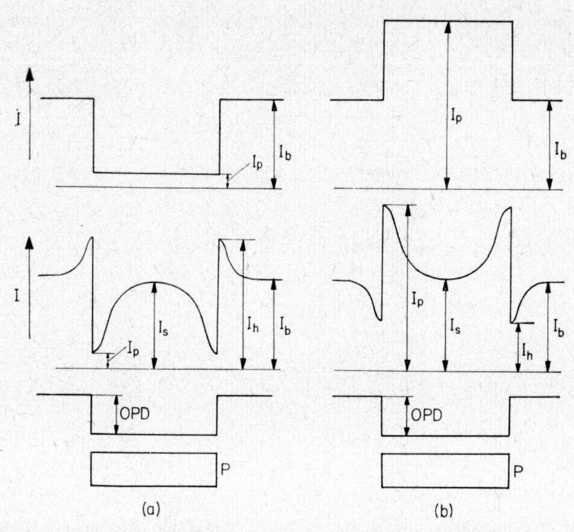

FIG. 7. "Halo" and "shading off" effects: (a) positive (dark) phase contrast image; (b) negative (bright) phase contrast image.

For phase objects of given OPD and dimensions the halo and shading-off effects depend on both the geometrical (width, diameter), and optical (absorption, phase shift) properties of the phase plate, as well as on the state of correction of the objective aberrations. In general, these effects are the more obtrusive the wider is the phase ring, and the greater are its diameter and light absorption, whereas moderate alterations in phase shift ψ are unimportant. For a given phase plate the halo and shading-off effects depend on both optical (refractive index, absorption) and geometrical (thickness, size, shape) properties of the objects. Moreover, these effects, and especially the shading-off, greatly depend on the magnifying power of the objective. This and other factors are illustrated by pictures and microdensitometric graphs in Figs 8–11. As seen, the faithfulness of imaging is better the smaller are the width and diameter of the phase ring, and the smaller in size is the object. In these conditions, the approximate vector theory of phase contrast can be applied.

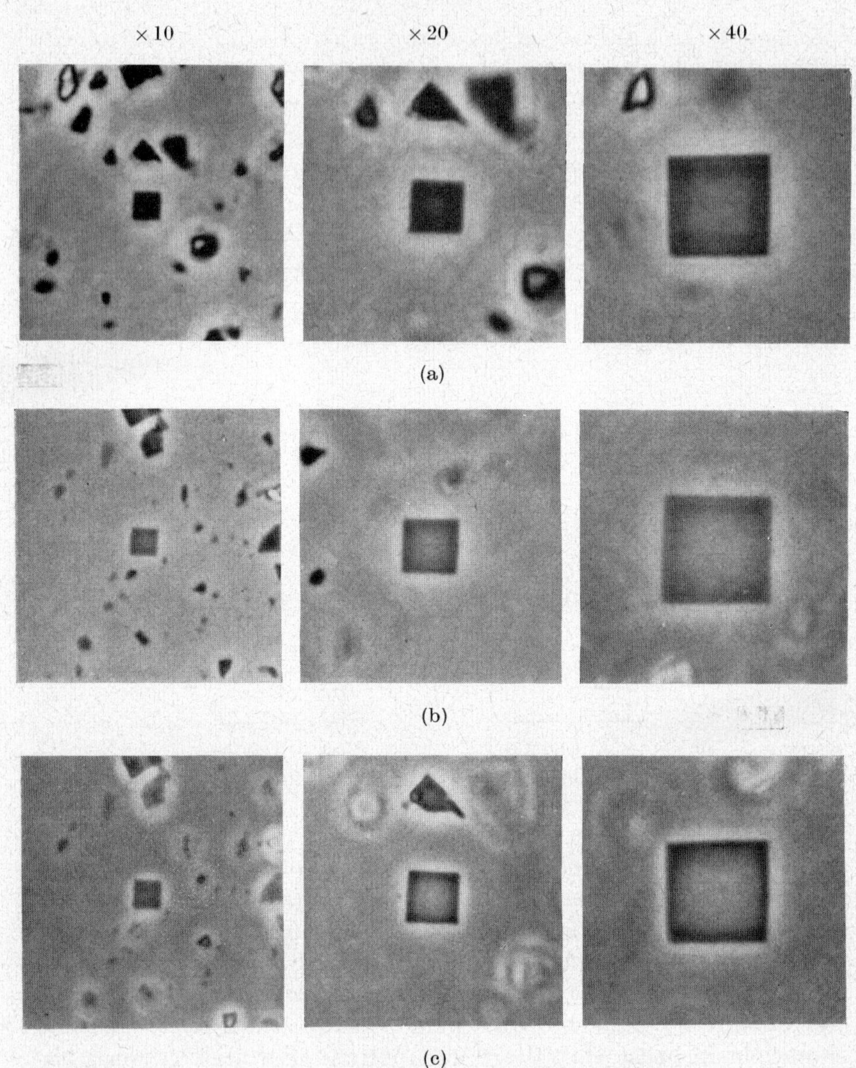

FIG. 8. A sequence of photomicrographs showing the dependence of the halo and shading-off effects on the geometrical and optical properties of the phase plate. The photomicrographs are taken with three different sets, (a), (b), and (c), of positive phase objectives of magnification ×10, ×20, and ×40. The phase rings in these objectives had the following characteristics: (a) $\psi = +90°$, $T = 15\%$, $d/D = 0{\cdot}28$, $w/D = 0{\cdot}045$, $a/A = 0{\cdot}055$; (b) $\psi = +90°$, $T = 25\%$, $d/D = 0{\cdot}55$, $w/D = 0{\cdot}07$, $a/A = 0{\cdot}14$; (c) $\psi = +90°$, $T = 15\%$, $d/D = 0{\cdot}5$, $w/D = 0{\cdot}1$, $a/A = 0.17$: where ψ, T, d, w, and a are the phase shift, transmittance, diameter, width, and area of the phase ring, respectively; D and A are the diameter and area of the exit pupil of objective, respectively. Square flat object in the centre of the photomicrographs has the dimensions $0{\cdot}01 \times 0{\cdot}01$ mm, and phase shift between light passing through this object and its surrounding medium is about $-8°$.

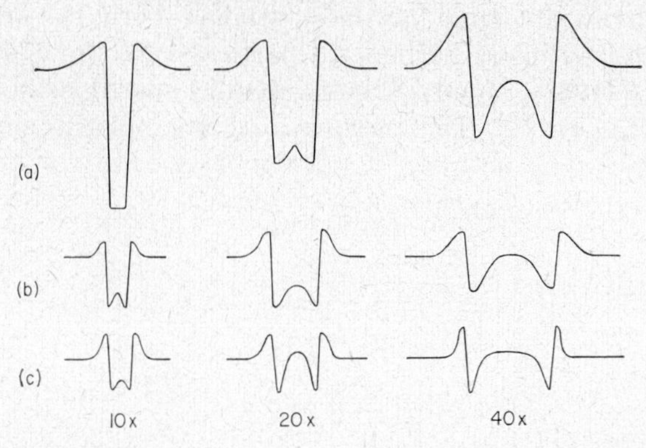

FIG. 9. Microdensitometric graphs across the images of the square object in Fig. 8. The curves (a), (b), and (c) relate to the photomicrographs (a), (b), and (c) in Fig. 8, respectively.

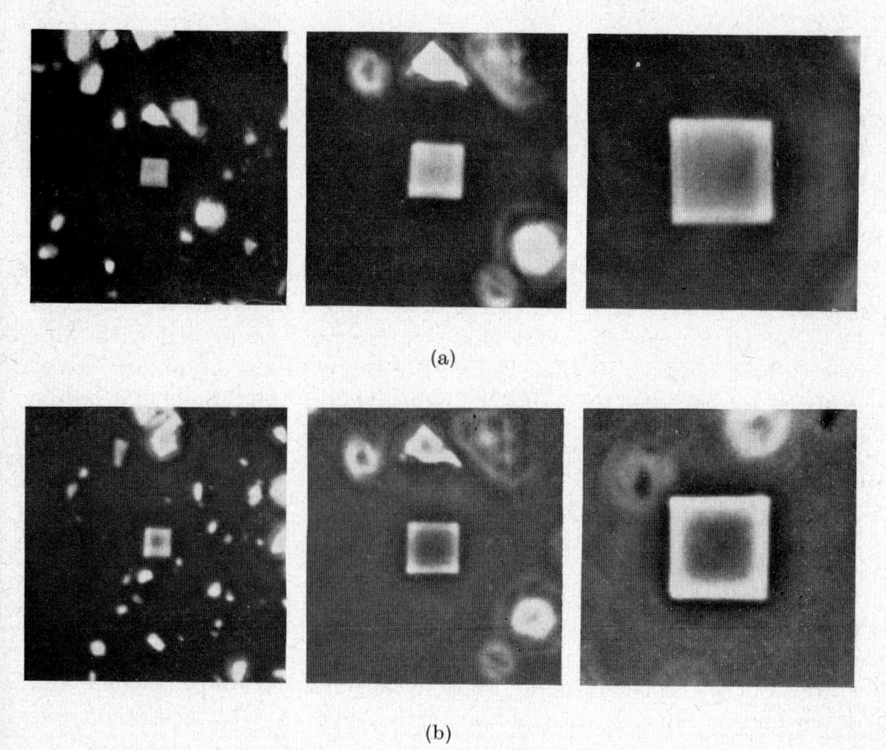

(a)

(b)

FIG. 10. As in Fig. 8, but photomicrographs are taken with objectives having negative phase rings made of soot, the parameters of which are as follows: (a) $\psi = -90°$, $T = 3$ to 4%, $d/D = 0{\cdot}55$, $w/D = 0{\cdot}065$, $a/A = 0{\cdot}13$; (b) $\psi = -90°$, $T = 3$ to 4%, $d/D = 0{\cdot}5$, $w/D = 0{\cdot}1$, $a/A = 0{\cdot}17$.

Phase contrast imaging has been studied both theoretically and experimentally by many authors, in particular Wolter (1950), Jupnik (1951), Beyer (1953), Mondal (1969), Mondal and Slansky (1970), and Khan and Rao (1972). The pictures and microdensitometric graphs

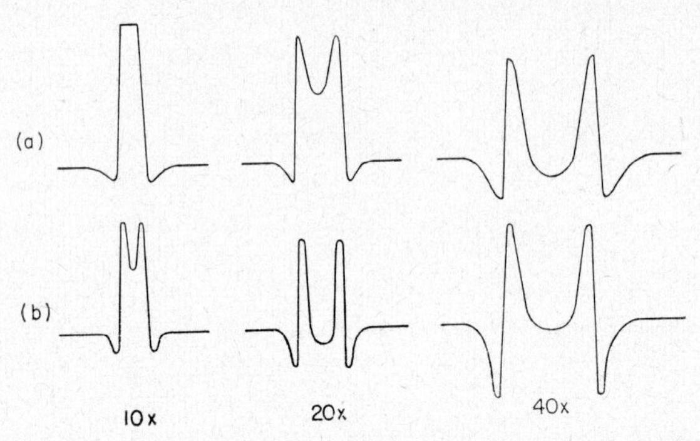

FIG. 11. Microdensitometric graphs across the images of the square object in Fig. 10. The curves (a) and (b) relate to the photomicrographs (a) and (b) in Fig. 10.

shown in Figs 8–11 are from experimental work by the present writer: the purpose was to study the dependence between image contrast and OPD for different commercially available phase contrast devices.

4. *Sensitivity*

One of the most important aspects of the phase contrast method is its sensitivity, defined as the smallest optical path difference δ_m that can be detected by the observer. Here we are only concerned with very small phase shifts ϕ so that Eqns (7) apply and the minimum image contrast (or contrast threshold) C_m perceived by the human eye can be expressed as

$$C_m = 2\phi_m N^{\frac{1}{2}} = 2\frac{2\pi}{\lambda}\delta_m N^{\frac{1}{2}},$$

and thus

$$\delta_m = \frac{C_m \lambda}{4\pi N^{\frac{1}{2}}}. \tag{12}$$

The contrast threshold C_m depends on the shape and size of the object, brightness of the background, state of accommodation of the eye, and other factors. In the case of two areas separated by a straight-line boundary $C_m = 0{\cdot}02$ is usually accepted. Then for a non-absorbing phase plate ($N = 1$) $\delta_m = 10$ Å, and for a phase plate which absorbs 96% ($N = 25$) of the direct light, $\delta_m = 2$ Å. In favourable circumstances these theoretical limits are achieved in practice.

D. *Nomenclature*

The nomenclature of phase contrast microscopy is not uniform and results in some confusion about the optical properties, capabilities and commercial specifications of phase contrast devices. This confusion results from (1) different definitions of image contrast, (2) various expressions for OPD or phase shifts, and (3) different descriptions of phase plates. The basic nomenclature used in this article is summarized in Table I. It is mostly in agreement with the nomenclature of many other authors, and in particular with that proposed by Zernike (1958).

All phase contrast devices having objectives with quarter wavelength phase plate decreasing the intensity of the direct light to the range of about 25% to 10% will be designated as standard phase contrast devices. They are commercially manufactured by many optical firms in Europe, USA and Japan as equipment for typical biological microscopes. Each comprises a set of phase objectives with fixed positive or negative phase rings, a phase condenser with annular diaphragms arranged in a turret (this condenser usually serves for bright-field illumination as well), and an auxiliary telescope for observing the objective exit pupil while centring the image of the condenser annulus on the phase ring. Standard phase contrast devices also include equipment for metallurgical microscopes which have quarter wavelength phase rings arranged in a slide or turret behind the objective, as well as devices for inverted biological microscopes which need long-working-distance phase condensers.

Consequently, non-standard phase contrast devices comprise all other special systems, and some which resemble standard ones but incorporate non-typical phase rings (e.g. highly or weakly absorbing, with small or zero phase shift), and phase plates termed by Bennett *et al.* (1951) B-type (intensity of diffracted light is reduced relative to that of the direct light), as well as some devices incorporating non-ring-shaped phase plates.

Many different non-standard phase contrast systems are described in scientific or technical journals, as well as in patents, and it is impossible to present all of them in an article of limited scope. Thus, in the first place a survey will be given of some non-standard devices which are, or have been, commercially available. Some were developed in the author's laboratory: these will be described more fully though devices developed by others will also be discussed.

TABLE I

Basic nomenclature and relations between optical properties of the object, type of phase plate, and image contrast

Object		Phase plate		Image contrast	
Relation between n_p and n_m	Sign of phase shift ϕ	Relation between n_c and n_r	Sign of phase plate	Relation between I_p and I_b	Sign of contrast
$n_p > n_m$	negative (−)	$n_c > n_r$	positive (+)	$I_p < I_b$	positive (+)
		$n_c < n_r$	negative (−)	$I_p > I_b$	negative (−)
$n_p < n_m$	positive (+)	$n_c > n_r$	positive (+)	$I_p > I_b$	negative (−)
		$n_c < n_r$	negative (−)	$I_p < I_b$	positive (+)

n_p and n_m, refractive indices of the object and its surrounding medium, respectively;
n_c and n_r, refractive indices of the complementary and conjugate areas of the phase plate;
I_p and I_b, light intensity (or brightness) of the object image and its background.

II. Phase Contrast Devices With Soot Phase Rings

A. *Historical Background*

In the late nineteenth century, when Abbe was establishing the diffraction theory of image formation in the microscope, Bratuscheck (1892), one of his co-workers, was on the verge of discovering phase contrast. For certain diffraction experiments he introduced into the back focal plane of the objective an absorbing diaphragm made of a soot layer, which reduced the intensity of the zero-order diffraction spectrum produced by a grating consisting of alternate clear and slightly absorbing strips. The strips were recessed in a very thin film of soot deposited on a glass slide, and they were almost invisible in the ordinary microscope. The visibility of these strips was greatly improved by inserting the absorbing objective diaphragm. Thus, Bratuscheck proved that by weakening the direct light the image contrast of this almost transparent specimen could be much increased.

Almost 50 years later Wilska carried out experiments on the application of the "Schlieren" method to the microscopy of living cells. He obtained good results by introducing an objective aperture made of a 50% transmission stearin soot layer, containing a clear annulus conjugate with an annular condenser aperture. In this way the intensity of the diffracted light was reduced by 50% relative to the direct light. Highly refractile living objects such as yeast cells gave clear, well defined images. Moreover these had a pleasing brownish tint against a bright background. The images were similar to those obtained with a Zeiss positive phase contrast microscope but they lacked the disturbing bright halo seen by phase contrast.

However, Wilska's first system was only suitable for observation of highly refracting objects.

In later experiments Wilska (1953a, 1953b) stated that if the coating of soot in the objective was ring-shaped, images with reversed contrast appeared, resembling those obtained with negative phase contrast. With such objectives the best results were obtained when the absorption of the soot rings was increased to about 90%. The images of slightly phase retarding objects were of good contrast with an agreeable golden brown tint.

A soot layer applied to a glass surface by exposing a plate or lens to a flame is not resistant to abrasion and can be easily damaged; as Wilska (1954) himself stated, objectives with soot rings are not suitable for commercial production. Wilska's experiments were, however, partially commercialized by C. Reichert of Vienna, as the so-called

"Anoptral Contrast Equipment" (Wilska, 1954). Instead of soot, highly-absorbing negative phase rings, made of dielectric and metallic substances evaporated in vacuum onto a lens surface, were applied. The ring lay between two antireflecting thin films (Françon, 1961). Such a non-reflecting (anoptral meaning non-reflecting) phase ring produces nearly the same effect as the soot ring in Wilska's second system. In reality, the Anoptral contrast equipment is a negative phase contrast device which produces high image contrast with golden-brownish background.

Wilska's experiments were successfully repeated and extended by Peschkov (1955a, 1955b). In 1956 the present writer undertook experiments to find a suitable industrial process for making soot phase rings. First it was necessary to find a method of hardening and fixing soot layers on glass surfaces. It was discovered that this is attained by moistening with a wetting liquid such as alcohol or ether. When the liquid evaporates, the soot layer becomes harder, adheres to the glass surface better and can be recessed with great accuracy by means of a suitable cutting tool; most important of all, a ring made of such a soot layer can be cemented without damage. Finally, as a result of this procedure and studies of the optical properties of soot layers, there were designed two highly sensitive phase contrast devices with positive and negative soot phase rings (Pluta, 1958, 1967b), and an amplitude contrast device (Pluta, 1968a), which will be described later.

B. *Optical Properties of Soot Layers*

A thin layer of soot obtained by exposing a glass slide to the flame of a stearin candle, changes both the amplitude and phase of the transmitted light. Light passing through a strip of a partially transparent layer of soot is considerably retarded in phase relative to light passing outside the strip through the surrounding air. On the analogy of a transparent object, we may write $\delta = (n - n_s)t$, where n is the refractive index of the medium, n_s is the effective refractive index of the soot strip, and t its thickness. The optical path difference δ depends, in general, on the light transmittance T of the soot layer. This dependence has been measured by using a double refracting interference microscope (Pluta, 1962, 1967a). The results for soot layers obtained from the flames of illuminating gas, stearin and kerosene are presented in Fig. 12. Curve 2 shows that for the stearin soot layer of transmittance about 10%, the optical path difference $\delta = -0{\cdot}25\lambda$. Hence, a ring made of stearin soot layer absorbing 90% of the light constitutes a negative phase plate, which shifts the phase of the transmitted light by $-90°$. This satisfies the condition of optimal negative phase contrast for slightly refractile phase objects, and this clearly explains Wilska's second method. Objectives with highly absorbing soot rings realized by Wilska were simply negative phase contrast objectives which shifted the phase of the direct

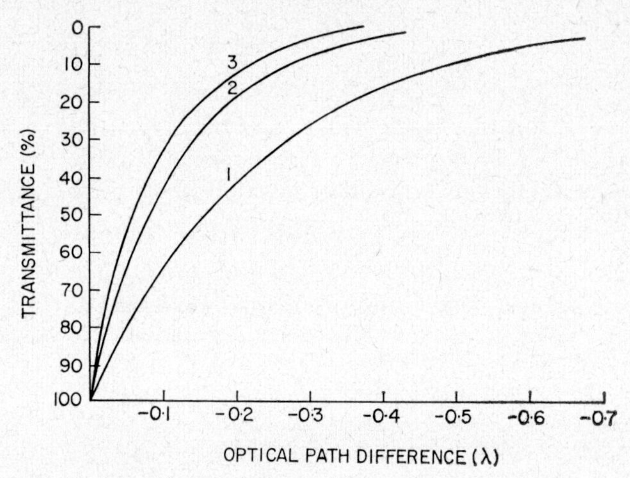

FIG. 12. Optical path difference (measured in relation to air medium) as a function of light transmittance for soot layers obtained by exposing a glass plate to flames of illuminating gas (curve 1), stearin (curve 2), and kerosene (curve 3).

light by about $-90°$. Graph 2 in Fig. 12 also shows that for the stearin soot layer of transmittance 50%, the optical path difference δ is equal to about $-0{\cdot}08\lambda$. Thus, the annular diaphragm opening in Wilska's first experiments changed the phase of the diffracted light relative to the direct light by about $-30°$, or advanced the phase of the direct light relative to the diffracted one by $+30°$. This was, of course, a positive phase contrast of type B, according to the notation proposed by Bennett *et al.* (1951). Thus, by measuring the optical path difference of soot layers versus the transmittance it was possible to explain precisely the principles of both Wilska's methods. These principles are the same as for Zernike's phase contrast method. Initially Wilska's method was interpreted mainly on the basis of amplitude or stop contrast (Barer, 1953a; Peschkov, 1955a, 1955b; Loveland, 1970). This was a half-truth, because the relative phase shift between the direct and diffracted light is of great importance in this method as well.

Using the interferometric technique of double immersion medium, the thickness t and effective refractive indices n_s of various soot layers differing in transmittance of light were determined. Then the dependence between the optical density D (or log $1/T$) and thickness t was examined, and it was established that the functions $t(D)$ and $\delta(D)$ are linear (at least over the density interval $D = 0{\cdot}1$ to $1{\cdot}1$) for different soot layers. From this statement it follows that Lambert's law, $T = e^{-kt}$ is obeyed by such layers. From these facts, the following relationship has been established

$$\delta = -2.3(n-n_s)\frac{\log T}{k}. \tag{13}$$

Knowing the absorption coefficient k, and refractive indices n_s and n, one can calculate δ for a given T. The determined refractive index n_s and absorption coefficient k for different soot layers are given in Table II.

TABLE II

Refractive index n_s and absorption coefficient k of soot obtained from the flames of stearin, kerosene and illuminating gas (for wavelength $\lambda = 546$ nm)

Type of soot	n_s	k 10^5 cm^{-1}
stearin	2·32	2·01
kerosene	2·26	2·37
illuminating gas	2·12	0·95

C. *Manufacture of Durable Soot Phase Rings*

Soot applied to a glass surface by exposing a plate or lens to a flame is not resistant to abrasion, adheres poorly and in the natural state is not suitable for practical application, particularly under production conditions. However, as already mentioned, the durability of the soot can be improved by wetting with alcohol or ether. When the alcohol has evaporated a soot ring is made using a special lathe. The most important element of this lathe is the cutting tool which is made of soft metal, e.g. copper, and mounted on an articulated joint. The cutting tool penetrates the soot layer under its own weight of a few grams only. The glass plate (or lens) with deposited soot layer is glued by a special plasticine to a small electric motor spindle rotated about a vertical axis. The worked soot layer is observed through a measuring microscope with a micrometer eyepiece. The lathe cuts soot rings of very regular contour and reproducible dimensions with an accuracy of 0·01 mm or better. The most important step in this procedure is hardening of the soot layer with alcohol. This makes it possible to cut soot rings with very sharp edges, and to cement them between two glass surfaces without damage.

Figure 13 shows the edges of two rings made of a hardened (a) and not hardened (b) stearin soot layer. In (b) the contour of the soot ring is irregular and some flakes of soot loosened by the cutting tool have fallen on the ring area and adhere there. In addition, there appears, in the vicinity of the principal irregular contour, an additional contour seen as

(a)

(b)

FIG. 13. Fragments of two soot ring contours: (a) soot hardened with alcohol, and (b) not hardened. The photomicrographs were taken with an objective $\times 20/0{\cdot}4$. Print magnification $\times 350$.

a fine line. None of these defects result when the soot layer is hardened with alcohol (Fig. 13a). The optical properties, refractive index and absorption coefficient, of the hardened soot layer remain the same as before hardening. In the hardened layer the structure of soot is, however, rougher and more porous (Fig. 14).

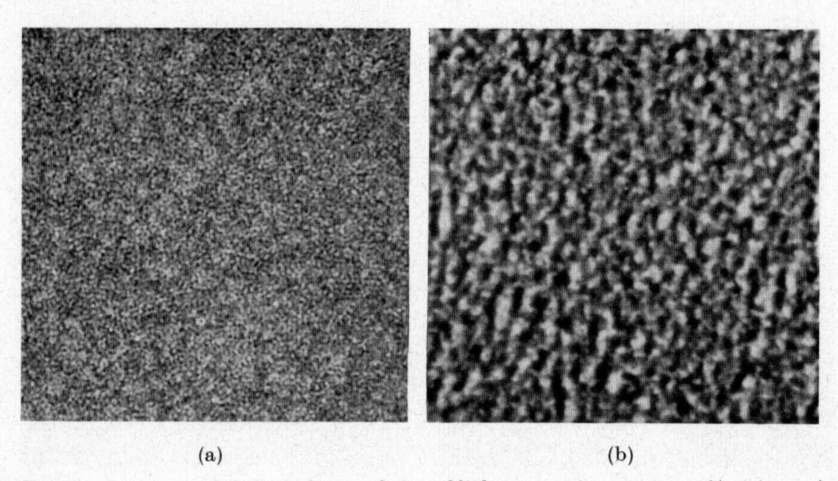

(a) (b)

Fig. 14. Structure of the stearin soot layer of light transmittance 3 to 4%: (a) soot in natural state; (b) after hardening with alcohol. The photomicrographs were taken with an objective $\times 40/0{\cdot}65$. Print magnification $\times 800$.

D. *Highly Sensitive Negative Phase Contrast Device*

1. *Technical specifications*

Soot layers have a high refractive index, greater than 2 (Table II), and are thus very convenient for making negative phase rings. For thin and slightly refracting phase objects such a ring must retard the phase of the direct light relative to the diffracted beams by $-90°$. From the values n_s and k given in Table II, and assuming that the soot phase ring

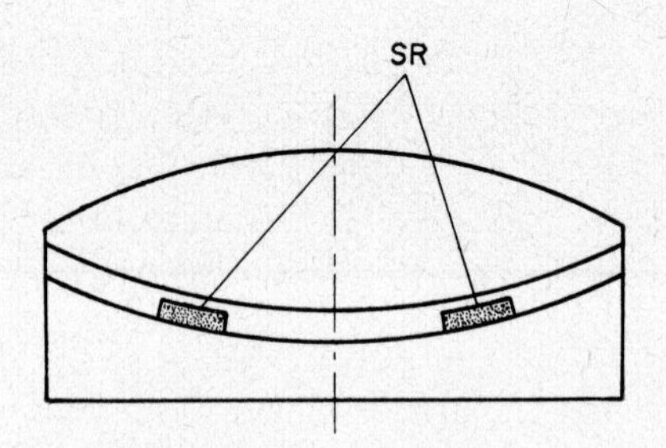

Fig. 15. Soot phase ring (SR) cemented between two lenses of an objective doublet.

(SR) (Fig. 15) is cemented by Canada balsam having a refractive index $n = 1{\cdot}53$, the light transmittance T corresponding to the phase shift $\psi = -90^\circ$ can be calculated from Eqn 13. This transmittance will be $3{\cdot}2\%$, $1{\cdot}2\%$ and 11% for the soot rings of stearin, kerosene and illuminating gas flames, respectively.

Thus, for highly sensitive phase contrast microscopy which needs a reduction of the direct light as great as possible, the soot of stearin or kerosene are most suitable. After some trials, one decided on the stearin soot. It has been found experimentally that the tolerance for the light transmittance T can be up to $\pm 1\%$, and finally $T = 3$ to 4% was chosen. For this transmittance the phase retardation ψ, according to Eqn 13, should remain within the range of 83° to 91°.

From this data a highly sensitive negative phase contrast equipment for biological microscopes has been constructed (Pluta, 1958, 1969a). It is designated "KFA", and is commercially available from Polish Optical Works (PZO), Warsaw.

The KFA equipment comprises four objectives $10\times/0{\cdot}25$, $20\times/0{\cdot}40$, $40\times/0{\cdot}65$ and $100\times/1{\cdot}3$ (oil immersion), a phase condenser with annular diaphragms arranged in a turret, and auxiliary centring telescope. Each objective incorporates a soot phase ring. This ring is located between two lenses of a doublet nearest the back focal plane of the objective (Fig. 15). The geometrical characteristics of the phase rings are as follows:

$$d/D = 0{\cdot}5, \quad w/D = 0{\cdot}1, \quad a/A = 0{\cdot}17,$$

where d, w and a are diameter, width and area of the phase ring, D and A are diameter and area of the exit pupil of the objectives, respectively.

The variation in optical properties of the soot rings, i.e. phase shift and light transmittance, has been measured for a number of different objectives taken from production. Measurements were made by means of a special shearing polarizing interferometer (Pluta, 1963). The results were within the range of manufacturing tolerance to the theoretical values.

2. *Imaging properties*

The sensitivity and range of unreversed contrast were determined using test objects consisting of narrow strips of thin silica films evaporated in vacuum on a glass slide. Applying different immersion liquids between a cover slip and the slide, various optical path differences were obtained. By this means it was established that the KFA device reveals OPD of about 2 Å. The range of unreversed image contrast was estimated to be 0 to -150° for phase-retarding objects, and 0 to $+20^\circ$ for phase-advancing objects.

The soot ring is practically achromatic and its phase shift is almost constant over the whole visible spectrum. Thus, the KFA device gives very good image contrast in both white light and monochromatic light of different wavelengths. When white light is used the field of view of the microscope presents an agreeable brownish colour. This colour is the result of a weak filtering effect of the soot layer, which transmits the longer wavelengths somewhat better than short wavelengths of the visible spectrum.

The soot phase ring reduces the intensity of the direct light only by absorption. This feature is a great advantage of the KFA system compared with devices which incorporate phase rings made of evaporated dielectric and metallic thin films. The metallic film decreases the light intensity not only by absorption but also to a high degree by reflection. The light reflected from the phase ring causes an increase of stray light in the field of view; thus the image contrast and sensitivity are decreased. None of these defects results when soot phase rings are used.

3. *Applications*

Due to its high sensitivity, the KFA device is suitable for observation of fine details, small particles, close-packed structures, and very thin

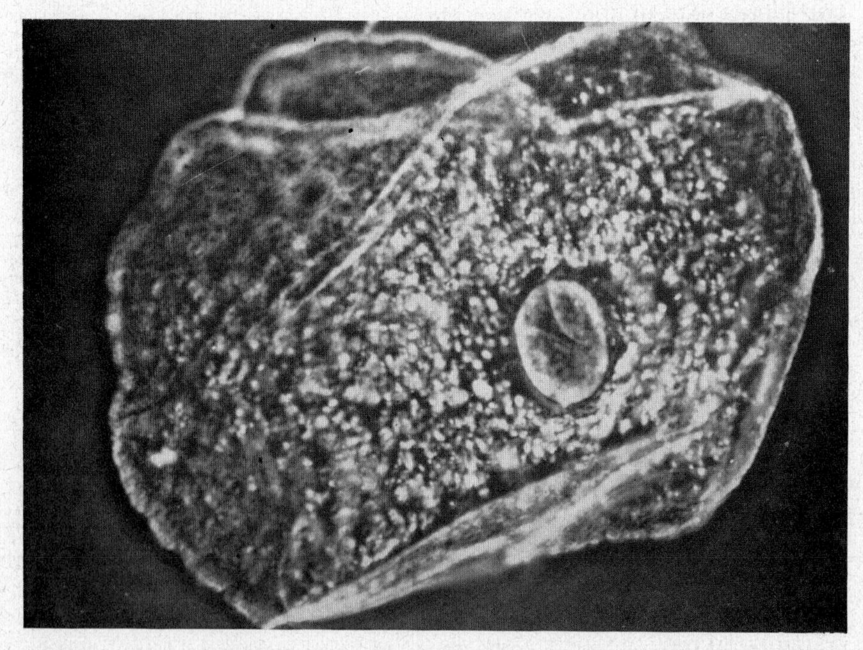

FIG. 16. Oral epitheliel cell immersed in saliva. Highly sensitive phase contrast device (KFA). Objective ×100/1·2 (oil immersion). Print magnification ×1500.

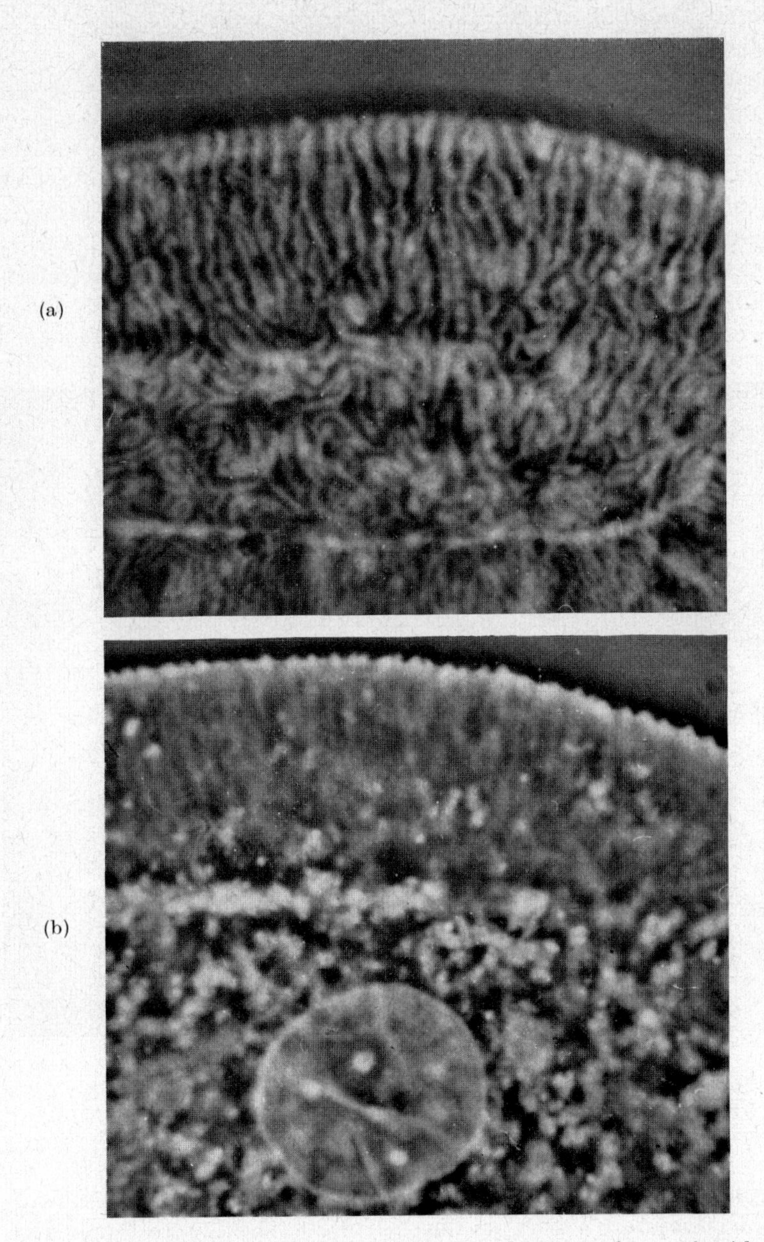

FIG. 17. Differently focused images of the same fragment of an oral epithelial cell embedded in saliva. (a) Upper ridged membrane surface focused; (b) nucleus and cell edge focused. Phase contrast device KFA. Objective ×100/1·2. Print magnification 2500×.

completely transparent objects, the refractive indices of which differ only slightly from that of the surrounding medium. In particular, it reveals very subtle structures, small granules and inclusions in living cells and tissues. Because of the large range of unreversed contrast for phase-retarding objects, the KFA device is also convenient for the examination of any other phase objects, the refractive index of which differs more markedly from that of the surround. The application of this device to observation of phase-advancing objects is, however, very limited, because the range of unreversed contrast for these objects is small. But the majority of microscopic specimens, and especially living tissues and cells and their inclusions, behave as phase-retarding objects.

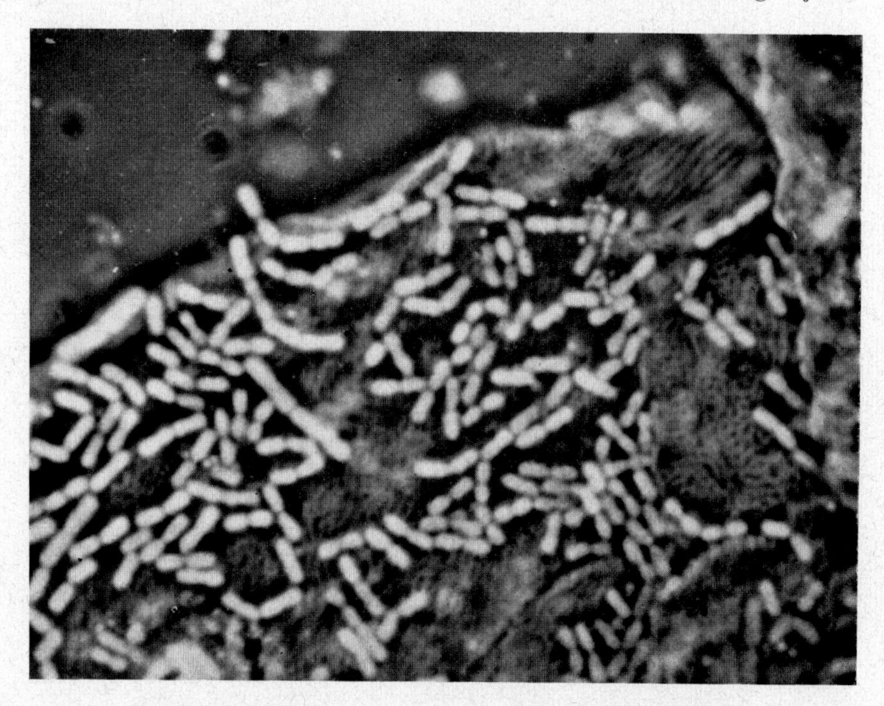

Fig. 18. A fragment of another oral epitheliel cell with fine membrane structure covered by a colony of bacteria. Phase contrast device KFA, objective ×100/1·2. Print magnification ×2700.

Photomicrographs of epithelial cells from oral mucosa, and human blood corpuscles are shown in Figs 16 to 20. These were taken with objectives of high magnification. Figure 16 shows an abundance of protoplasmic granules characteristic of most oral epithelial cells. There are some highly refractile particles. Another characteristic feature is a

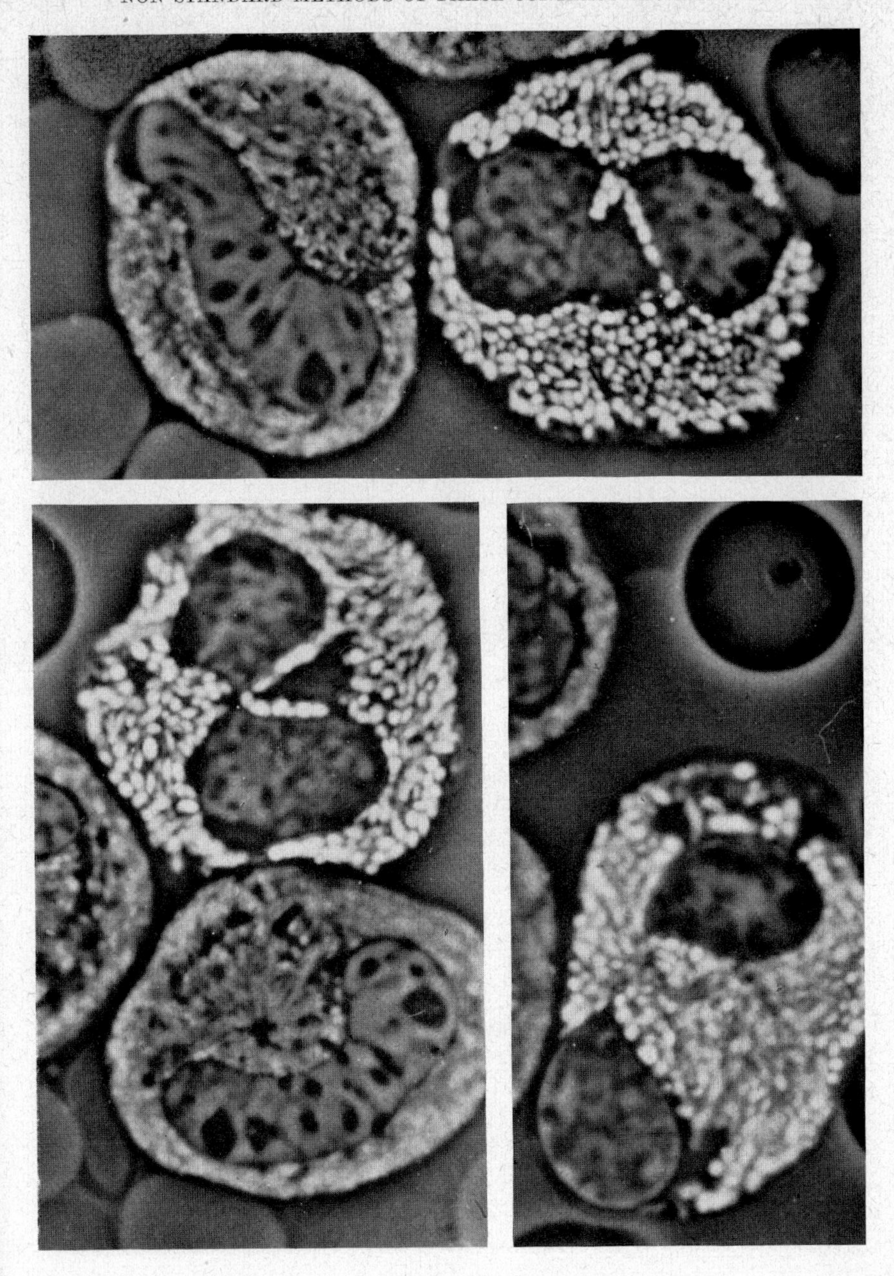

FIG. 19. Neutrophil and eosinophil granulocytes among red corpuscles "ghosts" in a fresh preparation of living human blood. Phase contrast device KFA, objective ×100/1·2. Print magnification ×2600.

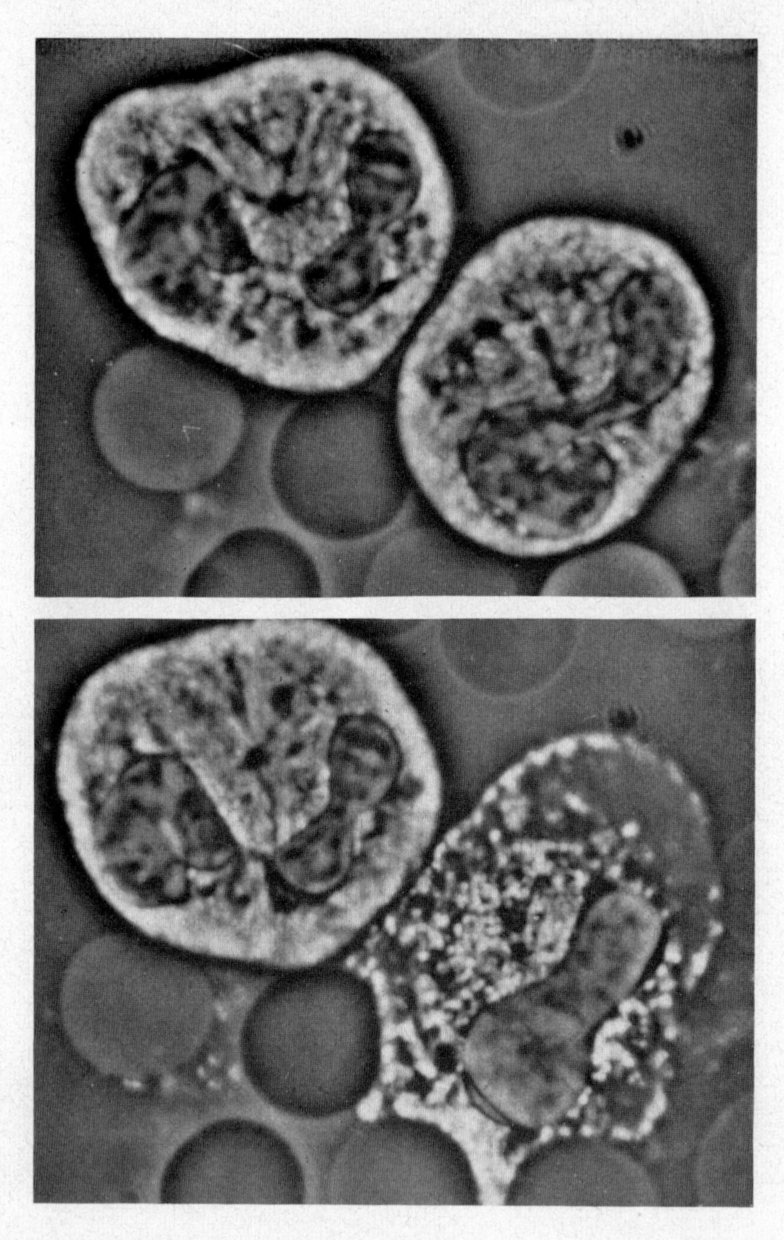

FIG. 20. Two of the same neutrophilic granulocytes photographed after a few minutes interval. Phase contrast device KFA, objective ×100/1·2. Print magnification ×2600.

peculiar ridged and microfolded structure of the membrane (Fig. 17). This structure, which is at the limit of the resolving power of the light microscope, can be particularly well seen with this device. Figure 18 shows a fragment of another epithelial cell with somewhat finer membrane structure covered by a cluster of bacteria. In this case the cell membrane appears as a delicate spongy network. This network and the highly refractile bacteria are both well seen. The ability to produce clear images of fine structures and of highly refractile phase-retarding particles at the same time is a basic advantage of this device.

This feature is seen in photographs of living leucocytes taken with the KFA device. Large structural elements such as nuclei and the small cytoplasmic granules are both well displayed. Figure 19 shows neutrophil and eosinophil granulocytes among some red cell "ghosts". Some of the granules undergo Brownian movement and do not always appear sharp in photographs taken with an exposure of a few seconds. Figure 20 shows two neutrophils, one of which burst after a few minutes (lower picture) when the Brownian movement ceased and the granules appeared sharp.

The KFA device is especially suitable for studies on cells grown in tissue culture and has been successfully used for this purpose by Veselý (1972). He studied the surface movements of cultured cells by means of high resolution cinemicrography with the KFA device fitted to an inverted microscope. Observations were made on the ruffling movement of the cell membrane, the behaviour of microvilli and the opening of pinocytotic channels onto the surface.

In addition, the KFA device can be used for very accurate immersion refractometry. Experiments showed that refractive index differences of about 0·0002 can be detected when small objects such as thin strips of transparent material 1 μm thick are successively immersed in a series of media of different refractive index.

E. *Highly Sensitive Positive Phase Contrast Device*

This device is designated "KFS". An essential difference between it and the KFA equipment is in the phase rings. Instead of the negative soot rings, quarter wavelength phase rings consisting of soot and dielectric films are used (Fig. 21). The dielectric film shifts the phase of the direct light by $+90°$, while a soot film is applied to reduce the intensity of this light by absorption only.

In order to make such a phase ring it is necessary to compensate the negative phase shift produced by the soot layer. This compensation cannot be achieved by embedding the soot ring in a suitably matched medium. The refractive index of soot layers is greater than 2 (Table II), and there is no optical cement having so high a refractive index. The only way to compensate the negative phase shift of light passing through the soot layer is to underlay the soot ring S (Fig. 21) with a transparent (dielectric) thin film D, whose refractive index n_D is considerably lower than the refractive index n_c of the optical cement C by means of which

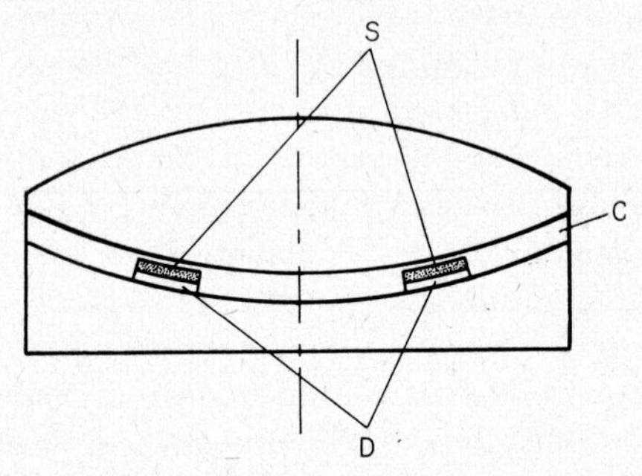

FIG. 21. Positive phase ring consisting of soot layer (S) and dielectric film (D) cemented between two lenses of an objective doublet.

the soot layer is glued between two lenses. In order to obtain the desired phase shift, equal to $+90°$, between the light passing through the ring S and light passing outside it, the thickness t_D of the dielectric film D should be chosen so that the following condition is fulfilled

$$t_D = \frac{|\delta|}{n_c - n_D} + \frac{\lambda}{4(n_c - n_D)} \tag{14}$$

where λ is the wavelength, and δ is the optical path difference produced by the soot layer.

The construction of such phase rings, as well as their properties, are described elsewhere (Pluta, 1967b). The phase ring (Fig. 21) consists of a cryolite film D and stearin soot layer S, and is cemented by Canada balsam C between two lenses of a doublet situated near the back focal plane of the microscope objective. In preliminary experiments it was established that the light transmittance T of this phase ring should be 10–12%. This is much higher than in the case of the KFA device, and it may be questioned whether a phase plate which transmits 10–12% of the direct light can be termed "highly sensitive". The answer is affirmative if the properties of positive phase contrast are taken into account. An essential objection, to $T < 10\%$ is, in this case, a small range of unreversed contrast for phase objects which exceed their surround in refractive index. No such objections exist for objects the refractive index of which is smaller than that of the surrounding medium, but such objects are in the minority.

The sensitivity of the KFS device is somewhat less than that of the KFA device, and, as measured experimentally, is equal to about 4 Å. When white light is used the field of view has an agreeable brownish tint, resembling that obtained with the KFA device. However, the achromatism of the phase plate is not as good as in KFA objectives, and the image contrast may be somewhat increased by using green light.

To illustrate the performance of this device some photomicrographs are presented in Figs 22–24. For comparison an equivalent picture taken with an objective with a typical quarter wavelength positive phase ring made of dielectric (cryolite) and metallic (chromium) substances is presented in Fig. 22(b). Geometrical parameters and light transmittance of the phase rings, as well as optical systems, were otherwise identical in the two objectives.

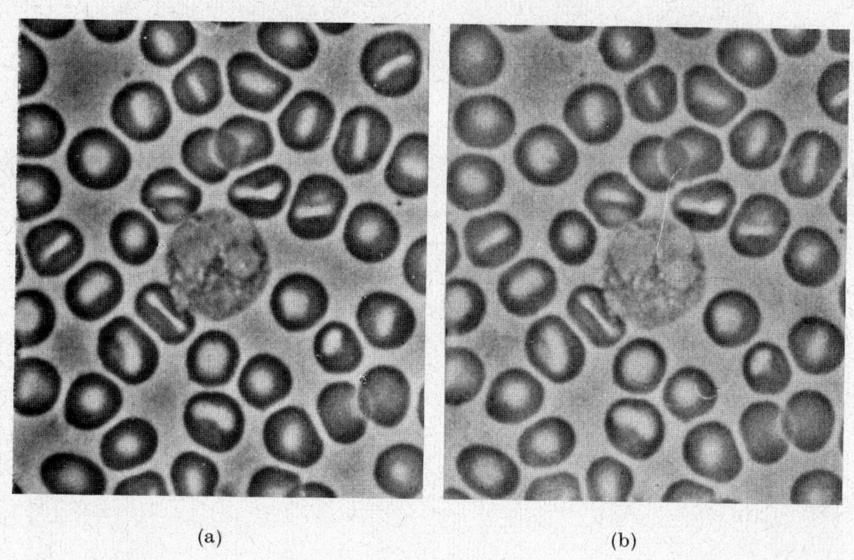

(a) (b)

FIG. 22. Unstained smear of human blood embedded in immersion oil and covered with a cover slip. Phase ring as shown in Fig. 21 gives greater image contrast (a) than a conventional phase ring of the same transmittance and phase shift (b). Objective ×40/0·65. Print magnification ×1200.

The image contrast in the case of the KFS objectives is significantly greater. This results exclusively from the non-reflecting properties of the soot which decreases the intensity of the direct light by absorption only. On the other hand, the metallic film reduces the light intensity not by absorption alone but also by reflection. The reflected light causes an increase of stray light in the field of view, decreasing the image contrast and sensitivity.

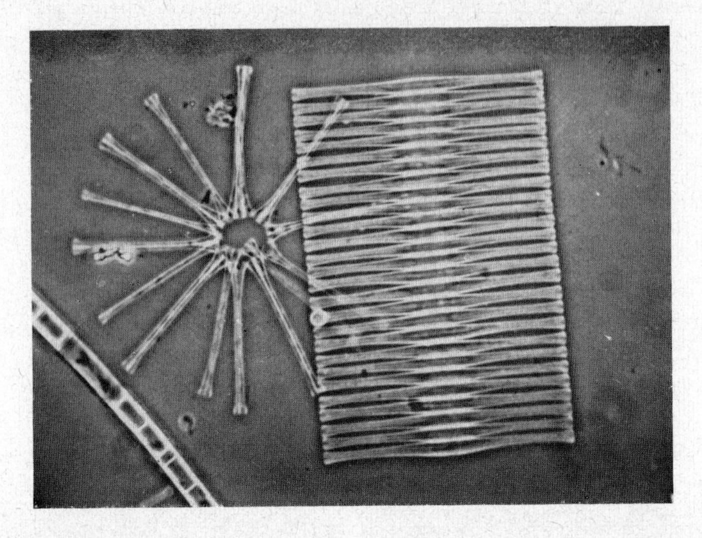

Fig. 23. Diatoms mounted in Canada balsam. Highly sensitive positive phase contrast device KFS. Objective $\times 20/0{\cdot}4$. Print magnification $\times 400$.

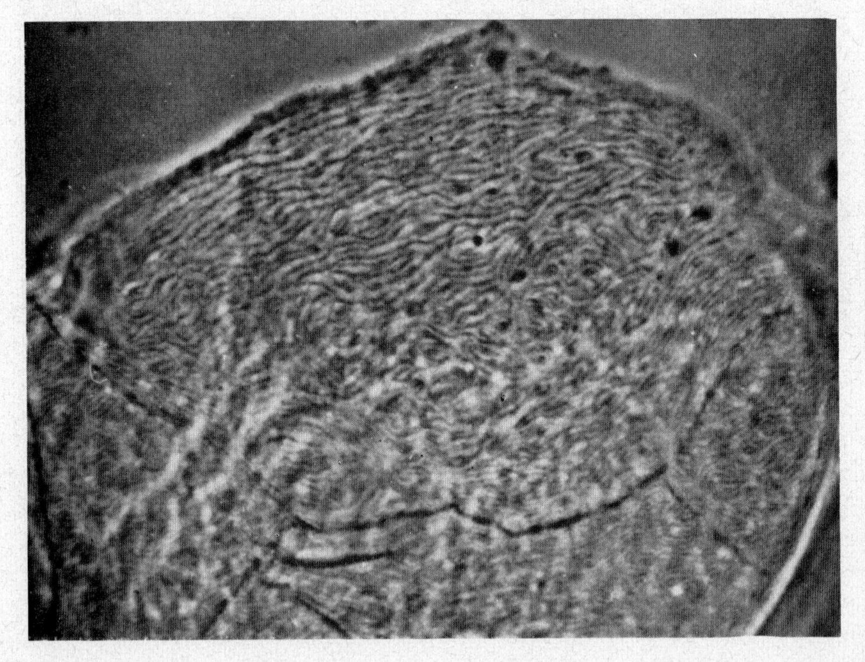

Fig. 24. Ridged membrane surface of an oral epithelial cell immersed in saliva. Phase contrast device KFS. Objective $\times 100/1{\cdot}2$ (oil immersion). Print magnification $\times 1800$.

Figure 22 shows the image of a specimen which belongs to the phase-retarding objects, whereas phase-advancing objects are shown in Fig. 23. In relation to these the KFS device presents almost the same advantages as does KFA equipment for phase-retarding objects.

Figure 24 shows an oral epithelial cell mounted in saliva. The picture was taken with a high power, $100\times/1{\cdot}2$ (oil immersion), objective focused at the top cell surface. The ridges and microfolds of the cell membrane are clearly visible.

In comparison with most standard commercial positive phase contrast systems, the KFS device is more sensitive, but has a somewhat smaller range of unreversed contrast; it yields more contrast images of minute and thin phase objects, and produces a brownish coloured background (when white light is used). This device is manufactured by PZO, Warsaw.

F. *Amplitude Contrast Device*

Up to now phase objects which transmit practically the same amount of light as their surrounding medium have been considered. However, it appears that using the phase contrast technique the image contrast of light absorbing objects can be controlled (increased, reduced or even reversed). These possibilities are predicted by both the vector representation (Barer 1952a,e, 1953b, 1954, 1955) and algebraic theory (Bennett *et al.*, 1951; Osterberg, 1955) of phase contrast. According to theory Richter's condition for optimum image contrast (Eqns (10) and (9)) takes the following, more general, form

$$T = \frac{1}{N} = 1 - T_p^{\frac{1}{2}} \cos \phi + T_p \tag{15}$$

$$\sin^2 \psi_{op} = \frac{T_p}{T} \sin^2 \phi \tag{16}$$

here T is the ratio of the light transmittance of the conjugate area to that of the complementary area of the phase plate, ψ_{op} is the phase difference between these two areas, T_p is the ratio of the light transmittance of the object to that of an equal thickness of the surrounding medium, and ϕ is the phase shift produced by the object.

Considering amplitude objects for which $\phi = 0$, one can easily show that Eqns (15) and (16) become

$$T = (1 - T_p^{\frac{1}{2}})^2 \tag{17}$$

$$\psi_{op} = 0 \quad \text{for } T_p < 1$$

$$\psi_{op} = 180^\circ \quad \text{for } T_p > 1 \tag{18}$$

$T_p > 1$ signifies that the object absorbs less than an equal thickness of its surrounding medium. This case is neglected, and only objects which absorb more light than their surround ($T_p < 1$) will be considered. When $T_p \to 1$ the light transmittance of the object and its surround becomes alike, and the required value T for optimum positive contrast approaches zero.

It follows that the image of weakly absorbing amplitude objects can be optimally increased by using a ring having zero phase shift and transmittance $T = (1-T_p^{\frac{1}{2}})^2$. Such a ring will be termed "amplitude ring". Consequently, a device which incorporates objectives with amplitude rings will be called "amplitude contrast device". Such a device has been developed (Pluta, 1968a, 1969b).

One reason for making this device was that in stained biological specimens there are always details and structures which are insufficiently stained and are therefore difficult to distinguish by bright-field microscopy. It is true that such details can be observed well enough with phase contrast, but the image of more intensely stained objects is then wrong.

The device is termed "KA". Its construction is not unlike that of the KFS equipment. The amplitude rings are made by the same process and they consist of a soot layer S (Fig. 21) and dielectric film D which cancels the negative phase shift of the soot layer. For this the thickness t_D of the dielectric film is given by

$$t_D = \frac{|\delta|}{n_c - n_D} \tag{19}$$

where δ, n_c and n_D are as in Eqn 14.

In initial experiments, typical achromatic objectives $10\times/0{\cdot}25$, $20\times/0{\cdot}4$, $40\times/0{\cdot}65$ and $100\times/1{\cdot}2$ (oil immersion) were used. These were changed into amplitude-contrast objectives by making, according to Fig. 21 and Eqn (19), an amplitude ring between the two lenses nearest the back focal plane. Amplitude rings having light absorption 75, 80, 85 and 90% were prepared. Their geometrical parameters were $d/D = 0{\cdot}5$, $w/D = 0{\cdot}1$ and $a/A = 0{\cdot}17$. For comparative studies the same achromatic objectives without the amplitude ring, as well as the same objectives fitted with positive phase rings having a phase shift of $+90°$ and a light absorption of about 80%, were used. The quality of each objective was tested by means of a shearing interferometer (Pluta, 1963). Only objectives having the same degree of aberration correction were selected. A condenser with annular aperture diaphragms conjugate with the amplitude and phase rings was used. The same condenser was used for all observations. Thus, the illumination of all specimens was identical for amplitude contrast, phase contrast and ordinary bright-field observation.

The experiments have shown that the best results are obtained with stained biological specimens when the soot amplitude ring absorbs 75–80% of the direct light. This value has been selected for the final version of the KA device. When the amplitude ring absorbs more than 85%, a decrease of image contrast is observed. A halo also appears, similar to the phase contrast halo. The decrease of image contrast, or even its reversal, and appearance of a bright halo are also observed with strongly absorbing objects.

The basic features and advantages of this device, as well as its possibilities for studies on stained specimens, in particular biological cells, are illustrated in Figs 25–28. These are photomicrographs of some rather lightly stained specimens. Each of the amplitude contrast photomicrographs (marked "a") is confronted with the bright-field image of the same specimen taken under comparable conditions. The exposure time was controlled automatically, using a photomicrographic attachment with an automatic exposure-meter.

It can be seen that the KA device yields, in comparison with the ordinary bright-field microscope, more image contrast for insufficiently or moderately stained specimens. An increase of image contrast is achieved both by using ring-shaped illumination and by reducing the light intensity of the zero order diffraction maximum. Thus more favourable conditions for effective interference of the direct and diffracted light beams are fulfilled. Another reason for the higher contrast is the elimination of the diffuse (stray) light from the field of view. In the ordinary bright-field microscope some haze (glare) always exists in the image plane. This is caused by instrumental factors such as light reflection at optical surfaces, light scattering by lens inhomogeneities and dust, aberrations of the objective, and so on. Often the specimen itself is a source of intense stray light. In the case of the amplitude contrast device the intensity of this injurious light is considerably reduced, images with more contrast are obtained, and the delineation of fine detail is improved.

This feature is a consequence of the enhancement of contrast, and finds theoretical support in the theory of the optical transfer function for partially coherent illumination. This advantage is particularly useful in the investigation of minute components and structures of biological cells.

When a specimen includes both large heavily stained objects and some fine low-contrast details the latter are more enhanced than the former, when amplitude contrast is used. This effect permits small

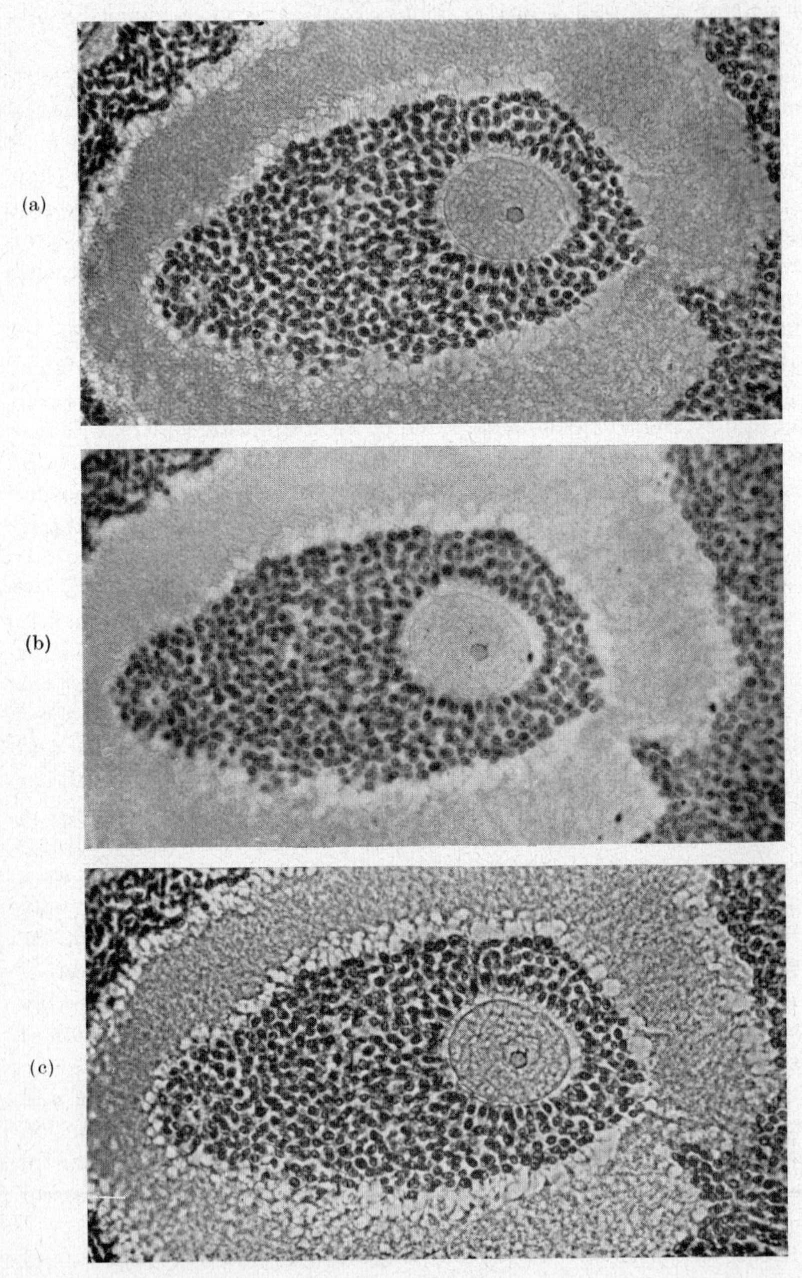

FIG. 25. Tissue section (ovary of rat) stained with haematoxylin and mounted in Canada balsam. (a) amplitude contrast (KA), transmittance of the amplitude ring $T = 20\%$, (b) bright-field, (c) positive phase contrast ($\psi = +90°$, $T = 20\%$). Objective $\times 20/0{\cdot}4$. Print magnification $\times 320$.

details on a background of large objects or in their vicinity to be observed better than in the bright-field microscope.

It is true that weakly stained objects can be observed quite well with a standard positive phase contrast microscope, but some undesirable

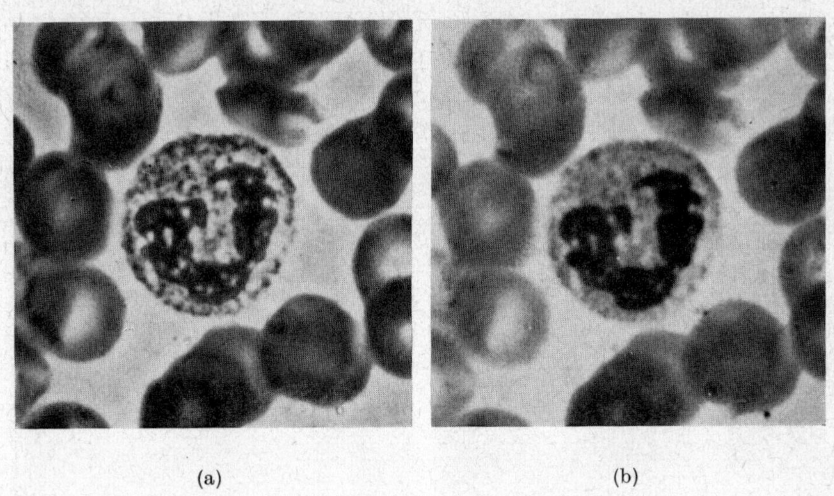

(a) (b)

FIG. 26. Routinely stained human blood smear embedded in immersion oil. (a) amplitude contrast (KA); (b) bright-field. Objective $\times 100/1 \cdot 25$ (oil immersion). Print magnification $\times 1750$.

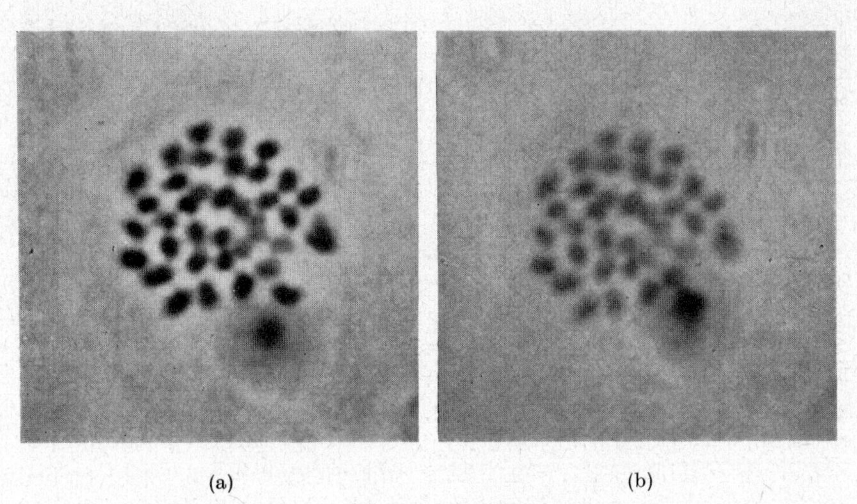

(a) (b)

FIG. 27. Mitosis of sugar beet cell stained by Feulgen method. (a) Amplitude contrast (KA); (b) bright-field. Objective magnification/aperture $\times 100/1 \cdot 25$. Print magnification $\times 3200$.

phase artefacts may then efface the amplitude details of interest. On the contrary, the amplitude contrast method tends to emphasize amplitude details and minimize phase details. However, the amplitude contrast device is useless for objects which are intensely stained. It tends to minimize or even reverse the contrast of the image of objects absorbing more than about 70%. A defect of the KA device is the bright halo similar to that in standard positive phase contrast. In general, however, this perturbing effect is small.

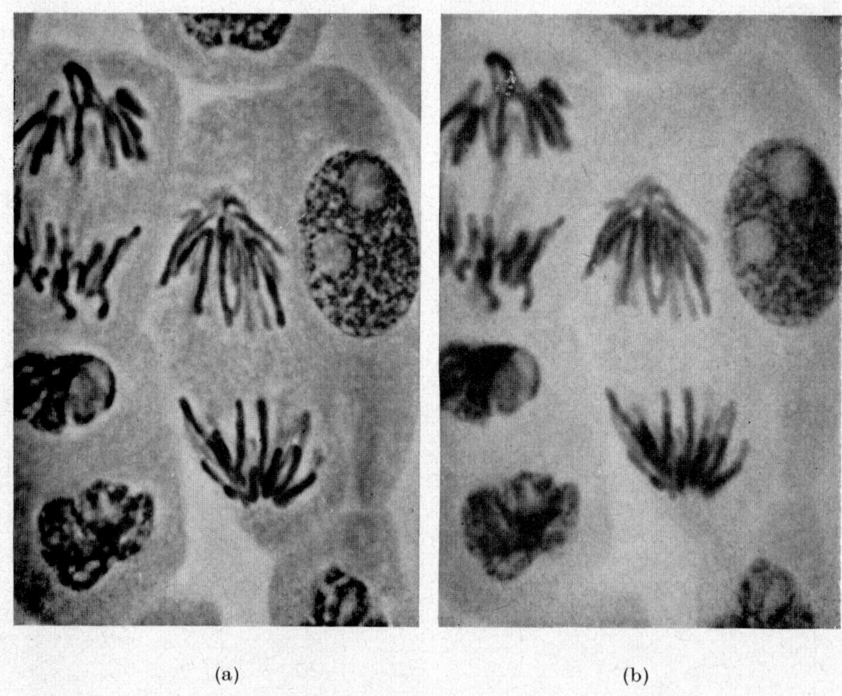

(a) (b)

Fig. 28. Chromosomes of onion cells stained by Feulgen method. (a) Amplitude contrast (KA); (b) bright-field. Objective ×100/1·25. Print magnification ×1050.

The KA device is commercially manufactured by PZO, Warsaw, for biological microscopes. Recently it was successfully adapted to reflected light microscopy by Popielas (1972), who demonstrated that the amplitude contrast method has many interesting advantages for studies of reflecting specimens as well. Using this method some finely etched structures appear more clearly than in standard metallurgical microscopes.

III. Variable Phase Contrast Systems with Isotropic Phase Plates

A. *Why Variable Phase Contrast?*

Phase contrast microscopy incorporates some contradictions. For instance, on increasing the image contrast and sensitivity of the method, the range of unreversed contrast decreases, and the halo and shading-off effects become more disturbing. In order to eliminate these and obtain a truer representation of the specimen, some microscopists prefer objectives with low absorption phase plates. Advantages of such objectives for studies of biological cells were discussed by Ross (1967) and other cytologists. Positive phase contrast objectives with 25% absorption of the direct light were manufactured by W. Watson & Sons, Ltd., England, whereas American Optical Company, Buffalo, USA, markets phase objectives with zero absorption phase plates. They are recommended for highly refractile objects and specimens including both phase and amplitude structures.

The halo and shading-off effects can be reduced, and the faithfulness of imaging can be improved without diminution of contrast by using a phase ring as narrow as possible. But, in this case a powerful source of illumination is needed. Also, injurious diffraction effects arise as a consequence of an over-narrow annular condenser diaphragm conjugate with the phase ring.

Another contradiction is that positive phase contrast is more acceptable for biologists than negative, since it gives, in general, dark images which resemble those of stained specimens in ordinary bright-field microscopy. However, from the physical point of view negative phase contrast appears to be better because its range of unreversed contrast for phase-retarding objects (these objects are in the majority) is much larger, dark haloes around bright object images are less disturbing than bright haloes around dark images and negative phase plates can be made more sensitive than positive ones. Advantages and disadvantages of positive and negative phase contrast have been discussed by Wilska (1954) and Gabler (1955).

In the writer's opinion both types of phase contrast are generally acceptable and they are complementary to one another. In practice, particularly in biological and medical research, there are usually very complicated specimens which contain small and large details, slightly or heavily refractile particles, structures of different spatial frequencies, and objects of diverse light absorption. Some are seen better in negative phase contrast, others in positive or even in bright field. Unquestionably, it is also advantageous to be able to analyze an unfamiliar image using different phase shifts and various intensity ratios between the direct and diffracted light beams. Thus it appeared necessary to design phase contrast devices with variable image contrast. Some of those which have found practical application will be described below.

B. *Beyer's Phase Contrast Device*

As has been mentioned, the halo and shading-off effects depend on the diameter and width of the phase ring. When these parameters diminish the halo spreads, and it becomes less intense in the immediate vicinity of the image. Simultaneously the shading-off abates and the image contrast becomes more uniform. Thus, the light distribution in the image corresponds more with variations of OPD in the specimen, and the general faithfulness of phase contrast imaging improves. These facts were carefully examined both theoretically and experimentally by Beyer (1953). He characterizes the faithfulness of phase contrast imaging by a parameter Γ defined as follows

$$\Gamma = d\frac{k}{f}\Delta r \tag{20}$$

where d is the diameter of the object (measured in the object plane of the microscope), $k = \pi/\lambda$ (λ is the wavelength), f is the focal length of the objective, and Δr is the width of the phase ring.

When the phase shift ϕ produced by a transparent object is small $|\phi| \ll 90°$, and $\Delta r \to 0$, the parameter Γ approaches zero, and the faithfulness of imaging is ideal. Such a situation is theoretical only, and in practice it is not attainable. If $0 < \Gamma \leqslant 1$ the image represents the object fairly well, but for $1 < \Gamma \leqslant 2{\cdot}5$ the halo and shading-off effects are quite noticeable, and become more evident the larger is the object. When $\Gamma > 2{\cdot}5$ the halo and shading-off effects are obtrusive and inadmissibly disturb the object image. The faithfulness of imaging of broader objects is then very poor.

In summary, for a given diameter d of the phase object, its image is a truer representation of OPD variations the smaller is the width Δr of the phase ring. But a narrow phase ring has some defects, i.e. it reduces the resolution and image definition. This relates, in particular, to fine details the dimensions of which are somewhat greater than the limit of resolution. For such objects a wider phase ring is preferable.

Taking these facts into account, Beyer developed a variable phase contrast device which permits both small and large objects to be observed with optimum contrast and improved image representation.

The basic feature (Fig. 29) is a combination (Ph) of two concentric quarter wavelength positive phase rings R_1 and R_2 situated in the back focal plane of the microscope objective O, and an aperture diaphragm D_1 consisting of two annuli A_1 and A_2, located in the front focal plane of the condenser C. The outer phase ring R_1 is of such a size as to cover the image of the outer condenser annulus A_1. The diameter and width

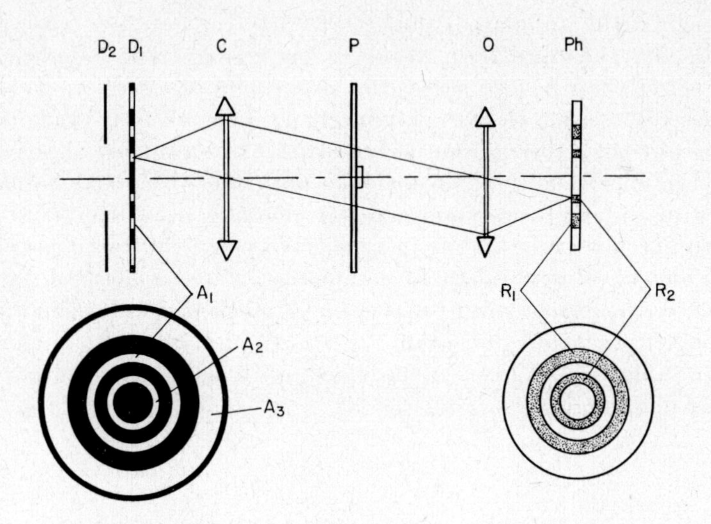

FIG. 29. Principle of Beyer's variable phase contrast device (see text).

of these elements are more or less as in a standard Zernike phase contrast device. The inner phase ring R_2 is arranged to cover the image of the inner condenser annulus A_2. These two elements are much smaller than the outer ones. The gap between the phase ring R_1 and R_2 is fixed so that the diffracted light beams, the primary sources of which are the condenser annuli A_1 and A_2, are not greatly affected by the phase rings R_2 and R_1, respectively. This is achieved when the following relation is fulfilled

$$G = d\frac{k}{f}\Delta g > 5 \tag{21}$$

where Δg is the desired gap between the two phase rings, and the remaining symbols are as in Eqn (20).

Just before the annular diaphragm D_1 there is an iris diaphragm D_2. When the latter is entirely open the light passes through both annuli A_1 and A_2. Consequently, both phase rings are simultaneously in action, but the inner ring R_2 contributes much less than the outer one to the production of the phase contrast image. In this case the image is similar to that obtained with any normal positive phase contrast device. If, however, the iris diaphragm D_2 is set so that the outer annulus is obscured, then the inner phase ring R_2 acts alone and an image with spread halo and reduced shading-off effect is obtained. According to Beyer's (1965) nomenclature this is strict phase contrast (strenger Phasenkontrast).

In reality the annular diaphragm D_1 has three annuli, two as described, and the third one (A_3) lying outside A_1. The annulus A_3 is much larger than A_1 and serves for bright-field observation. Then, by progressively closing the iris diaphragm D_2 one can pass continuously from bright-field, through normal to strict phase contrast observation. Thus, the specimen P which includes both the phase and amplitude objects of different sizes and various characteristics can be studied more comprehensively than with a standard phase contrast microscope.

This device is manufactured commercially by C. Zeiss, Jena, and supplied as an attachment to conventional microscopes. It includes four Phv objectives of magnification 10, 20, 40 and 90× (oil immersion) and an aplanatic condenser provided with a turret incorporating annular diaphragms.

C. *Device with both Positive and Negative Phase Contrast*

An ordinary microscope equipped with standard positive and negative phase contrast devices of Zernike-type with quarter-wave phase plates having a fixed light absorption assures sufficiently wide possibilities for investigating most phase objects. However, changing from positive phase contrast to negative, and vice versa, is inconvenient and slow. A device which enables the user to change rapidly from one to the other, using a single objective, has been developed (Pluta, 1965, 1969c).

The basic elements of this device (Fig. 30) are two concentric phase rings, the negative R_1 and the positive R_2, sandwiched between the two cemented lenses L_1 and L_2 of a doublet situated in (or near) the back focal plane of the microscope objective O. An aperture diaphragm D with two transparent annuli A_1 and A_2 covered by polarized films P_1 and P_2 is inserted in (or near) the front focal plane of the substage condenser C. The transparent annuli A_1 and A_2 and the phase rings R_1 and R_2 are in conjugate planes, so that the image of A_1 is formed on R_1 and the image of A_2 on R_2. The ring-shaped polarizing film P_1 covers the annulus A_1 and the disc-shaped polarizing film P_2 covers the annulus A_2. The polarization planes of these films are perpendicular to each other. The third polarizer P_3 is placed below the aperture diaphragm D and it covers both transparent annuli A_1 and A_2 (this polarizer can also be placed above the objective O or eyepiece). By rotating the polarizer P_3 around the axis of the microscope, it is possible to control the ratio of the intensity of the light passing through the transparent annuli A_1 and A_2. When the polarizer P_3 is crossed with the polarizer P_2, the light passes through the annulus A_1 alone and not through A_2. Then the phase contrast defined by the phase ring R_1 (negative) is obtained. If, however, the polarizer P_3 is crossed with polarizer P_1, the light does not pass through the annulus A_1 but through A_2 only, and the phase contrast determined by the phase ring R_2 (positive) is obtained. In intermediate positions of the polarizer

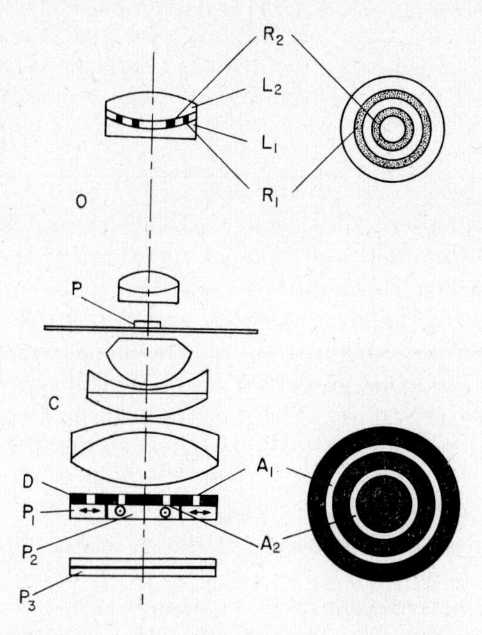

FIG. 30. Optical system of variable phase contrast device with both negative (R_1) and positive (R_2) phase rings (see text).

P_3, when the direct light passes through both annuli, a "mixed" phase contrast appears in which either bright or dark images prevail. Thus, by rotating the polarizer P_3 it is possible to change the image contrast rapidly and continuously and to pass from positive phase contrast to negative (and vice versa) without interrupting the observation of the specimen P. The dimensions of the phase rings as well as diameters and widths of the condenser annuli are so designed that there is no change in image sharpness during the transition from negative to positive phase contrast, and the diffracted light, the primary source of which is the annulus A_1, is not greatly disturbed by the phase ring R_2, and conversely the diffracted light derived from the annulus A_2, is not greatly affected by the phase ring R_1. These requirements can be fulfilled when the diameters of the phase rings R_1 and R_2 are equal to 1/2 and 1/4 of the exit pupil diameter of objective and the widths of these rings are equal to 1/6 to 1/9 of their diameters, respectively. The principal geometrical characteristics of the phase rings are given in Table III.

TABLE III

Diameters (d), widths (w) and areas (a) of phase rings R_1 and R_2 relative to the diameter (D) and area (A) of the exit pupil of objective

Phase ring	d/D	w/D	a/A
negative R_1	0·55	0·065	0·13
positive R_2	0·28	0·045	0·055

The positive phase ring R_2 is made of dielectric and metallic materials (cryolite and chromium) and changes the phase of the direct light by about $+90°$, whilst the negative phase ring R_1 is made of soot and alters the phase of the direct light by about $-90°$. The phase ring R_2 reduces the light intensity to 15% and the phase ring R_1 to about 4%. This latter has the same properties as the soot phase rings in the KFA device described previously. The procedure for making two such phase rings on one surface of a lens is described in detail elsewhere (Pluta, 1965).

The polarizing films P_1 and P_2 are cemented between the diaphragm D and an additional glass plate (not shown in Fig. 30). The quality of the polarizing films P_1, P_2 and polarizer P_3 need not be excellent. A degree of light polarization of 99·8% is quite sufficient. There are four ring diaphragms D with polarizing films P_1 and P_2 in the constructed device (one for each of the four objectives of magnification 10, 20, 40, and 100×). They are placed in a revolving disc attached to an aplantic-achromatic condenser.

Apart from the ability to change rapidly from one type of contrast to another, this device gives relatively good image faithfulness because narrow phase rings are used. The sensitivity is high, being about 2 Å for the negative phase ring and about 6 Å for the positive one. The photomicrographs in Figs 31 and 32 illustrate the performance of the system.

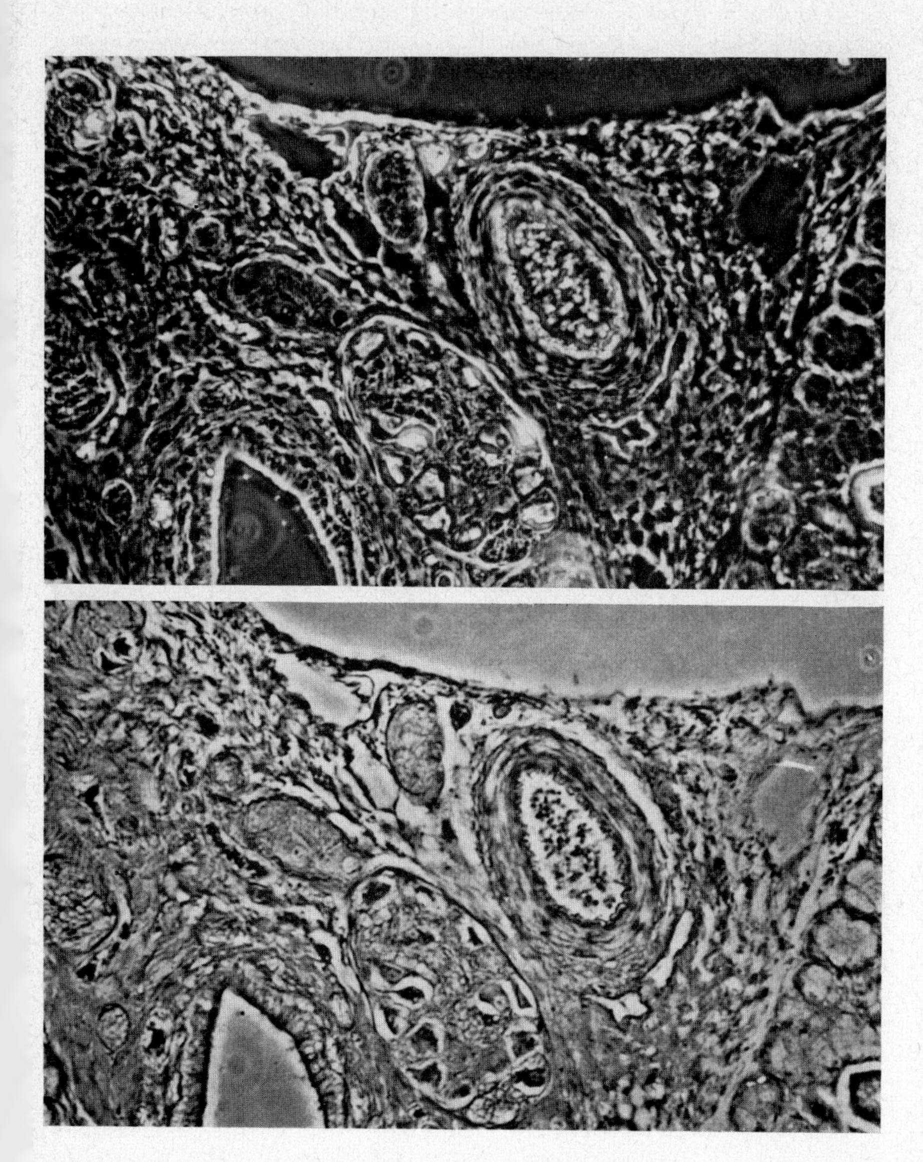

FIG. 31. Unstained tissue section (salivary gland of cat) mounted in Canada balsam. Variable phase contrast device (Fig. 30) set at negative (top) and positive (bottom) image contrast. Objective $\times 10/0{\cdot}25$. Print magnification $\times 180$.

Some experiments with objectives in which the negative phase ring R_1 (Fig. 30) was made of dielectric and metallic materials, instead of soot, have also been carried out. However, their performance was not as good, because of reflection by the thin metallic film. The reflected

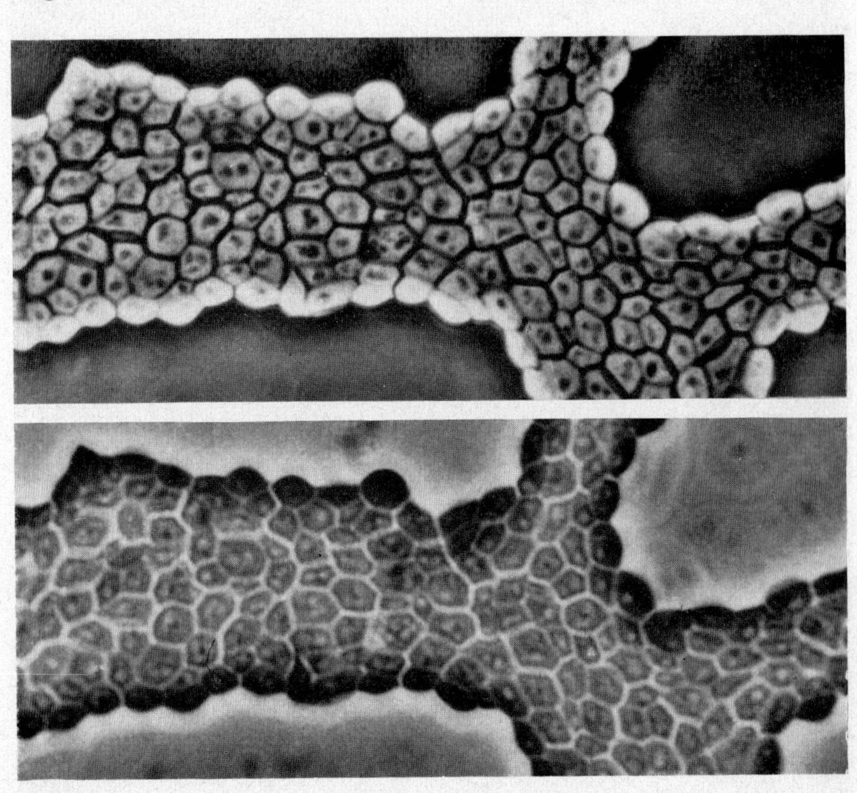

FIG. 32. Monolayer of dry yeast cells obtained by evaporation from an aqueous suspension of yeast between a slide and cover slip. Variable phase contrast device (Fig. 30) set at negative (top) and positive (bottom) image contrast. Objective $\times 40/0{\cdot}65$. Print magnification $\times 1100$.

light is particularly injurious for negative phase contrast as the background is dark and should not be brightened by stray light. In the case of positive phase contrast the reflected light is less injurious, especially as the ring R_2 has a small area (three times less than the area of the negative phase ring R_1), so it does not reflect as much light.

Objectives in which the inner phase ring R_2 is negative and the outer R_1 is positive have also been constructed, but less satisfactory results were obtained. Experimentally, both the positive and negative contrast gave rather poor quality. Only the first device has been manufactured by PZO Warsaw.

D. "*Varicolor*" *and other Systems*

Several variable phase contrast devices with isotropic phase plates have been developed. Most of them belong to so-called *colour phase contrast* systems. They have found little practical application and will only be discussed briefly.

The first colour phase contrast system was proposed by Zernike and investigated by his co-workers (Saylor *et al.*, 1950). It makes use of the dispersive properties of two transparent substances. A phase ring is made of a thin layer of one substance and the complementary area is coated with another. The refractive index of these two materials is the same for the middle of the spectrum (green light) only, whereas in the violet the phase ring is more refractile than the adjacent complementary layer and less refractile in the red. If the thickness of the phase ring and of the complementary layer are suitably chosen, one can obtain a negative or positive phase shift between the direct and diffracted beams, equal, e.g. to $-90°$ and $+90°$ in violet and red light, respectively. Thus, a variable colour phase contrast is obtained when the specimen is illuminated with monochromatic light of different wavelengths.

An interesting system, so-called "Varicolor", has been developed by Dufour and Locquin (1951; Dufour, 1952; Locquin, 1952). In this system the conjugate area is covered by a ring-shaped interference filter of the narrow band-pass type. This filter gives different phase shifts and different amplitude ratios between the direct and diffracted light beams when monochromatic light of various wavelengths is used. In particular, the interference filter can be so designed that for short wavelengths (blue-green light) it produces a phase shift ψ of $+90°$ and $\psi = -90°$ for long wavelengths. In the first case one obtains a positive phase contrast with blue-violet colour and in the other case negative phase contrast with red colour. For the intermediate colours one obtains lesser phase shifts and greater transmittances of such an interference phase ring. Thus, by varying the wavelength of the light one can pass continuously between bright and dark image contrast. Simultaneously the colour of the object image and that of the background change. An interference filter, progressively inclined to the objective axis, is used as a monochromator. The Varicolor device was formerly available from Wild Heerbrugg (Switzerland).

Grigg (1950) developed a colour phase contrast device by using an ordinary phase plate in the objectives and a Rheinberg-type condenser annular diaphragm. The annulus is made of a filter of one colour and a complementary colour filter fills the remaining area of the condenser diaphragm.

Barham and Taylor (1948) suggested using two phase rings and two condenser annuli covered by filters of different colours one of which gives positive and the other negative phase contrast.

A simple way of producing variable phase contrast depends on using variable-characteristic phase plates, with step-shaped or wedged distributions of refracting and absorbing materials (Lyot, 1946).

The optimum position for the phase plates is the back focal plane of the microscope objective. Unfortunately, this plane, except for low-magnification objectives, is located in the objective itself and is not accessible. To overcome this drawback re-imaging or transferred-image attachments for use in phase contrast microscopy have been developed (Payne, 1952a 1954; Françon and Nomarski, 1950, 1952). These re-image the back focal plane of the objective in an easily accessible position where different phase plates can be introduced and exchanged.

Other systems for variable and/or colour phase contrast microscopy have been realized by using anisotropic phase plates. These are described in the next section.

IV. Variable Phase Contrast Systems using Anisotropic Elements

A. "*Polanret*" *Method*

Polarizing, birefringent and optically active elements are particularly suitable for variable phase contrast microscopy. They enable systems with continuous variation of the phase ψ and/or transmittance ratio T between the direct and diffracted light to be constructed. Thus, it is possible to adjust the microscope for optimum image contrast according to Richter's conditions (Eqns 9 and 10), and in principle quantitative measurements of OPD in the specimen can be made.

A great number of phase contrast systems with anisotropic phase plates has been described in the literature. One of the best known is the "Polanret" system developed by Osterberg (1946, 1947a). The most general version is shown in Fig. 33. Its basic element is a disc MD placed in the back focal plane of the objective O. This is composed of a quarter-wave retarding plate Q made, e.g. of mica, and of two zonal polarizers, Z_1 and Z_2, made of polarizing film. The polarizer Z_1 covers the conjugate area, defined by the image of a condenser annulus A, whereas Z_2 covers the complementary areas. The polarization planes of these polarizers are perpendicular to each other and their directions of light vibration V_1 and V_2 make an angle of 45° with the slow (or fast) axis X of the quarter-wave plate Q. A linear polarizer P_1 placed before the disc MD, and an analyzer P_2 inserted behind it complete the basic Polanret system. Both P_1 and P_2 are rotatable around the

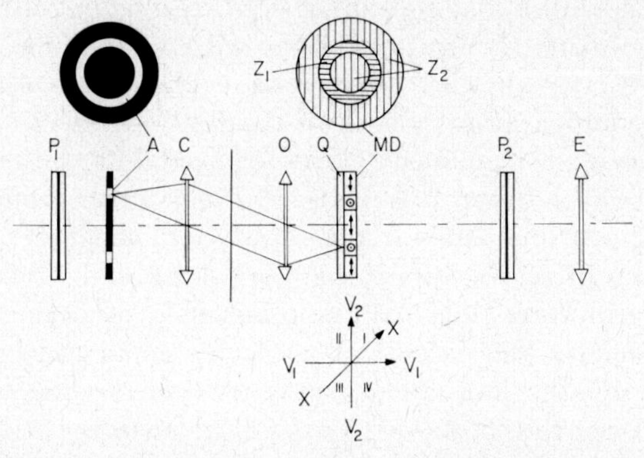

FIG. 33. Osterberg's Polanret phase contrast system (see text).

objective axis. Rotation of the polarizer changes the phase ψ between the conjugate and complementary areas, whereas rotation of the analyzer varies the light transmittance ratio T of these two areas. Let the initial (zero) orientation of P_1 and P_2 be defined as follows: direction of light vibration transmitted by P_1 is parallel to the X axis of the quarter-wave plate Q, and direction of light vibration transmitted by P_2 is parallel to the light vector V_2 transmitted by the zonal polarizer Z_2. Then, the rotation of P_1 through an angle θ introduces a phase difference $\psi = 2\theta$ between the direct and diffracted beams, whereas rotation of the analyzer through an angle α varies the transmittance ratio T of the conjugate area relative to the complementary area such that $T = \tan^2\alpha$. Thus, in general, ψ can be varied from $-180°$ to $+180°$, and T from zero (conjugate area completely dark) to infinity (complementary area completely dark).

The polarizer P_1 and quarter-wave plate Q constitute, in fact, a Sénarmont compensator. The transmission direction of the polarizer P_1 together with vectors V_1 and V_2 determine four quadrants (I, II, III and IV) of which two (I and III) are neutral and two anti-neutral (II and IV). The neutral quadrants contain the transmission direction of the polarizer P_1. When the transmission direction of the analyzer P_2 passes from the neutral to anti-neutral quadrants, the phase shift ψ

between the diffracted and direct light beams jumps suddenly by 180°. For example, if $\psi = 90°$ in the neutral quadrant then in anti-neutral quadrants ψ is equal to 270° (or $-90°$). Thus, by rotating the analyzer one can simultaneously vary the transmittance ratio T and pass from positive to negative phase contrast. This important feature of the Polanret method is true both with and without quarter-wave plate Q. When this plate is omitted a required phase difference ψ, e.g. equal to 90°, between the diffracted and direct light, can be introduced in other ways, e.g. by coating the conjugate area with a thin dielectric film.

The Polanret system with quarter-wave plate (or with other birefringent base compensators) constitutes a highly versatile system which is capable of giving continuously variable phase shift and transmittance ratio T, whereas without this plate it constitutes a phase contrast system having continuously variable T and two constant, positive and negative, values of phase shift ψ, e.g. $\psi = +90°$ and $-90°$.

The Polanret system (Fig. 33) can sometimes be used for measuring OPD in specimens. This is only possible for small objects which are not disturbed by the halo and shading-off effects. The measurement depends on setting the rotatable polarizer and analyzer at two positions for which first the image and then its surrounding background are made maximally dark. Another interesting measuring feature of this system is the possibility of determinating the area or effective radius of small, unresolvable, opaque particles (Osterberg, 1947b; Osterberg and Pride, 1950; Osterberg, 1952).

Despite its versatility for qualitative observation and some measurements, Osterberg's Polanret system has found little practical application. Its manufacture is difficult and expensive. Besides, as Osterberg (1965) himself stated, this system has some optical defects. In particular it incorporates anisotropic elements in the image space of the objective, where converging light beams pass. Thus the anisotropic elements produce non-uniform phase shifts for individual rays traversing the objective at different angles. This causes a blurring of the focused image. Moreover, due to the elliptical polarization introduced by lenses the system fails to achieve the maximum possible image contrast. The problem of injurious halo effects exists in these systems as well.

Taking account of these defects, Osterberg (1965) has proposed another system, based on novel principles. A characteristic feature is a plurality of spatially separated light beams illuminating the specimen, as well as localization of phase-shifting and amplitude controlling elements in the substage of the microscope. This system is, however, very complicated, and its practical application seems doubtful.

The most recent commercial version of the Polanret system has been described by Richards (1973). It uses improved phase plates with the conjugate and complementary areas uniformly polarized at 90° to each other. A wide range of phase and amplitude variation is provided by two separate controls. Richards discussed the use of this instrument for studying a range of specimens.

Despite its drawbacks, the Polanret method has stimulated some practical solutions of the problem of highly flexible variable phase contrast microscopy; some of these are described in the following sections.

B. *Nomarski's Variable Achromatic Phase Contrast System*

In the Polanret method, as well as in some other variable phase contrast systems, a basic phase difference ψ_b, usually equal to 90°, is introduced between the direct and diffracted light, to which a variable component is added or subtracted by means of a compensator. In the system shown in Fig. 33 the basic phase difference $\psi_b = 90°$ is produced by the quarter-wave plate Q. This plate as well as all the other standard phase plates made of transparent refracting materials are chromatic, i.e. ψ_b varies with wavelength.

Another defect of the Polanret method is that optimum contrast is achieved by two separate manipulations. One step is matching the phase shift by means of polarizer P_1 (Fig. 33), the other is the variation of amplitude by using analyzer P_2. This is an inconvenience which can introduce inaccuracies in phase matching.

These defects are eliminated in a variable phase contrast microscope developed by Nomarski (1968a, b). This microscope is similar to the Polanret system, but the arrangement of its individual polarizing and birefringent elements is basically different. First, zonal polarizers Z_1 and Z_2 (Fig. 33) are set up so that their directions of light vibration (vectors V_1 and V_2) make an angle of 45° with each other. Secondly, the direction of light vibration transmitted by the polarizer P_1 is parallel to the light vibrations from the conjugate area (zonal polarizer Z_1). Thirdly, quarter-wave plate Q is placed between the zonal polarizers and analyzer P_2 and is oriented so that its slow (or fast) axis is parallel to the light vibrations of the conjugate area (zonal polarizer Z_1). Now, the quarter-wave plate and analyzer constitute a Sénarmont compensator. Other birefringent compensators can be used equally well.

The theory formulated by Nomarski (1968b) shows that in these conditions the compensator alters both the phase ψ and intensity ratio T of the direct light simultaneously. The alteration is, moreover, in accordance with Richter's condition (Eqns (9) and (10)). Thus the image of a phase object becomes darkest (or brightest) by performing only one action, and so one has the possibility of making measurements of OPD, or other properties of the object, more exactly.

Returning to the vector representation of phase contrast (Fig. 4), the principle of Nomarski's system is based on the fact that the maximally dark image of a phase object can be obtained by adding to the vector of the direct light **u** a vector **−p** being opposed to the vector **p** which represents the light deviated in phase by the object under study. The sum of vectors $\mathbf{u}+(-\mathbf{p})$ gives a resultant vector $\mathbf{u}'$ which now represents the modified background light. As can be seen, the vector $\mathbf{u}'$ has the same length as vector **d**, which represents the diffracted light, but it is anti-parallel to **d**. Now, the brightness of the image results from the sum of vectors $\mathbf{u}'$ and **d**, but $\mathbf{u}'+\mathbf{d} = 0$, so the image is maximally dark. As can easily be proved, the angle between the vector $\mathbf{u}'$ and **u** is the optimum phase shift, equal to $\psi_{op} = -90° + \phi/2$, whereas the length of the vector $\mathbf{u}'$ is equal to $2\sin(\phi/2)$; thus the transmittance ratio $T = 4\sin^2(\phi/2)$. This is in agreement with Eqns (9) and (10).

The zonal polarizers in this system cannot produce any additional phase shift between the diffracted and direct light. Thus, they must be made very carefully; their thickness should be exactly identical. If this condition is fulfilled, the achromatic error $\Delta\psi$ of phase matching is very small and given by

$$\Delta\psi = \frac{\psi(\lambda) - \phi(\lambda)}{2} \tag{22}$$

where $\psi(\lambda)$ and $\phi(\lambda)$ are the dispersions of the phase shift in the compensator and specimen, respectively.

A prototype of Nomarski's variable achromatic phase contrast microscope was built in 1968 and exhibited at 62 Exposition de Physique, Paris. However, this microscope is not yet commercially available as its industrial manufacture is difficult and expensive.

C. *Nikon Interference-phase Device*

The Nikon interference-phase device is a continuously variable phase contrast device similar to the Polanret system shown in Fig. 33. An essential difference lies in the design of the polarization phase plate, which in the Nikon device consists of the zonal polarizer Z_2 only; the polarizer Z_1 is omitted. Thus, only the complementary area is covered by the polarizing films and the conjugate area is clear. The remaining polarizing and birefringent elements are identical in both systems, but their arrangement is somewhat different. In the Nikon system these elements are placed in a transferred-image unit attached to the microscope between the objective O and ocular head. The polarization phase plate is located in the reimaged back focal plane of the objective, and immediately in front of this plate a rotatable polarizer and quarter-wave plate are placed. The quarter-wave plate is orientated so that its slow (or fast) axis makes an angle of 45° with the direction of light vibration transmitted by the polarization phase plate. Thus, rotating the polarizer changes the phase shift ψ between the diffracted and direct light waves, whereas rotating the analyzer varies the intensity ratio T of the direct beam. This is nearly the same effect as in the Polanret system.

When white light illumination is used every small phase detail shows different interference colours produced by the mutual interference of the direct and diffracted light (Fig. 34). The colours depend on phase differences ϕ in the specimen, and the OPD between the object and its surrounding medium can be evaluated if any difference in colour appears between the object image and its background.* This is, of course, only an approximate evaluation of OPD. A more accurate method of measurement is to use monochromatic (green) light and set the rotatable polarizer and analyzer so that the object image appears maximally dark, and then rotate these polarizing elements until the background appears the darkest. The optical path difference δ is calculated from the formula $\delta = (\theta_1 - \theta_2)\lambda/180°$, where θ_1 and θ_2 are the readings (in degrees) of setting of the polarizer at maximum darkness of the object image and background, respectively.

It must be stressed that this device does not have all the measuring possibilities of a true interference microscope. In reality, this is a variable phase contrast accessory and it has defects typical of the phase contrast method. In particular, the halo and shading-off effects appear, therefore measurements can only be made on very small objects, the dimensions of which are less than 0·01 mm. In the case of 2 μm objects or smaller, an accuracy of OPD measurement of about $\lambda/50$ can be attained.

The Nikon system is available commercially in the form of an attachment for Nikon transmitted light microscopes. It consists of a special body tube inserted between the nose piece and ocular head. This carries a slider with four polarizing phase rings corresponding

* The most differentiated colours occur in the image when the background is red-purple. The Nikon technical descriptions do not explain how that is done. Probably, a first order red plate is added to the quarter-wave plate.

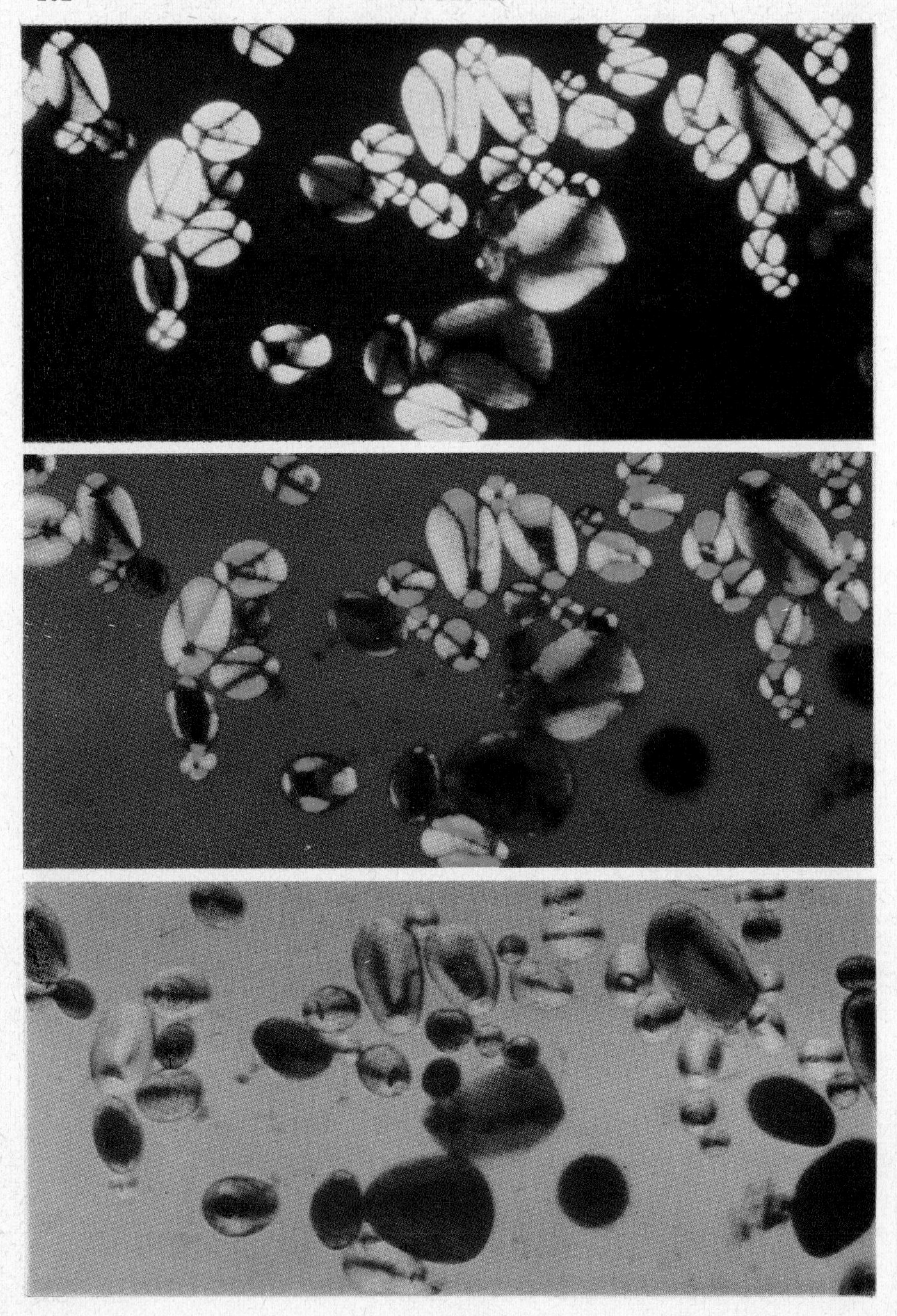

FIG. 34. Phase-interference images observed with the Nikon interference-phase device set at black, purple and light green background. Reproduced by permission of Nikon Europe B.V., the Netherlands.

to 10, 20, 40 and 100× standard objectives. A conventional turret condenser, centring telescope and filters are provided (Fig. 35).

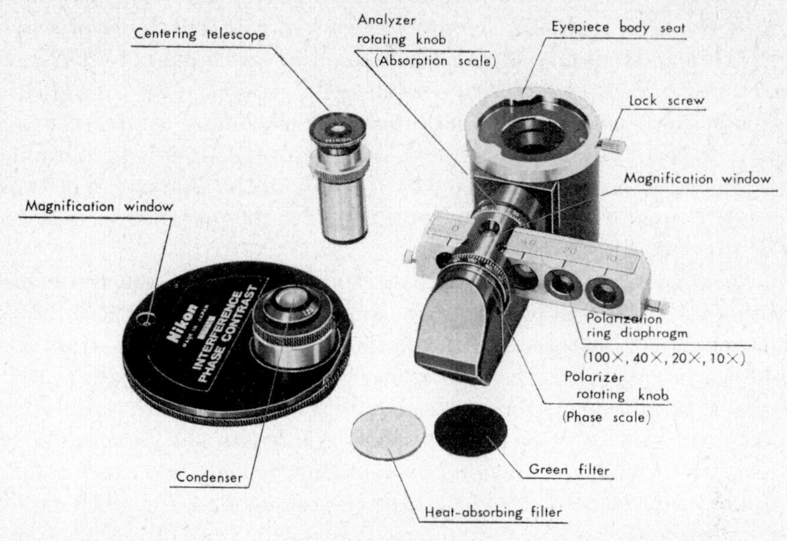

FIG. 35. Nikon interference-phase attachment. Reproduced by permission of Nippon Kogaku K.K., Japan.

D. *Other Systems*

Alternative methods related to the Polanret system were described by Taylor (1947), Hartley (1947), Kastler and Montarnal (1948), Locquin (1948) and Wolter (1955). Payne (1950, 1954) described commercial instruments which incorporated birefringent phase plates made of half wave quartz. Oettlé (1950a, b) used one of these instruments to investigate the effects of variation of both phase and amplitude on the appearance of images of a number of test objects. Some of these special systems are dealt with adequately in a number of standard works (see Bennett *et al.*, 1951; Françon, 1954, 1961; Beyer, 1965) and will not be discussed here.

It is, however, worth mentioning an interesting variable colour-amplitude phase system developed by Barer (1949), who used a type of dichroic filter the spectral transmission of which changes with plane of polarization of the incident light. This filter combined with a linear polarizer transmits, e.g. green light maximally for one orientation of the plane of polarization, and red light when the polarizer is rotated through 90°. If in a phase contrast system of the Polanret-type (Fig. 33) the polarizer P_1 is replaced by such a dichroic filter then one can obtain mainly green light from the conjugate area and mainly red from the complementary area, or vice versa. The phase contrast image is produced by interference of light of the common wavelengths passed by both the conjugate and complementary areas. The light of wavelengths transmitted only by the complementary area cannot effectively interfere with light of different wavelengths passing only through the conjugate area, and these superimpose on the phase contrast image giving a type of coloured darkfield illumination (Rheinberg illumination). The image contrast is varied by rotation of the analyzer. The same effect is obtained when the analyzer, instead of the polarizer, is replaced by the dichroic filter.

Another interesting device for continuously variable phase contrast microscopy was developed by Françon and Nomarski (1950, 1952). A characteristic feature, which distinguished this device from the Polanret system, is a reflection phase plate set at Brewster's angle between rotatable polarizer and analyzer. This phase plate has a strip-shaped form and is conjugated with a condenser slit diaphragm by means of an auxiliary re-imaging attachment. This device was formerly available from Nachet of Paris, and recommended for both trans-illumination and incident-light phase contrast microscopy (Lamarche, 1952).

V. Systems for both Phase Contrast and Interference Microscopy

A. *Why Phase Contrast together with Interference Microscopy?*

Both phase contrast and interference microscopy display phase objects and structures of different refractive indices or different thicknesses. The phase contrast method provides a higher sensitivity to minute changes in optical path in the specimen, but, on the other hand, produces disturbing halo and shading-off effects which make it difficult to carry out accurate studies of large objects. This defect is not present in some types of interference microscopy, and a true interference method does not discriminate between large and small objects, so that all phase objects can be imaged with equal faithfulness regardless of their shape and dimensions. Furthermore, a true interference microscope is capable of giving both the qualitative and quantitative information on the distribution of optical path differences. It is true that by using some variable phase contrast devices one can measure OPD, but this relates to very small objects and fine structures, whereas some interference microscopes permit measurements of OPD on large objects, such as biological cells and tissues. Interferometric measurements are, in particular, very useful for dry mass determination in biological cells and their organelles (Barer 1966; Ross 1967).

Recently, the differential interference contrast (DIC) method has become very popular. The most useful is Nomarski's (1955) system which is now manufactured by several firms. An important feature of DIC images is a relief appearance or shadow-cast effect. For transparent biological specimens this effect results from both the gradient of OPD in the interior of cells and from surface irregularities. To avoid misinterpretation, it should be established whether the relief appearance does or does not represent the object surface topography (Padawer, 1968). For this purpose a combination of DIC and phase contrast is useful.

Another interesting feature of the DIC method is that it produces effectively optical sectioning (Allen *et al.*, 1969). This is particularly obvious when high-aperture objectives are used together with high condenser illumination apertures. Then the depth of field is extremely shallow, and it is possible to obtain sharp and contrasting images of structures in a focused plane without appreciable image impairment by out-of-focus details. This is an advantage of the system which may be important for morphological studies of living cells (David *et al.*, 1966; Bajer and Allen, 1966; Padawer 1968).

Contrary to the DIC technique phase contrast is characterized by a rather large depth of field, and a focused phase contrast image shows up a thicker layer of the specimen than the DIC image.

These examples show that a combination of phase contrast and true interference technique could be very useful to reveal more completely the nature and properties of the specimen.

B. "*Interphako*"

The "Interphako" system offers an approach to microscopy with both phase contrast and interference methods. It has been developed by Beyer, and is commercially available from C. Zeiss, Jena, together with the "Amplival" microscope for transparent objects (Beyer and Schöppe, 1965), or the "Epival" microscope for reflecting specimens (Beyer, 1971).

The essential part of this system is a compact Mach-Zehnder interferometer which is composed of two identical rhomboidal prisms R_1 and R_2 (Fig. 36) arranged behind the objective. The skew surfaces S_1 and S_3 of these prisms constitute semi-transparent mirrors, whereas surfaces S_2 and S_4 are fully-reflecting mirrors. Before the interferometer an image-transfer system, T_1 and T_2, is inserted, which re-images the back focal plane F′ of the objective O to an easily accessible plane F″ inside the interferometer. For maintaining this plane in a fixed position a reflecting prism RP is used, which is suitably translated in the direction p parallel to the optical axis of the re-imaging system T_1 when one objective is replaced by another of different magnification.

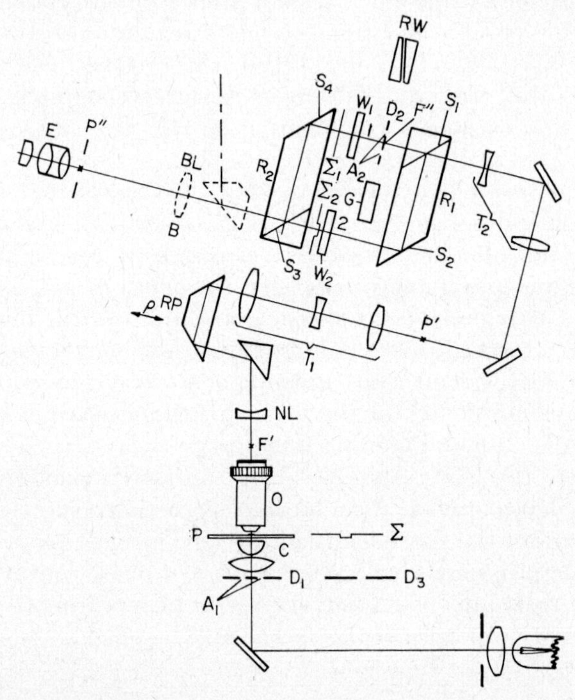

Fig. 36. Optical diagram of the "interphako" microscope. Reproduced with permission from Beyer and Schöppe (1965).

The geometrical image of the object P is projected to infinity by the negative lens NL, and is then focused by the image-transfer system T_1 to an intermediate image plane P′. This latter is then transferred by lenses T_2 into the front focal plane of the eyepiece E. Before that, however, the light is divided into two beams by the semi-transparent surface S_1 of the rhomboidal prism R_1. One beam follows path 1 and the other path 2. They are then recombined by the semi-transparent mirror S_3 of the rhomboidal prism R_2 and interfere with each other.

This system is capable of giving either uniform or fringe interference images with either large or differential image separation, as well as three different kinds of phase contrast: "interphako" (i.e. phase-interference contrast), normal positive and negative phase contrast, and coloured phase contrast. The most interesting is the interphako method. It is achieved by inserting an annular diaphragm D_2 between the rhomboidal prisms in the focal plane F''. This diaphragm is conjugate with another diaphragm D_1 located in the front focal plane of the condenser C, so that the direct light, the primary source of which is the condenser annulus A_1, passes through the annular opening A_2, whereas the diffracted light is stopped. Thus an essential difference is established between the light beams 1 and 2. Beam 1 consists of the direct rays only, and it can be regarded as the reference beam, the wavefront Σ_1 of which is not deformed by the object (this is true for fine objects only), whereas beam 2 includes both the direct and diffracted rays, and its wavefront Σ_2 is the same as the original object waveform Σ. Recombining the wavefronts Σ_1 and Σ_2 produces a phase-interference image (interphako image) of the object under examination (Fig. 37 top and middle). The brightness (Fig. 38) or colours can be varied continuously by means of a phase compensator consisting of two glass wedges W_1 and W_2. Sliding one of these wedges changes the phase difference between the reference wavefront Σ_1 and deformed image wavefront Σ_2. When these wedges are suitably calibrated OPD can be measured. An exact measurement may, however, be performed only on very small objects. In the case of large objects the diffracted rays are insufficiently deviated from the direct beam and cannot be effectively separated by the annular diaphragm D_2, and the wavefront Σ_1 is deformed by the object. This deformation is the greater the larger the object, the smaller the effective focal length of the imaging system, and the greater the width of the annuli A_1 and A_2. This is, in fact, the problem of halo and shading-off effects.

Typical positive or negative phase contrast is realized by using turret-mounted phase rings instead of the annular diaphragm D_2. Simultaneously, light path 2 is blocked. For coloured phase contrast a highly dispersive phase plate is used. It produces greatly differing phase shifts ψ between the direct and diffracted light. Over the range of wavelengths of the visible spectrum this phase shift varies from about 360° to 540°. According to their individual phase difference, phase details are thus variously coloured. Striking colour phase contrast is especially obtained when the objects are embedded according to the so-called colour-dispersion method.

These phase contrast techniques can be rapidly replaced by shearing interference methods. For this a slit condenser diaphragm D_3 is used instead of the annular diaphragm, and the upper annular diaphragm

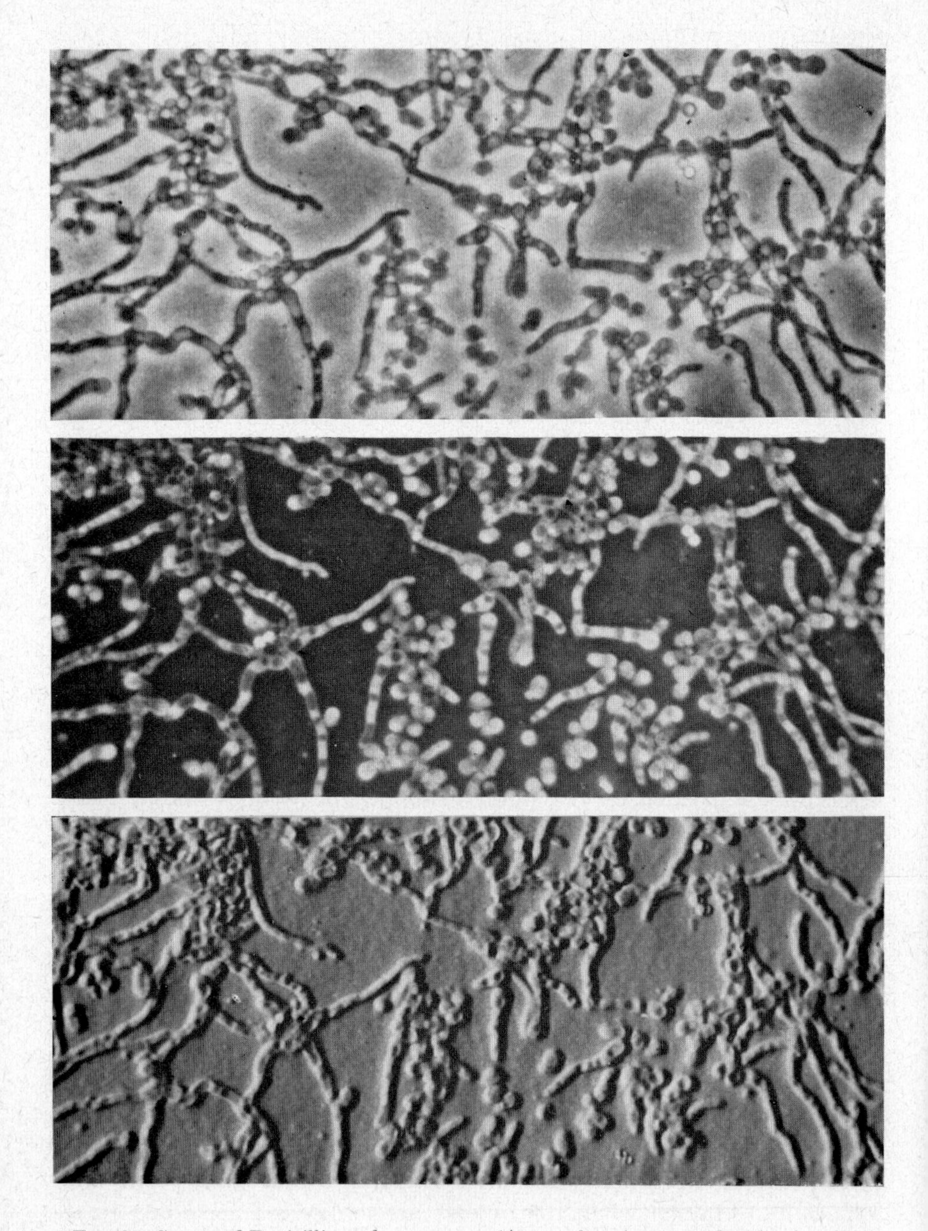

FIG. 37. Spores of Penicillium chrysogenum (Agar culture) seen under the Amplival-interphako microscope set at positive (top) and negative (middle) phase-interference (interphako) image contrast, and differential interference contrast (bottom). Planachromatic objective ×12·5/0·25. Print magnification ×240. Reproduced by permission of VEB Carl Zeiss JENA.

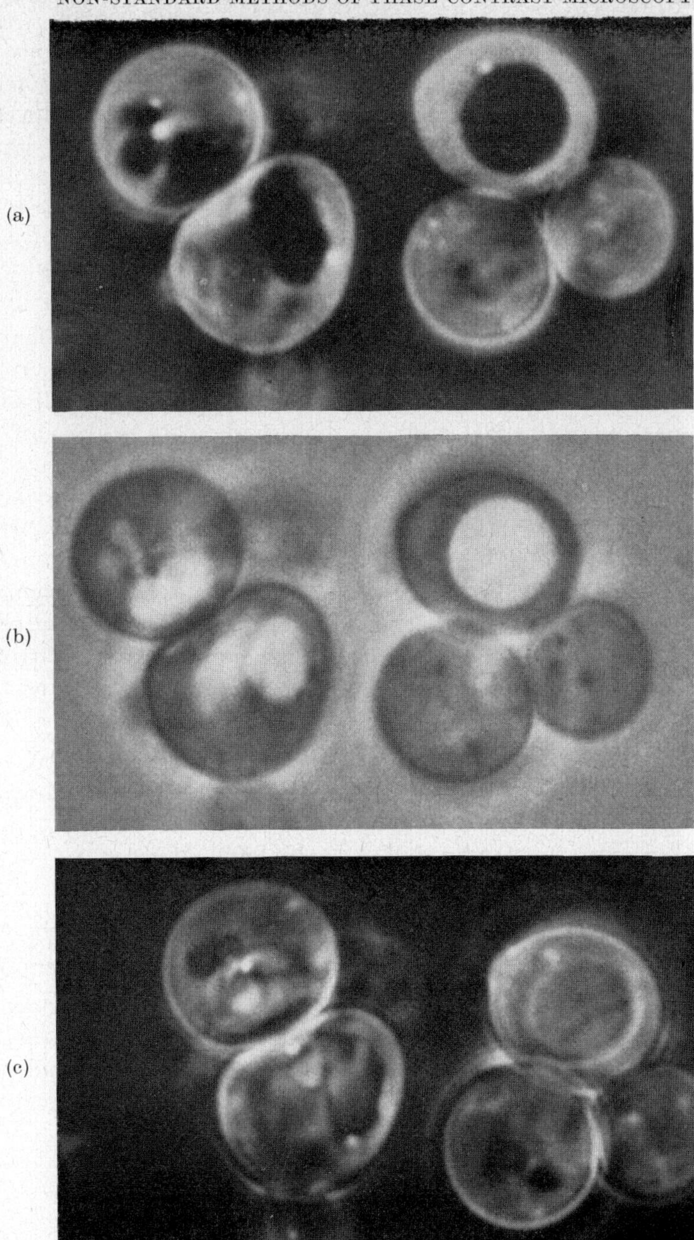

FIG. 38. Interference-phase contrast (interphako) images of living yeast cells immersed in sugar solution. (a) optimum dark contrast set for vacuoles; (b) optimum dark contrast set for granules; (c) setting for maximum dark background. Planachromatic objective ×100/1·25. Print magnification ×4000. Reproduced by permission of VEB Carl Zeiss JENA.

(D_2) is replaced by a pair of rotatable wedges RW. Simultaneously, a compensating glass plate G is included in path 2 of the interferometer system. Now, both interfering light components are identical. Rotating the wedges RW about the axis of light beam 1 enables the lateral wavefront shear (image duplication) to be continuously varied from zero to about 4 mm (this value relates to the image plane P″). If the wavefront shear is comparable to the limit of the resolving power of the objective, differential interference contrast is obtained (Fig. 37, bottom). When the wavefront shear is great, two adjacent interference images result from one object. Both images are of identical quality, and measurements of OPD can be made from one image to the other, using the minimum intensity (extinction) method or half-shade method. To employ this last method a half-shade plate is placed in the image plane P′.

By suitably adjusting the wedges W_1 and W_2 uniform field interference ("even field" or "flat-tint" interference) as well as fringe interference is realized. To obtain a good interference effect in the image plane and well contrasty images, the condenser diaphragm slit should be sufficiently narrow. The width w, of the slit should be adjusted so that its image in the back focal plane F″ of the objective is narrower than one-quarter of the inter-fringe spacing of the interference pattern appearing in that plane. This plane is observed by inserting the Bertrand lens BL. It can be shown that the width w of the condenser slit is defined by the relation

$$w < \frac{\lambda}{4} \frac{f_c}{d} \tag{23}$$

where λ is the wavelength of the light, f_c is the focal length of the condenser, d is the wavefront shear (image duplication) related to the object plane.

The narrow slit is, of course, a defect because it reduces the condenser aperture, and the resolution in the direction perpendicular to the slit is decreased. But, when monochromatic light is used, the condenser diaphragm slit can be replaced by a combination of linear gratings with different intervals i defined as follows

$$i = \frac{f_c}{d} n\lambda \tag{24}$$

where $n = 1, 2, 3, \ldots$ For a grating with a given i and w (Eqn 23) one cannot obtain an arbitrary image duplication d and contrasty interface image. Such an image appears for $d_1 = f_c\lambda/i$, $d_2 = 2f_c\lambda/i$, $d_3 = 3f_c\lambda/i, \ldots$, and so on.

C. *Phase Contrast and Interference Microscope based on the Michelson Interferometer*

This microscope is a result of work on simultaneous variable phase contrast and interference microscopy performed by the present writer in 1966–68.

The basic optical elements of the system finally developed are shown in Fig. 39. The most important is the polarizing interferometer of the Michelson type which incorporates an interference polarizer IP used as the beam splitter, two quarter-wavelength plates Q_1 and Q_2, two mirrors M_1 and M_2, and a Sénarmont compensator consisting of a quarter-wavelength plate Q_3 and linear polarizer (analyzer) P_2. This interference system is located between the objective (O) and ocular head (RP and E). The lenses, L_1 and L_2, act as a re-imaging system, and in reality they are more complicated than shown.

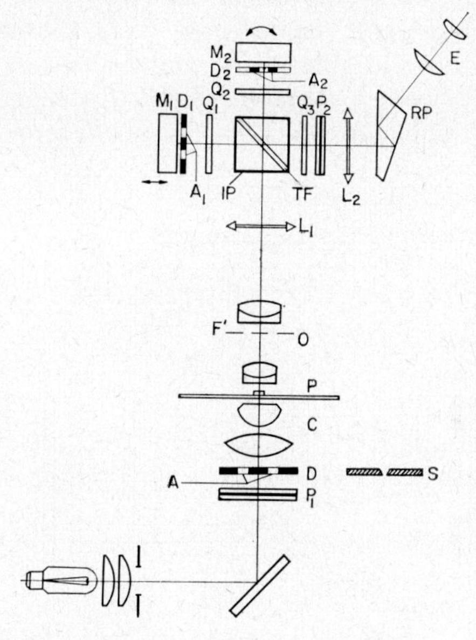

FIG. 39. Optical diagram of the variable phase contrast and interference microscope based on a polarization interferometer of the Michelson type.

The interference polarizer IP consists of two right-angle prisms, the hypotenuse surfaces of which are covered, by vacuum evaporation, by a stack of dielectric thin films TF alternately of high and low refractive index, and cemented together. This interference polarizer is made, e.g. by the procedure developed by Banning (1947). A light beam incident upon the interference polarizer IP is split into two parts: a reflected component and transmitted component, polarized at right angles to each other. The reflected component vibrates perpendicularly to the plane of light incident on IP and the transmitted component vibrates in this plane. The quarter-wave plates, Q_1 and Q_2, rotate the directions of light vibrations through 90° by the double passage of the light, so the beam reflected by the mirror M_1 passes through the interference polarizer IP, whereas the other beam reflected by the mirror M_2 is totally reflected from the stack of thin films TF and recombines with the first beam. Both beams, after passing through the quarter-wave plate Q_3 and analyzer P_2, vibrate in the same direction and thus can interfere with each other. The quarter-wave plates Q_1, Q_2 and Q_3 are set in such a manner that their axes (slow and fast) make an angle of 45° with the vibration directions of light beams split by the interference polarizer IP.

This system is used for both variable phase contrast microscopy and shearing interference microscopy with continuously variable wavefront shear. For variable phase contrast an annular diaphragm D is located in the front focal plane of the condenser C, and two diaphragms D_1 and D_2, one of which has an annular opening A_1 and the

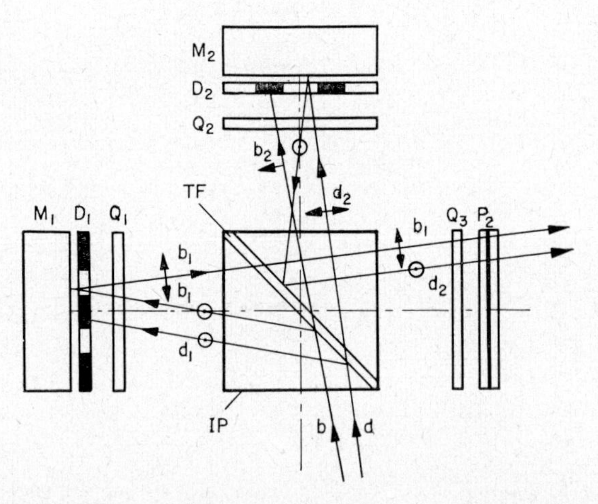

FIG. 40. A more detailed ray diagram of the interference system shown in Fig. 39.

other has an absorbing ring A_2, are located close to or in the mirrors M_1 and M_2 in such a manner that the ring A_2 and annulus A_1 are conjugate to the condenser annulus A. The image of this latter is formed in the back focal plane F′ of the objective O and projected by means of the re-imaging lens L_1 into the planes of the diaphragms D_1 and D_2.

In these conditions the direct light emerging from the condenser annulus A is totally absorbed by the ring A_2, whereas it is transmitted by the annular opening A_1, then reflected by the mirror M_1, and passes through the interference polarizer IP to reach the eyepiece E. The diffracted light is, in contrast with the direct light, stopped by the diaphragm D_1 and reflected by the mirror M_2 only. It is then reflected by the interference polarizer IP, and on passing through the quarter-wave plate Q_3 and analyzer P_2, can interfere with the direct light reflected from the mirror M_1. This is illustrated in detail in Fig. 40 which shows the paths of two rays: direct b and diffracted d, arbitrarily selected from the light beam leaving the object. Each of these rays is split into two components: reflected b_1, d_1 and transmitted b_2, d_2. The reflected component d_1 of the diffracted ray and the transmitted component b_2 of the direct ray are absorbed by the diaphragm D_1 and D_2, respectively, whereas the reflected component b_1 of the direct ray and transmitted component d_2 of the diffracted ray are reflected by the mirrors M_1 and M_2, and then directed to the eyepiece E by the interference polarizer IP.

Rotating the analyzer P_2 varies the phase difference ψ between the direct and diffracted light vibrations. This phase difference is $\psi = 2\theta$, where θ is the angle between the direction of light vibrations in the analyzer and one of the axes, e.g. slow axis, of the quarter-wave plate Q_3. Thus a continuous variation of the phase shift ψ between the direct and diffracted light beams is achieved. The amplitude ratio of the direct and diffracted beams can be varied by rotating the quarter-wave plates Q_1 and Q_2, or by using an additional linear polarizer P_1 located before the interference polarizer IP, e.g. below the substage condenser C. Rotating this polarizer alters the amplitude ratio of the light vibrations of the direct and diffracted beams, while their phase difference remains constant within the range of a given quadrant. When the vibration direction of the polarizer P_1 passes from the neutral to anti-neutral quadrants the phase shift jumps suddenly by 180°, similarly to the Polanret procedure. If α denotes the angle between the direction of light vibration in the polarizer P_1 and principal section of the interference polarizing splitter IP (this section is the plane of the paper in Fig. 39 or 40), then for $\alpha = 0$ only the diffracted light reaches the eyepiece E, and for $\alpha = 90°$ only the direct light passes to the eyepiece. For intermediate positions the intensity ratio T varies as $\tan^2 \alpha$. Thus, a phase contrast system is achieved wherein both the phase difference ψ and intensity ratio T of the direct and diffracted light beams are continuously variable.

This system is operated as follows. It is first aligned for Köhler illumination and the object P is brought into focus. Then the mirrors M_1 and M_2 are so adjusted that zero order uniform field interference without any lateral shearing in the exit pupil of the objective is observed. For this the diaphragms D_1 and D_2 are decentred or removed, and the eyepiece E is replaced by an auxiliary lens focused at the image of the condenser annulus A. In setting for zero order uniform field interference, the diaphragms D_1 and D_2 are returned, and the annular opening A of the condenser diaphragm is exactly centred with the annulus A_1 and absorbing ring A_2 of the diaphragms D_1 and D_2. The image contrast is then varied by rotating the analyzer P_2 and polarizer P_1.

This system can easily be transformed into a polarization interference microscope with variable wavefront shear. For this a slit diaphragm S is used instead of the annular diaphragm D, and the diaphragms D_1 and D_2 are removed. Then, in general, straight interference fringes are observed in the exit pupil of the objective O. If, simultaneously, the image of the condenser slit S is brought parallel with these fringes, and the width w of the slit is smaller than given by Eqn (23), then uniform field interference is produced in the image plane of the microscope, and the object P is seen as two identical images laterally duplicated to a greater or lesser degree and coloured according to the Newtonian scale, if white light is used. If monochromatic light is used, the background appears with uniform intensity and the duplicated images show up as differences in relative brightness. The amount of image duplication and direction of wavefront shear can be controlled by tilting one of the interferometer mirrors, or by means of two additional plane parallel birefringent plates located in the split light beams between the interference polarizer IP and quarter-wave plates Q_1 and Q_2 (Pluta, 1969d, 1970).

With uniform field interference, OPD is measured by the Sénarmont method. This can be done in two ways. The easier way is to use monochromatic light and first set the analyzer P_2 so that one of the separated images appears at its darkest, and then rotate the analyzer through an angle θ until the second image appears at its darkest. This is the minimum intensity or extinction method. Optical path difference δ is then calculated from the formula $\delta = \lambda\theta°/360°$. Another way is to start with the analyzer setting at which one image appears with exactly the same brightness as the background immediately adjacent to the object and then rotate the analyzer through an angle γ until the other image appears with the same brightness as the background. This is the matching or minimum visibility method. OPD is calculated from the formula $\delta = \lambda\gamma°/180°$. With these methods values of OPD equal to fractions of λ can be determined. OPDs greater than λ are measured by examination of interference colours or from displacements of interference fringes. Fringe field interference is achieved when the image of the condenser slit S is not in coincidence with the interference fringes in the exit pupil of the objective O. Then the interfering wavefronts are inclined to each other in the image plane and interference fringes appear in the field of view of the eyepiece E. This kind of interference is, in general, obtained by tilting one of the mirrors M_1 and M_2, or by shifting the slit diaphragm in the direction of the condenser axis.

Excluding one of the diaphragms D_1 or D_2 yields other possibilities for investigating phase objects. In particular, the phase-interference contrast ("interphako" method) can easily be realized. For this, only the diaphragm D_2 need be excluded from the interference system shown in Fig. 39.

D. *Other Systems*

The microscopes that have just been described enable the diffracted and direct light beams to be split, modulated in phase and/or amplitude, then recombined and brought into interference. Other systems based on this principle have been described in the literature. Among them is Meyer-Arendt's (1960) system in which the separation of the direct and diffracted light is produced by means of a fully reflecting ring conjugate with a condenser annulus.

The phase-interference systems considered so far are capable of giving either variable phase contrast or true interference images by the use of one and the same objective. Many authors have, however, attempted to construct systems capable of giving a phase contrast image mixed with a fringe field or uniform field interference image. Systems which produce, strictly speaking, a simultaneous phase contrast and interference image have been proposed and studied by Barer (1952d) and Nomarski (1952). In Barer's system the phase contrast method is combined with the transmission Fabry-Perot interferometer as modified by Merton (1947). In this case, a phase object is placed between a pair of glass plates (slide and cover slip) having half-silvered surfaces set parallel to and opposite one another, and light beams undergoing successive partial reflections at the two reflecting surfaces contribute to the interference pattern which superimposes on a phase contrast image. This procedure has been theoretically discussed by Nomarski (1952) and extended to reflecting specimens by combining phase contrast with Tolansky's multiple beam interference. Such a combination of methods enables the nature of some phase specimens to be imaged more completely, since the mixed interference–phase contrast image reveals both fine and broad structures. Small objects and minute details are revealed better in phase contrast than interference images, whereas broad phase structures are shown more faithfully by interference procedures.

Other rather complicated systems for simultaneous phase contrast and interference microscopy have been devised by Bennett (1949), Räntsch (1949), Lohmann (1954), and others. These systems are described in more detail in the book by Krug *et al.* (1961).

VI. Phase Contrast Combined with other Microscopic Techniques

A. *Phase Contrast with both Bright-Field and Dark-Field Illumination*

Any phase contrast device can normally be used for bright-field microscopy as well. The most popular phase condensers are provided with a rotatable disc containing three to five annuli for phase contrast and one or two clear openings for bright-field observation. The phase ring in objectives causes, of course, some disturbance of bright-field images, but in many cases this is insignificant, and one and the same phase objective can be used for both purposes.

A turret phase condenser can also be used for dark-field, or rather quasi dark-field, microscopy with low- and medium-power objectives. In this case the condenser annulus is chosen so that its image in the back focal plane of the objective lies completely beyond the edge of the objective exit pupil.

For simultaneous phase contrast, bright-field and dark-field microscopy so called "universal phase condensers" with continuously variable diameter of the annulus are more suitable. To such systems belong, e.g. the Heine condenser manufactured by E. Leitz, Wetzler, the "Polyphos" condenser produced by C. Reichert, Vienna, and the pancratic condenser available from C. Zeiss, Jena.

It may be pointed out that variable phase contrast systems with continuously controlled transmittance ratio T of the conjugate area give a type of dark-ground image when $T = 0$ (direct light is completely extinguished).

The "ultra-anoptral" condenser devised by Soran and Diaconeasa (1957) gives a dark-field image superimposed on an anoptral contrast image. The dark-field illumination is achieved by an additional condenser annulus of high diameter. This annulus is covered by a short-wavelength (blue or violet) monochromatic filter. Such a combination of anoptral contrast with dark-field illumination reduces some of the injurious diffraction and light-refraction effects that occur at highly refractile objects.

B. *Simultaneous Phase Contrast and Fluorescence Microscopy*

Fluorescence microscopy has recently become very important for the examination of some biological specimens. A combination of this technique with phase contrast is often useful in biological and medical research. Alternating or simultaneous phase contrast and fluorescence image display enables an exact localization of fluorescing substances, e.g. specifically fluorochromed antigens, in cell cultures and tissue sections to be made. Such a combination poses no problem if the fluorescence excitation is by incident light from a vertical illuminator. The use of any standard phase contrast equipment for transmitted light is then possible. In the case of a transmitted light fluorescence microscope this simple procedure is useless as the condenser annular diaphragm greatly reduces the intensity of the exciting light. To overcome this limitation several systems have been devised. A well known one is that developed by Gabler and Herzog (1965), and manufactured by C. Reichert. This consists of a double illuminator and special phase-fluorescence condenser which incorporates annular diaphragms made of dark glass (e.g. Schott UG1 filter) transparent to the exciting light (near u.v. and blue-violet). The phase contrast image is produced by visible light from a separate low-voltage illuminator, which passes through a clear condenser annulus conjugate with the phase ring. The fluorescence image is excited by short-wavelength light emitted by a high-pressure mercury lamp and transmitted to the specimen by both the clear annulus and remaining filter-covered area of the condenser diaphragm. A 45° mirror in the light path partially reflects the light beam from the low-voltage illuminator. In order to increase the colour contrast between fluorescing and non-fluorescing parts of the image, a red filter is sometimes inserted into the beam producing the phase contrast image. An example of the performance of this system is shown by the photomicrographs in Fig. 41.

The writer has combined fluorescence microscopy with phase contrast by using annular condenser diaphragms made of interference thin films instead of the dark glass filter. The use of such diaphragms has some advantages because the spectral transmission curve of an interference band filter is sharper than that of a glass filter. The fluorescence light of wavelengths much closer to the exciting wavelengths can be observed. Experiments with this system show that for simultaneous phase

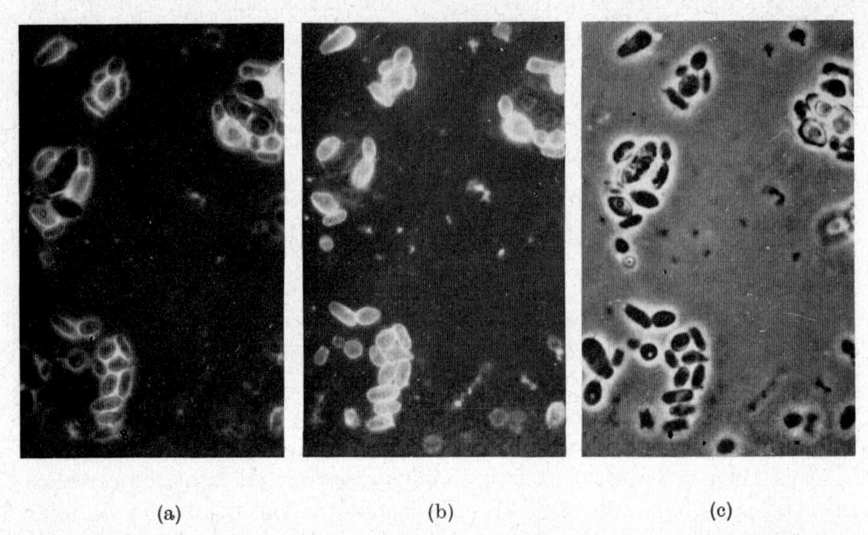

Fig. 41. Two different types of yeast cells one of which is stained with FITC, seen under the Reichert phase-fluorescence microscope. (a) Simultaneous fluorescence and positive phase contrast image; (b) fluorescence image only; (c) positive phase contrast image only. Objective ×40/0·65. Print magnification about ×500. Photomicrographs courtesy of F. Herzog (C. Reichert Vienna).

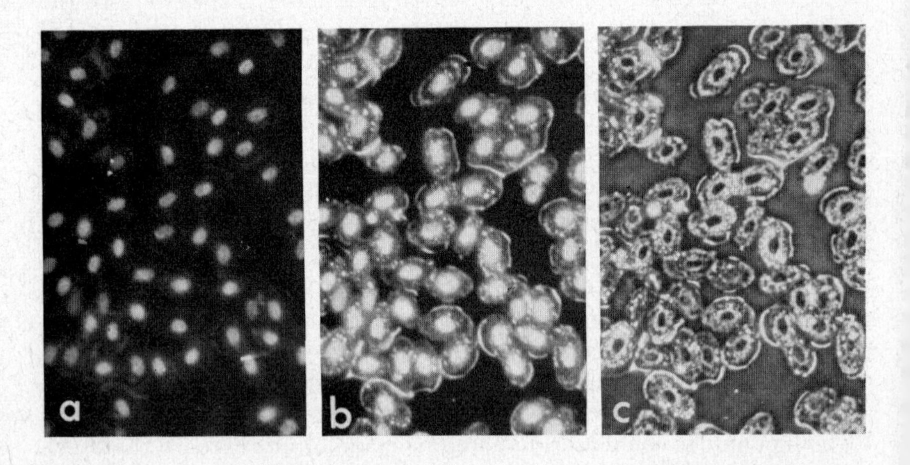

Fig. 42. Smear of frog blood stained with acridine orange seen under a fluorescence microscope (Polish Optical Works) combined with phase contrast by using annular diaphragms made of interference thin films. (a) Fluorescence image only; (b) simultaneous fluorescence and negative phase contrast image; (c) negative phase contrast image only. Objective ×20/0·4. Print magnification ×300.

contrast and fluorescence negative phase objectives are preferred (Pluta, 1971). Three typical photomicrographs are presented in Fig. 42.

Essentially different equipment is manufactured by C. Zeiss, Jena. In this an image-transfer sub-condenser system is applied which projects the image of a clear annulus through the partially transparent 45° mirror into the front focal plane of the condenser. This equipment, as well as some other systems, is described in more detail by Beyer (1965).

C. *Phase Contrast in Polarized Light*

The combination of phase contrast with a polarizing microscope presents no difficulties. Any standard phase contrast device can easily be adapted to the polarizing microscope, or any conventional phase contrast microscope can be fitted with a polarizer and analyzer. Only a single illuminating beam is needed. Nevertheless, some systems with two illuminating beams (one for producing a phase contrast image, and the other for observation in polarized light) have been designed, and even patented (e.g. Austrian Pat. 245283). The advantages of such systems are, however, doubtful.

An interesting phase contrast system with polarized light has been devised by Wolter (1951). This gives a mixed phase contrast, bright-field and polarization image in which phase, amplitude and birefringent objects are displayed in different colours. An essential element is a condenser diaphragm made of polarizing film and containing a clear annulus or slit. The annulus is conjugate with a typical phase plate followed by an optically active plane-parallel plate cut at an angle of 90° to the optical axis of a quartz crystal. A rotatable polarizer and analyzer are placed before the condenser and behind the quartz plate, respectively. When the polarizer is crossed with the polarizing film of the condenser diaphragm, and the analyzer is set so that its vibration direction is parallel to that of the polarizer, a phase contrast image is obtained. In any other case a bright-field image is superimposed on the phase contrast image. Light vibrations in these two images are parallel to the vibration directions of the polarizer and polarizing film of the diaphragm, respectively. The optically active plate rotates the planes of polarization of these two images and, because the angle of this rotation varies rapidly with the wavelength, a variously coloured image is obtained depending on the properties of different parts of the specimen. With all its interesting features, this system has not found much practical application. This is understandable, because a polarizer and analyzer added to a typical phase contrast microscope give sufficient scope for distinguishing birefringent objects.

Experiments performed by Barer (1952b) showed that the use of only a single rotatable polarizer (placed before the phase condenser or above the phase objective) yields many interesting possibilities for the examination of different birefringent objects, such as cellulose fibres, microcrystals, liquid crystals, and plant tissues. Some birefringent properties and polarization phenomena which are not revealed by either phase contrast or normal interference methods become clearly visible if plane polarized light is added to the phase contrast or interference system.

The polarization microscope, provided that its instrumental extinction factor is high, enables minute birefringent details in cells and tissues to be detected. Since the background and isotropic parts of the phase specimen are dark (between crossed polars), it is sometimes difficult to localize exactly the birefringent details in relation to the rest of the specimen. This difficulty is the same as in fluorescence microscopy, and it is easily overcome by the simultaneous use of phase contrast.

The variable phase contrast device described in section III.D, p. 95, as well as Polanret systems, can be used for distinguishing birefringent and polarizing objects.

D. *U.v. and i.r. Phase Contrast Microscopy*

Ultraviolet and infrared phase contrast systems, known to date, are not far removed from standard phase contrast devices for visible light. Only the phase plate is optimized for u.v. or i.r. radiations, and the optical systems are adapted to transmit and display the ultraviolet or infrared image.

Bennett *et al.* (1948) were the first to have experimented in the field of ultraviolet phase contrast microscopy but they used only glass objectives. Next, Taylor (1950) constructed a true u.v. phase-contrast attachment for the Cooke, Troughton and Simms u.v. microscope. In this a suitable phase ring was placed in the back focal plane of ultraviolet glycerine immersion quartz objectives corrected for wavelengths 2563 or 2750 Å. The phase ring was conjugate with a transparent annulus located in the front focal plane of a quartz condenser. The ultra-violet phase contrast image was observed on a fluorescent screen placed in the focal plane of the eyepiece. This attachment has been used successfully at the National Institute for Medical Research in London (Taylor, 1950; Payne, 1952b).

Some components of living cells, especially those containing nucleic acids, absorb u.v. light strongly at certain wavelengths (*c.* 2600 Å for nucleic acids and *c.* 2800 Å for proteins). Thus heavily absorbing structures such as chromosomes may give mixed phase and amplitude images that are hard to interpret. In such cases a visible light phase contrast attachment for ultraviolet microscopes may be more useful and helps in the location of intracellular structures as a prelude to quantitative microphotometry.

Barer (1952c) has devised an infrared phase contrast system using an electronic image converter and standard phase objectives or a special objective with phase plate optimized for infrared radiation. This system appeared to be useful for the examination of some specimens opaque in the visible spectrum, but transparent in near infrared ($\lambda \approx 1\ \mu m$). Most routinely stained histological specimens are of this nature. In particular, tissue sections stained with haematoxylin and eosin are quite transparent ("unstained") in infrared and can be visualized with phase contrast. This procedure of optical destaining has, however, found little practical application.

In recent years there has been much interest in developing infrared phase contrast microscopes for investigating semiconducting materials, which are, as a rule, transparent in infrared, but opaque in visible light.

E. *Stereoscopic Phase Contrast Microscope*

Stereoscopic imagery has become increasingly important for both optical and electron microscopy, as it gives a means of studying the three-dimensional form and characteristics of irregular objects. A mono-objective stereoscopic phase contrast system has therefore been developed (Pluta, 1968b) with cytological applications in mind.

In this microscope a wellknown technique of stereoscopic vision, based on dividing the aperture of an optical instrument into two halves by means of crossed polarizers, has been used. Several variations of this technique have been investigated and, as a result, a versatile instrument for effectively combining phase contrast and stereoscopic observation has been constructed. Its basic optical elements are shown in Fig. 43. The aperture diaphragm D, containing a transparent annulus A, is covered by semi-circular polarizers P_1 and P_2. These polarizers are crossed and arranged in such a manner that the light passing through vibrates 45° from the bisecting line B–B, which is perpendicular to the line joining the eyepieces E_1 and E_2. The annulus A of the condenser diaphragm is conjugate with the phase ring R of a phase objective O. Two other polarizers P_3 and P_4 are placed in the ocular tubes of a typical binocular head BH. Rotating the polarizers P_3 and P_4 around the axis of eyepieces E_1 and E_2 extinguishes either the left or the right half of the annulus image A′ formed on the phase ring R. Setting the polarizer P_3 so as to extinguish the right half of the annulus image A′ (Fig. 43 (b)), and the polarizer P_4 so as to extinguish the left half of this image (Fig. 43 (c)), the left eye LE of the observer receives only the light passing through the left half of the exit pupil of the objective O, and the right eye RE, only the light passing through the right half of

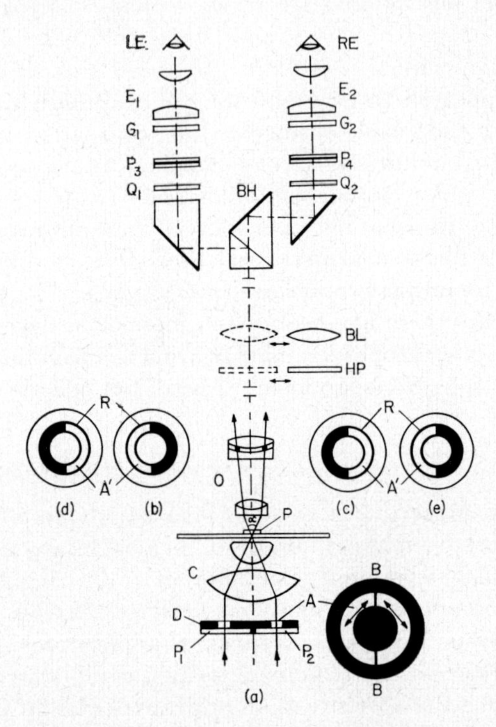

FIG. 43. Schematic diagram of the stereoscopic phase contrast microscope (see text).

this pupil. The object P then appears to be observed with the left eye from the left side and with the right eye from the right side only. The condition for stereoscopic observation is therefore satisfied. The phase details of the specimen located nearer the objective O are really seen nearer and the more distant details further away. The stereoscopic effect is greater the greater is the angular aperture α of the microscope system. The angle α is equal to 15°, 22°, 31° and 62° for objectives of magnification 10×, 20×, 40× and 100× (oil immersion), respectively.

The conditions for three-dimensional observation are also satisfied when the polarizers P_3 and P_4 are set so that P_3 extinguishes the left and P_4 the right half of the annulus image A′, as shown in Fig. 43 (d) and (e). But in this case the stereoscopic effect is reversed; the specimen details nearer the objective are seen further away and vice versa. This is a pseudoscopic image. The change from stereoscopy to pseudoscopy enables one to obtain a lot of additional information concerning the relative distribution of individual elements in complex specimens and to demonstrate small differences in their spatial location. The possibility of instantaneous interchange between the stereoscopic and pseudoscopic image is very useful, particularly in biological research. For quickly changing the stereoscopic image into a pseudoscopic one, a half-wave plate HP is inserted in the light path in such a manner that one of its vibration directions makes an angle of 45° with the polarization planes of the polarizers P_1 and P_2. In this position the half-wave plate HP rotates the vibration direction of the polarized light (coming from each half of the exit pupil of the objective) through 90°. Setting the polarizers P_3 and P_4 in the position of stereoscopic vision and then inserting the half-wave plate HP immediately changes the image to pseudoscopic.

Light incident on the inclined prism surfaces in the binocular head BH is subject to some depolarization. Consequently the light passing through the prism system becomes elliptically polarized, and the polarizers P_3 and P_4 do not extinguish completely the left and right halves of the annulus image A′. To overcome this defect, the quarter-wave plates Q_1 and Q_2 are placed before the polarizers P_3 and P_4. By suitable adjustment of these quarter-wave plates elliptically polarized light again becomes linearly polarized. For precise setting of the plates Q_1 and Q_2 and polarizers P_3 and P_4, as well as for matching the annulus image A′ and the phase ring R, the Bertrand lens BL is inserted.

For quantitative evaluation of depth in the specimen two identical glass plates G_1 and G_2 with stereoscopic marks can be placed in the front focal planes of the eyepieces. If these plates are properly adjusted an observer with normal vision sees a three-dimensional image of the specimen together with a single (stereoscopic) image of the stereometric marks. The latter image appears to be "immersed" in the specimen image. A series of stereometric marks can be arranged in such a manner as to constitute a stereoscopic scale of depth. Another procedure for the stereoscopic measurement of depth is based on the method of "floating image" (Pluta, 1973). This makes it possible to measure depth or vertical distances under the microscope with an accuracy equal to about 0·1–0·05 of visual depth of focus.

Highly sensitive negative phase objectives as described in section

II.D are especially suitable for this stereoscopic phase contrast system. With these objectives the stereoscopic effect and depth perception are, in general, better than with positive phase objectives. This has been shown with specimens of living cells cultured *in vitro* (Veselý and Pluta, 1972).

It is obvious that using ordinary rather than phase objectives permits amplitude specimens to be observed stereoscopically in bright field. In this case, as found experimentally, annular condenser apertures rather than full cone illumination are advantageous; the stereoscopic effect and resolution are thereby much improved. The amplitude objectives described in section II.F, p. 81, can be used, and a further increase in the stereoscopic effect and image contrast can thus be achieved. In practice, the stereoscopic effect is generally the greater the higher the image contrast.

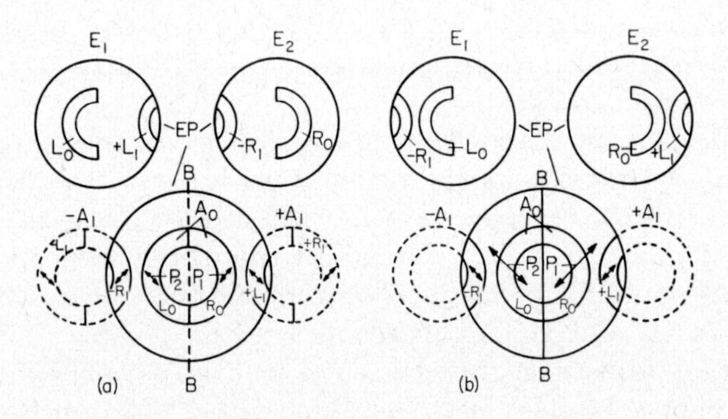

FIG. 44. Auxiliary diagram for discussion of resolution problem of the stereoscopic phase-contrast microscope shown in Fig. 43 (see text).

Stereoscopic vision is also obtained by placing the bisecting polarizers P_1 and P_2 (Fig. 43) behind the objective O. This position is however less satisfactory. Above all, the resolving power of the objective is halved in the direction perpendicular to the bisecting line, whereas in the case of the system presented in Fig. 43, where the aperture of condenser is bisected, the resolving power of the objective is the same in all directions and, moreover, it is unchanged in comparison with a normal phase contrast microscope. This can be explained on the basis of the diffraction theory of image formation. Suppose the object is a linear grating, the period of which is only slightly greater than the resolution limit of the objective. This grating is orientated so that its grooves are parallel to the bisecting line of the condenser (or objective). Then the diffraction pattern in the back focal plane of objective appears as illustrated in Fig. 44, where diagram (a) relates to the stereoscopic system presented in Fig. 43, and diagram (b) represents the case when the bisecting polarizers P_1 and P_2 are placed behind the objective. In these diagrams EP denotes the objective exit pupil; P_1 and P_2 are the directions of the light vibrations in the semi-circular polarizers which bisect the condenser aperture (or exit pupil of objective); B–B is the bisecting line; A_0, $+A_1$ and $-A_1$ are the diffraction images of the condenser annulus, produced by the object grating; E_1 and E_2 are the images of the exit pupil of objective observed through the left-hand and right-hand eyepiece, respectively, when the microscope is adjusted for stereoscopic vision. The eyepiece polarizers P_3 and P_4 (Fig. 43) are set so that they extinguish the right half R_0 (Fig. 44) and left half L_0 of the direct (or zero diffraction order) image A_0 of the condenser annulus, respectively. In such a situation P_3 extinguishes the right half $-R_1$ of the first order diffraction image $-A_1$ as well, and P_4 extinguishes the left half $+L_1$ of $+A_1$. Thus, the observer views the left images L_0 and $+L_1$ through the left-hand eyepiece E_1, and right images R_0 and $-R_1$ through the right-hand eyepiece E_2. The halves L_0 and $+L_1$ are mutually coherent, and light waves issuing from them can interfere with each other and produce a resolved image of the object grating, which is observed by the left-hand eyepiece E_1. Equally, light waves emerging from the right halves R_0 and $-R_1$, which are also mutually coherent, give the resolved image of the grating in the right-hand eyepiece E_2.

Quite a different situation occurs when P_1 and P_2 are inserted behind the objective (Fig. 44 (b)). If the polarizer P_3 of the left-hand eyepiece E_1 extinguishes the right half R_0 of the direct annulus image A_0, the left half $+L_1$ of the first diffraction order image $+A_1$ becomes simultaneously extinguished, and, vice versa, the polarizer P_4 of the right-hand eyepiece E_2 extinguishes simultaneously the left half L_0 and right half $-R_1$ of the zero and first diffraction order images A_0 and $-A_1$. In this case the observer views (if the Bertrand lens BL is inserted) the left half L_0 of the direct image A_0 and right half $-R_1$ of the diffracted image $-A_1$ through the left-hand eyepiece E_1, whereas through the other eyepiece E_2 he views the right half R_0 of the direct image A_0 and left half $+L_1$ of the diffracted image $+A_1$. The halves L_0 and $-R_1$ are apposed, their mutual coherence is poor, and light waves issuing from them cannot effectively interfere to produce a resolved grating image in the left-hand eyepiece E_1. Equally, light waves emerging from the halves R_0 and $+L_1$ cannot produce a resolved image of the object grating in the right-hand eyepiece E_2. The result is as if the objective exit pupil were half-obscured by a light absorbing diaphragm.

The stereoscopic phase contrast microscope as presented in Fig. 43 is manufactured by PZO, Warsaw, under the name "Stereophase" microscope.

F. *Phase Contrast in Holographic Microscopy*

Holography or wavefront reconstruction imaging, invented by Gabor (1948), is a process in which both the amplitude and phase changes of an object wavefront are recorded on a photo-sensitive material together with a reference wavefront passing outside the object under examination. Photographic plates of high resolution are usually employed and after exposure and development a specific diffraction pattern (hologram), from which the original object wavefront can be reconstructed, is obtained. This two step process (hologram record and image reconstruction) is carried out by using a monochromatic and highly coherent light source such as a laser.

Different techniques of holographic imaging have been developed. In principle, they differ in the way by which the reference wavefront is introduced to the recording process. The best known is the two beam "off-axis" technique (also called two-beam carrier frequency technique), developed chiefly by Leith and Upatnieks (1962). A beam of coherent light is split into two parts, one of which serves as the reference beam passing outside the object, and the other passes through or is reflected by the object. Both parts are then recombined and brought into mutual interference at the photo-sensitive plate. The reconstruction process is accomplished by illuminating the developed photographic plate, usually located in its original or conjugate position, with a coherent light beam identical with or similar to the reference beam. In contrast to Gabor's "in-line" holographic process, the off-axis technique enables the twin images (real and virtual) to be effectively separated.

Gabor's original concept was formulated in an effort to improve the resolution capabilities of the electron microscope. Although no practical results have been achieved to date by this means, many variations of Gabor's idea have been successfully implanted into different fields of microscopy. Using their own off-axis technique, Leith and Upatnieks (1965) have devised a "lens-less" holographic microscope enabling enlarged images of good quality to be obtained without the use of lenses. A microholographic system with well-corrected microscope optics, located between the object and holographic plate, has been developed by van Ligten and Osterberg (1966). This system makes holograms of the magnified image (for this procedure the term "image holography" is sometimes employed). Another image holography system which does not require high quality microscope lenses, has been suggested by several authors and realized practically by Toth and Collins (1968). This system gives a reconstructed holographic image of magnification +1, which is then observed by means of a common microscope. Quite recently, Stroke and Halioua (1972) have described a method of *a posteriori* holographic image deblurring, enabling the theoretical diffraction limit of resolution (equal to about 1–2 Å) in high-resolution electron microscopy to be obtained. An essential element of this method is an amplitude and phase filter made holographically.

The phase contrast method can be combined with these systems in different ways based on the fact that the hologram contains the complete information about the object, and all the image-processing techniques which are normally applied to the original object can also be applied to the holographic reconstruction. In particular, phase contrast holographic imaging consists of the use of a phase plate to modulate the phase and/or amplitude of the zero-order diffraction term (direct light) at the time the object image is reconstructed from the hologram. This was suggested by Gabor (1951) and first practically realized by Ellis (1966). A somewhat different procedure was proposed by Anderson (1969). An essential feature of this is the suppression of the zero-order diffraction light at the time the hologram is recorded. A suitable spatial filter is used and the object beam, which transilluminates a phase specimen, is focused on a holographic plate by means of a well-corrected objective. The spatial filter is placed in front of this plate, and it blocks the zero-order diffraction light. After development, the hologram is located in its original position, and spatial filter as well as specimen are removed from the apparatus. Reconstruction of the phase contrast image is accomplished by using two coherent beams: one, the so-called restoration beam, is identical with the beam that formerly transilluminated the specimen, and the other, the reconstruction beam, is identical with the reference beam that was previously used for making the hologram. Introducing a continuous phase compensator and/or variable attenuator in the restoration beam permits variable phase contrast to be produced during the reconstruction process.

A somewhat different holographic phase contrast system has been described by Tsuruta and Itoh (1969). A hologram of unit magnification is made with the off-axis technique similar to Anderson's method, but the phase specimen is transilluminated with a diffused beam and the spatial filter is omitted. After passing through the specimen, the object beam is focused on a holographic plate, and it interferes with a reference plane wave falling obliquely upon the holographic plate. After development, the hologram is placed exactly in its original position, the specimen together with the reference beam is removed, and the hologram is illuminated with a convergent beam, identical with that formerly used to transilluminate the specimen. In this situation a bright point image appears at the focus of an additional objective placed coaxially with the previous reference beam behind the hologram. Placing a phase plate across this point image enables phase variations in the specimen to be clearly observed by the eye located close behind the phase plate. With this system Tsuruta and Itoh have examined phase phenomena (striations) in liquids. These were rather macroscopic objects, but there are no essential objections to employing this procedure in microscopy.

A major problem that retards progress in holographic microscopy is the granular (speckle) appearance of the reconstituted image. This injurious effect is caused by the coherence of the laser light used in both the hologram record and image reconstruction, and it causes serious difficulties in the observation of fine structure, especially of biological cells and tissues. Recently several systems have been developed to reduce the speckle pattern (Close, 1972; van Ligten, 1973; Mróz *et al.*, 1971; Pawluczyk, 1973), and it appears that the problem will be satisfactorily solved before long. Actually, for holographic examination of phase objects, the size of which is much greater than the resolution limit of the microscope system, holographic interference techniques are more useful. Several ingenious holographic interference microscopes have been constructed, and some of them are commercially available (van Ligten and Lawton, 1970; Pawluczyk, 1972; Snow and Vandewarker, 1968). Small phase objects and minute structures cannot however be studied effectively with these microscopes.

VII. Conclusions

In considering the current status of phase contrast microscopy, a wide dissemination of this technique in practice, especially in biology and medicine, has already been accomplished. For many years, standard phase contrast devices have been in routine use in different fields of cell biology and clinical medicine. Phase contrast microscopes are part of the basic equipment of most microscopical laboratories. The technique constitutes one of the principal points in the programme of microscopy courses in biology, medicine, biophysics, veterinary medicine, and other branches of life sciences, and, in order to aid the assimilation of the principles, some teaching aids have even been constructed (Marsden *et al.*, 1971).

Although the phase contrast method for routine microscopy has already been developed to a satisfactory state, nevertheless further development is necessary for many fundamental and detailed biological, especially cytological, studies.

The major problem is that of elimination of halo and shading-off effects. This problem is, however, in the essential nature of the phase contrast technique and can never be solved in a satisfactory manner. Nevertheless, in the case of continuously variable phase contrast systems these undesirable effects can be controlled and reduced to some extent to optimize the quality and fidelity of phase contrast imaging.

A development trend for combination of the phase contrast method with other techniques, in particular with shearing interference methods, is also most useful. A versatile microscope comprising both a variable phase contrast and a shearing interference system with continuously variable wave-front shear is capable of giving the greatest possibilities for extensive studies of phase specimens.

Finally, it is also interesting to note that several authors have become interested in studying the problem of phase contrast imaging on the basis of the theory of image formation in partially coherent light formulated twenty years ago by Hopkins (1953). A number of theoretical works dealing with different aspects of this problem have been published recently. They are listed, for example, in the paper by Khan and Rao (1972). Results in this field are both interesting and stimulating for further progress in phase contrast microscopy.

Acknowledgements

Many thanks are due to the following colleagues and firms for providing material concerning their own non-standard phase contrast systems: Dr. H. Beyer of VEB Carl Zeiss JENA, Dr. G. Nomarski of the Institut d'Optique (Paris), Dr. F. Herzog of C. Reichert Optische Werke A. G. (Vienna), Nikon Europe B.V. (the Netherlands), Nippon Kogaku K.K. (Tokyo), and Polish Optical Works P.Z.O. (Warsaw).

It is also a pleasure to acknowledge the valuable assistance of Mr. A. Kuc with photomicrography.

References

Allen, R. D., David, G. B. and Nomarski, G. (1969). *Z. wiss. Mikrosk.* **69**, 193–221.
Anderson, W. L. (1969). *J. opt. Soc. Am.* **59**, 224–226.

Bajer, A. and Allen, R. D. (1966). *Science* **151**, 572–574.

Banning, M. (1947). *J. opt. Soc. Am.* **37**, 792.

Barer, R. (1949). *Nature* **164**, 1087–1088.

Barer, R. (1952a). *In* "Contraste de phase et contraste par interférences". Colloques de la Commission International d'Optique, (M. Françon, ed.), pp. 56–64. Éditions "Revue d'Optique", Paris.

Barer, R. (1952b). *In* "Contraste de phase et contraste par interférences". Colloques de la Commission International d'Optique, (M. Françon, ed.), pp. 196–204.

Barer, R. (1952c). *In* "Contraste de phase et contraste par interférences". Colloques de la Commission International d'Optique, (M. Françon, ed.), pp. 133–135.

Barer, R. (1952d). *Nature* **169,** 108.

Barer, R. (1952e). *Jl. R. microsc. Soc.* **72**, 10–38; 81–98.

Barer, R. (1953a). *Nature* **171**, 697–698.

Barer, R. (1953b). *Jl. R. microsc. Soc.* **73**, 30–39.

Barer, R. (1954). *Jl. R. microsc. Soc.* **73**, 206–215.

Barer, R. (1955). *Jl. R. microsc. Soc.* **75**, 23–37.

Barer, R. (1966). *In* "Physical Techniques in Biological Research", (A. W. Pollister, ed.) 2nd edn. pp. 1–56. Academic Press, New York.

Barham, P. M. and Taylor, E. W. (1948). British Patent 648801.

Bennett, A. H. (1949). U.S. Pat. 2655077.

Bennett, A. H., Woernley, D. L. and Kavanagh, A. J. (1948). *J. opt. Soc. Am.* **38**, 739–740.

Bennett, A. H., Jupnik, H., Osterberg, H. and Richards, O. W. (1951). "Phase Microscopy". John Wiley & Sons, New York; Chapman & Hall, London.

Beyer, H. (1953). *Jenaer Jb. 1953*, 162–209.

Beyer, H. (1965). "Theorie und Praxis des Phasenhontrastverfahrens". Akademische Verlagsgesellschaft Geest Portig K.-G., Leipzig.

Beyer, H. (1971). *Jenaer Rundschau Sonderheft* **16**, 82–88.

Beyer, H. and Schöppe, G. (1965). *Jenaer Rundschau* **10**, 99–105.

Bratuscheck, K. (1892). *Z. wiss. Mikrosk.* **9**, 145–160.

Close, D. H. (1972). *Appl. Opt.* **11**, 376–382.

David, G. B., Allen, R. D., Hirsh, L. F. and Watters, C. D. (1966). *Proc. R. Micr. Soc.* **1**, 142.

Dufour, Ch. (1952). *In* "Contraste de phase et contraste par interférences". Colloques de la Commission Internationale d'Optique (M. Françon, ed.), pp. 108–113. Éditions "Revue d'Optique", Paris.

Dufour, Ch. and Locquin, M. (1951). *C. r. hebd. Séanc. Acad. Sci., Paris* **232,** 2087–2089.

Ellis, G. W. (1966). *Science* **154,** 1195–1197.

Françon, M. (1954). "Le microscope à contraste de phase et le microscope interférentiel". Éditions du Centre National de la Recherche Scientifique, Paris.

Françon, M. (1961). "Progress in Microscopy". Pergamon Press, Oxford.

Françon, M. and Nomarski, G. (1950). *C. r. hebd. Séanc. Acad. Sci., Paris* **230,** 1392–1394.

Françon, M. and Nomarski, G. (1952). *In* "Contraste de phase et contraste par interférences". Colloques de la Commission International d'Optique (M. Françon, ed.), pp. 136–141. Éditions "Revue d'Optique", Paris.

Gabler, F. (1955). *Mikroskopie* **10,** 119–124.
Gabler, F. and Herzog, F. (1965). *Appl. Opt.* **4,** 469–472.
Gabor, D. (1948). *Nature* **161,** 777–779.
Gabor, D. (1951). *Proc. phys. Soc.* **B64,** 449–487.
Grigg, F. C. (1950). *Nature* **165,** 368–369.
Hartley, W. G. (1947). *Nature* **159,** 880–881.
Hopkins, H. H. (1953). *Proc. R. Soc.* **A217,** 408.
Jupnik, H. (1951). *In* "Phase Microscopy" by A. H. Bennett, H. Jupnik, H. Osterberg and O. W. Richards. Chapter III. John Wiley & Sons, New York.
Kastler, A. and Montarnal, R. (1948). *Nature* **161,** 357.
Khan, M. A. W. and Rao, V. V. (1972). *Atti.Fond. Giorgio Ronchi.* **27,** 521–527.
Krug, W., Rienitz, J. and Schulz, G. (1961). "Beiträge zur Interferenzmikroskipie". Akademie-Verlag, Berlin.
Lamarche, J. (1952). *In* "Contraste de phase et contraste par interférences". Colloques de la Commission Internationale d'Optique (M. Françon, ed.) pp. 122–124. Éditions "Revue d'Optique", Paris.
Leith, E. N. and Upatnieks, J. (1962). *J. opt. Soc. Am.* **52,** 1123–1130.
Leith, E. N. and Upatnieks, J. (1965). *J. opt. Soc. Am.* **55,** 569–570.
Locquin, M. (1948). *Microscopie* **1,** M47–M48.
Locquin, M. (1952). *In* "Contraste de phase et contraste par interférences". Colloques de la Commission Internationale d'Optique (M. Françon, ed.), pp. 114–122. Éditions "Revue d'Optique", Paris.
Lohmann, A. (1954). *Optik* **11,** 478.
Loveland, R. P. (1970). "Photomicrography" Vol. 1. John Wiley & Sons, New York.
Lyot, B. (1946). *C. r. hebd. Séanc. Acad. Sci., Paris* **222,** 765–768.
Marsden, N. V. B., Zade-Oppen, A. M. M. and Öberg, P. A. (1971). *Optik* **33,** 423–436.
Merton, T. (1947). *Proc. R. Soc.* **A189,** 307.
Meyer-Arendt, J. R. (1960). *J. opt. Soc. Am.* **50,** 163–165.
Mondal, P. K. (1969). *Optica Acta* **16,** 85–93.
Mondal, P. K. and Slansky, S. (1970). *Appl. Opt.* **9,** 1879–1882.
Mróz, E., Pawluczyk, R. and Pluta, M. (1971). *Optica Applicata* (Wrocław, Poland) **1,** 9–16.
Nomarski, G. (1952). *In* "Contraste de phase et contraste par interférences". Colloques de la Commission Internationale d'Optique (M. Françon, ed.), pp. 65–86. Éditions "Revue d'Optique", Paris.
Nomarski, G. (1955). *J. Phys. Radium, Paris* **16,** 9S–11S.
Nomarski, G. (1968a). *J. opt. Soc. Am.* **58,** 1568.
Nomarski, G. (1968b). French Pat. 1591113.
Oettlé, A. G. (1950a). *Jl. R. microsc. Soc.* **70,** 232–254.
Oettlé, A. G. (1950b). *Jl. R. microsc. Soc.* **70,** 255–265.
Osterberg, H. (1946). *J. opt. Soc. Am.* **36,** 710.
Osterberg, H. (1947a). *J. opt. Soc. Am.* **37,** 726–730.
Osterberg, H. (1947b). *J. opt. Soc. Am.* **37,** 523–524.
Osterberg, H. (1952). *In* "Contraste de phase et contraste par interférences". Colloques de la Commission Internationale d'Optique (M. Françon, ed.), pp. 227–234. Éditions de la "Revue d'Optique", Paris.
Osterberg, H. (1955). *In* "Physical Techniques in Biological Research" (G. Oster and A. W. Pollister, eds.) Vol. 1, pp. 378–437. Academic Press, New York.

Osterberg, H. (1965). U.S. Pat. 3180216.
Osterberg, H. and Pride, G. E. (1950). *J. opt. Soc. Am.* **40**, 64–73.
Padawer, J. (1968). *Jl. R. microsc. Soc.* **88**, 305–349.
Pawluczyk, R. (1972). *Optica Applicata* (Wrocław, Poland) **2**, 27–34.
Pawluczyk, R. (1973). *Opt. Commun.* **7**, 366–370.
Payne, B. O. (1950). *Jl. R. microsc. Soc.* **70**, 225–231.
Payne, B. O. (1952a). *In* "Contraste de phase et contraste par interférences". Colloques de la Commission Internationale d'Optique (M. Françon, ed.), pp. 130–132. Éditions "Revue d'Optique", Paris.
Payne, B. O. (1952b). *In* "Contraste de phase et contraste par interférences". Colloques de la Commission Internationale d'Optique (M. Françon, ed.), pp. 124–130. Éditions "Revue d'Optique", Paris.
Payne, B. O. (1954). *Jl. R. microsc. Soc.* **74**, 108–112.
Peschkov, M. A. (1955a). *Usp. Sov. Biol.* **39**, 253–256 (in Russian).
Peschkov, M. A. (1955b). *Usp. Sov. Biol.* **40**, 372–376 (in Russian).
Pluta, M. (1958). *Kosmos A. Biologia* **7**, 587–593 (in Polish).
Pluta, M. (1962). *Pomiary Automatyka Kontrola* **8**, 634–639 (in Polish).
Pluta, M. (1963). *Pomiary Automatyka Kontrola* **9**, 292–295 (in Polish).
Pluta, M. (1965). *Pomiary Automatyka Kontrola* **11**, 33–35 (in Polish).
Pluta, M. (1967a). Proceedings of the Second Colloquium on Thin Films, Budapest, pp. 257–266.
Pluta, M. (1967b). *Mikroskopie* **22**, 326–336.
Pluta, M. (1968a). *Microscope* **16**, 211–226.
Pluta, M. (1968b). *Microscope* **16**, 32–36.
Pluta, M. (1969a). *Microscope* **17**, 235–248.
Pluta, M. (1969b). *Folia Histochemica et Cytochemica* (Cracow, Poland) **7**, 269–280.
Pluta, M. (1969c). *J. Microscopy* **89**, 205–216.
Pluta, M. (1969d). U.S. Pat. 3658405.
Pluta, M. (1970). *Microscope* **18**, 113–122.
Pluta, M. (1971). *Mikroskopie* **27**, 121–128.
Pluta, M. (1973). *Optica Applicata* (Wrocław, Poland) **3**, No. 4, 31–35.
Popielas, M. (1972). *Microscope* **20**, 101–110.
Räntsch, K. (1949). W. German Pat. 822023.
Richards, O. W. (1973). *J. Microscopy* **98**, 67–77.
Richter, R. (1947). *Optik* **2**, 342–345.
Ross, K. F. A. (1967). "Phase Contrast and Interference Microscopy for Cell Biologists". Edward Arnold, London.
Saylor, C. P., Brice, A. T. and Zernike, F. (1950). *J. opt. Soc. Am.* **40**, 329–334.
Snow, K. and Vandewarker, R. (1968). *Appl. Opt.* **7**, 549–554.
Soran, V. and Diaconeasa, S. (1957). *Naturwiss.* **44**, 465.
Stroke, G. W. and Halioua, M. (1972). *Optik* **35**, 50–65.
Taylor, E. W. (1947). *Proc. R. Soc.* **A190**, 422–426.
Taylor, E. W. (1950). *Proc. R. Soc.* **B137**, 332–339.
Toth, L. and Collins, S. A. (1968). *Appl. Phys. Lett.* **13**, 7–9.
Tsuruta, T. and Itoh, Y. (1969). *Jap. J. Appl. Phys.* **8**, 96–103.
van Ligten, R. F. (1973), *Appl. Opt.* **12**, 255–265.
van Ligten, R. F. and Lawton, K. C. (1970). *Ann. NY Acad. Sci.* **168**, 510–535.
van Ligten, R. F. and Osterberg, H. (1966). *Nature* **211**, 282–283.
Veselý, P. (1972). *Folia Biologica* (Praha) **18**, 395–401.

Veselý, P. and Pluta, M. (1972). *Folia Bilogica* (Praha) **18,** 374–375.
Wilska, A. (1953a). *Nature* **171,** 353.
Wilska, A. (1953b). *Nature* **171,** 697–698.
Wilska, A. (1954). *Mikroskopie* **9,** 1–80.
Wolter, H. (1950). *Ann. Phys.* **7,** 33–53.
Wolter, H. (1951). *Ann. Phys.* **9,** 57.
Wolter, H. (1955). *Z. Phys.* **140,** 57–74.
Zernike, F. (1935). *Z. tech. Phys.* **16,** 454–457.
Zernike, F. (1958). *In* "Concepts of Classical Optics" by J. Strong, pp. 255–536. W. H. Freeman and Co., San Francisco.

Development of the Vickers M85 Integrating Micro-densitometer

F. H. SMITH, D. S. MOORE

Vickers Limited, Vickers Instruments, Haxby Road, York, England

and

D. J. GOLDSTEIN

Department of Human Biology and Anatomy, University of Sheffield, England

I. Introduction

The decision to initiate a research programme for the development of a scanning photoelectric microscope was taken by Vickers in 1963. The primary aim was to provide an instrument having an optical system which would be equally appropriate for integrating micro-densitometry and for dry-mass determination by the integration of optical phase differences. The latter function was to be included because dry-mass measurement by commercially available interference microscopes had proved excessively slow and arduous, resulting in a temporary set-back in the application of this initially promising technique. However, marketing considerations indicated that it would be prudent to plan the project in such a way that the density integrating function should be achieved first, and the present paper will be mainly concerned with this aspect of the project which ultimately culminated in the production of the Vickers M85 instrument.

II. General Optical Design Considerations

The first stage of the project was directed towards the development of a scanning system which would be suitable not only for the optical density function but also for the interferometric mode of operation when subsequently included. At this stage our conception of the interferometric retardation measuring system was necessarily somewhat tentative but the already-established possibility of using electro-optic light-modulating crystals for a similar purpose seemed promising (Dyson, 1964). Now it happens that commercially available electro-optic crystals of adequate optical homogeneity are birefringent and require that the direction of light propagation be confined to the optic axis of the crystal. This, of course, means that they require a substantially parallel beam of light. Also they are so small that it is not practicable to traverse the parallel beam across the aperture. Consequently, our wish to keep an open option on the possibility of ultimately using an electro-optic crystal in the retardation-measuring system imposed a far reaching design constraint which strongly influenced the future development of the optical system, particularly with respect to the essential scanning function.

This specific function could have been obtained by imparting a scanning motion to the specimen relative to a stationary photometric beam, using a motorized stage. This is a familar approach, offering the advantage of optical simplification and reduction of veiling glare errors by the possibility of close field diaphragming (Piller, 1973). But these advantages had to be weighed against the very restricted rate at which it is feasible to operate a scanning stage. It was considered that this inherent restriction might unacceptably limit the range of application of the envisaged instrument, particularly bearing in mind the future possibility of dry-mass measurements on living material and also the numerous repeated measurements which are often needed to achieve statistical significance when cell populations are being classified.

We therefore had to face up to the implication that the scanning system would have to be of the beam-deflecting type, since we needed a photometric beam which would be stationary and collimated in one section of the instrument and scanning in another. This requirement was seen to eliminate devices which, like the Nipkow disc, scan a real image of the specimen with a small moving aperture (Deeley, 1955). The beam emerging from the moving aperture necessarily executes an angular motion which immediately precludes the possibility of confinement to the optic axis of a crystal. Moreover, the inherently invariant raster format associated with scanners of this type prevents the attainment of transmission-profile traces along a pre-selected line across the specimen, a peripheral facility which was envisaged at an early stage of the project.

Scanners of the television camera type were not very seriously considered because they operate upon a complete, real-time image of the specimen, so that the imaging light beam cannot be confined within the parallel pencil needed for electro-optic crystal modulation. Also, cameras which were then available were known to be unsuitable for photometry on account of inaccuracy of grey-scale response.

Flying-spot systems in which the specimen is scanned by a microscopic image of a raster generated on the face of a cathode-ray oscilloscope were obviously inapplicable because beam-motion is generated at source and available levels of illumination were too low for acceptable signal-to-noise conditions to be obtained (Box and Freund, 1959).

The only beam-deflecting type of scanner of which we were then aware was the mirror-drum, used in the pioneer days of television, in which a stationary beam of light is deflected into a repeated linear scan by sequential reflection at plane mirrors located around the periphery of a rotating drum. However, during the course of a theoretical design study we became concerned on account of the inconveniently close tolerances arising from the unavoidable doubling of errors in the angular locations of the mirrors. Even had it been feasible to achieve the required accuracy of location of the mirrors relative to the drum there remained the problem of preventing small angular displacements of the drum axis.

This tolerance problem, seemingly inherent in reflecting scanners of the drum variety, might have been avoided by recourse to an electrically vibrated single mirror to achieve the line-scan function; but forcing the naturally sinusoidal vibration of the mirror into the required linear, saw-tooth motion seemed to present possibly intractable problems which we were not then in a position to solve (Engle and Freed, 1968).

III. Prismatic Scanner Design

Thus discouraged by difficulties associated with reflecting scanners we turned our attention to refraction as a means of beam deflection, mainly because Snell's law implies that the direction of a refracted beam is less sensitive to variation of angle of incidence than is the corresponding direction of a beam reflected from a mirror. It was hoped that this reduction in sensitivity might facilitate the development of a refracting version of the mirror-drum. A known example of such a device is a glass cube rotated about an axis normal to the incident beam. The principle of operation is illustrated in Fig. 1. As the glass cube rotates, the refracted beam suffers variable transverse displacement due to the changing angular position of the cube, in accordance with the expression

$$D = t \sin \theta \left(1 - \sqrt{\frac{1-\sin^2 \theta}{n^2-\sin^2 \theta}}\right),$$

where D is the displacement of the beam due to refraction in a glass cube of thickness t and refractive index n at a changing angle of incidence θ. Unfortunately we had to reject this attractively simple device because the necessarily convergent transmission of the beam through the tilted glass block inevitably introduces both coma and astigmatism.

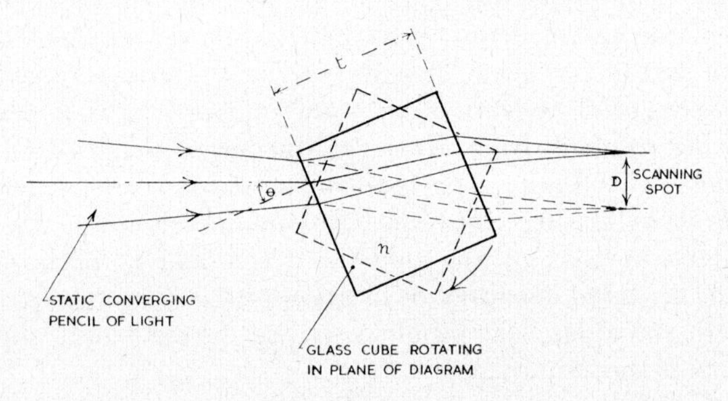

FIG. 1. Refracting glass-cube scanner considered at an early stage.

The first constructive step towards the development of the refracting scanning system was based upon the realization that variable deflection of a beam can be obtained by rotating a refracting prism or wedge about an axis parallel with the incident beam. The angle of deviation is then defined by the angle and refractive index of the prism but the plane containing the incident and deviated emerging beam rotates with the prism. The emerging beam therefore traces out a conical surface, having the prism as apex and the incident beam as axis. Since both the angle and refractive index of the prism are clearly immune from the effects of accidental mechanical dislocations this concept seemed to offer the inherent stability we sought.

There remains, however, a higher order effect whereby the angle of beam deviation is slightly dependent upon the angular attitude of the prism relative to the coincident beam, implying that small angular displacements of the axis of rotation could deflect the emerging, scanning beam. Fortunately, this effect can be reduced to vanishing point by maintaining the prism close to the minimum deviation condition in which the entrance and exit faces make equal angles respectively with the incident and emerging beams. Since this condition can, by definition, hold only for rays which are parallel with the general direction of the incident beam it follows that this beam must consist exclusively of mutually parallel rays, i.e. it must be collimated.

A cross-sectional diagram of a prism in this preferred position is depicted in Fig. 2.

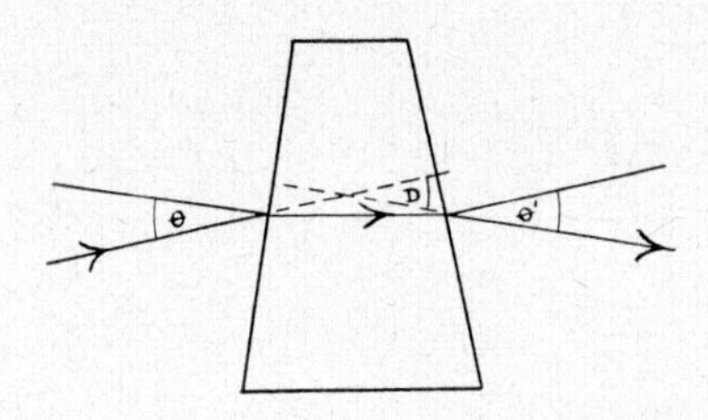

FIG. 2. Prismatic element of refracting scanner. When $\theta' = \theta$, the deviation angle, D, is minimally changed by small angular displacements of the prism in the plane of incidence.

The next step was to devise an assembly of elemental prisms such that each, in sequence, would provide a single scan line in the course of rotating across the path of the incident beam, in a manner analogous to the corresponding function performed by mirror-drum scanners. This was achieved by mounting the elemental prisms with their bases locating tangentially around the cylindrical periphery of a rotated drum, so that the principal section of each prism is continuously turned at constant angular velocity.

Figure 3 shows section views of the complete prismatic assembly. As each prism sequentially crosses the incident beam, the latter is deviated through a substantially constant angle, but in a plane which turns with the prism. In other words, the beam deflection is in the form of part of a cone of constant apex angle.

In order to avoid the instrumental inconvenience associated with the constant overall angular deviation we introduced a compensating, fixed prism of the same angle and glass-type as employed for the rotating prisms. Then, the output beam is not subjected to angular deflection in any plane when the rotating prism reaches its central position relative to the beam. This arrangement generates a scan consisting of a repeated sequence of arc-shaped lines—the line frequency being the product of the number of elemental prisms and the speed at which the assembly is rotated.

Although this looked promising, we wondered whether or not the widespread familiarity with the rectilinear scan format presented by television systems would prejudice prospective users against an instrument employing the curvilinear format associated with our prismatic system. Retrospectively such misgivings may seem naïve in view of the proven densitometric validity of the inherently curvilinear format produced by Nipkow scanners. Such was our concern,

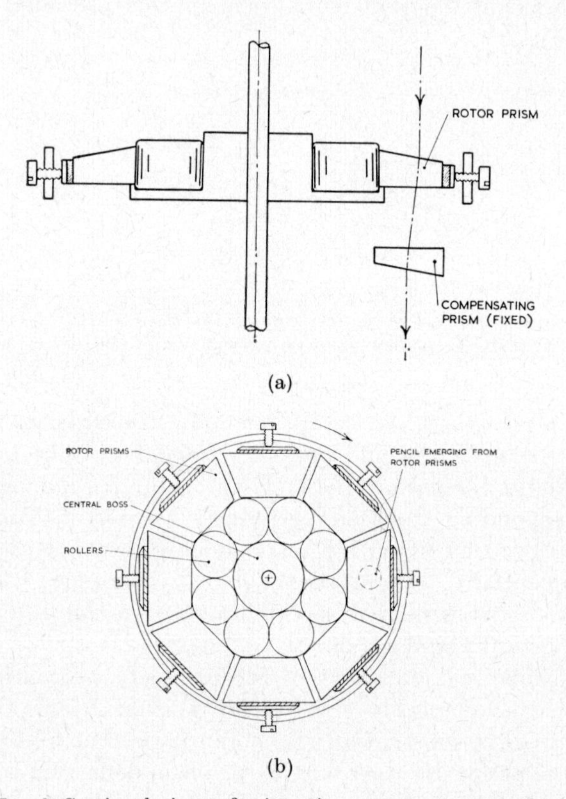

FIG. 3. Sectional views of prismatic scanner rotor assembly.

however, that we actually made up a working model of a more complicated version of this system in which the curved component was compensated by returning the beam through the moving prism by means of a Porro prism reflector. This system was finally abandoned on account of the complication and the difficult setting-up procedure it entailed.

Having agreed to accept the curved format for the fast, "x", component of the raster we next had to turn our attention to means for obtaining the slow, "y", component. In the first working model of the complete system this was achieved by duplicating the fast rotor assembly but rotated at an appropriately slow speed, so that each prism provided a complete raster. The necessary 90-degree displacement of the slow scan component was achieved by suitable orientation of the two rotor assemblies.

Before finally committing ourselves to this form of scanner we had to consider an apparent shortcoming which is more or less inherent in any form of refracting scanner—namely dispersion. Unlike a mirror drum, the deflection produced by a refracting scanner is a rather complex function of wavelength. For the simple, unachromatized prismatic system under consideration the deviation increases with decreasing wavelength, resulting in a format which is slightly larger for blue than for red light. It turns out, however, that this effect is closely equivalent to the familiar phenomenon known to opticians as chromatic difference of magnification and it proved possible to choose the glass type of the prisms so as to compensate quite nicely for the chromatic difference of magnification which is present in most otherwise well-corrected microscope objectives. It is rather analogous to the effect of using a compensating eyepiece. Subsequent experience, however, has shown this consideration to be somewhat academic, in so far as measurements are almost invariably made with monochromatic light.

The possible importance of static, "spot", measurements next claimed attention. It was seen to be desirable to provide for this facility for initial setting-up purposes with the photometric spot located at a clear reference region near the object to be measured. This implied the need to inhibit the scanning mode of operation and manually to steer the spot to the required point in the field. Clearly this facility could also be useful for making static absorption measurements on localized structural features of particular interest, such as nuclei and cytoplasmic inclusions of various kinds. The simplest way of instrumenting this requirement proved to be the provision of individual clutches whereby the drive to either the fast or slow scanner could be inhibited at will. This simple solution yielded the additional bonus of providing either fast or slow repeated linear scans for display of absorption profile traces taken along a pre-selected line across the specimen, the fast scan being suitable for oscilloscope displays and the slow one for pen-recorders.

The rotational speeds of the rotor assemblies obviously demanded serious consideration. Since it seemed important that the scan velocity should not vary during a single raster it was thought necessary to drive the rotors by a synchronous motor, thus sacrificing the feasibility of speed adjustment by the user. In the interests of optimum signal-to-noise conditions it was thought prudent to decide upon the lowest line frequency compatible with envisaged applications of the complete instrument. The facility for oscilloscope displays of light-absorption profiles indicated a line frequency of 16 Hz as being comfortably close to the fusion frequency of the eye. This was readily accomplished by rotating the eight-prism rotor assembly twice per second.

The rotational speed of the "slow" rotor assembly had to be selected with a view to what users might consider to be an acceptable integration time, bearing in mind the numerous determinations required for such statistical techniques as the preparation of frequency histograms in terms of integrated optical density. The range of ratios offered by available gears, in as simplified a train as possible, finally led us to accept a reduction ratio of 1: 180 yielding a complete raster of 180 lines over a period of 11·25 seconds.

IV. General Configuration of the Optical System

Having finally decided upon the detailed design of the scanner we turned our attention to the layout of the complete optical system. This was seen to depend largely upon the most suitable form of illumination for the photometric system. Our beam-deflecting type of scanner seemed to favour locating the scanner between the objective lens and the photometric light source, since the latter then has the easier task of merely producing an intense parallel beam instead of having to fill the microscope field. In other words the illuminating system has only to illuminate a single point of the field, because that point is moved over the field by the scanner. This offers the advantage that the dimensions of the light source need to be little larger than is necessary to fill the scanning aperture, the effective level of illumination then being determined by the source's intrinsic brightness, or radiated energy per unit area. This obviates the need for high wattage and the associated large and cumbersome power supplies. It was also thought that this "flying-spot" type of configuration would lend itself to the parallel beams generated by most laser sources, though a specific application for such a source was not then foreseen.

The foregoing considerations led to the formulation of a complete optical system which is illustrated in Fig. 4. This served as a useful basis for the first working model, built around a Vickers Patholux microscope stand. As can be seen the essential additional components comprised an intensely illuminated star collimator whose parallel output beam is reflected downwards through the conventional microscope system via the specially-designed prismatic scanner. The scanning illumination from the microscopic raster at the specimen plane is collected by a standard substage condenser and finally reflected into a photomultiplier at the base of the instrument.

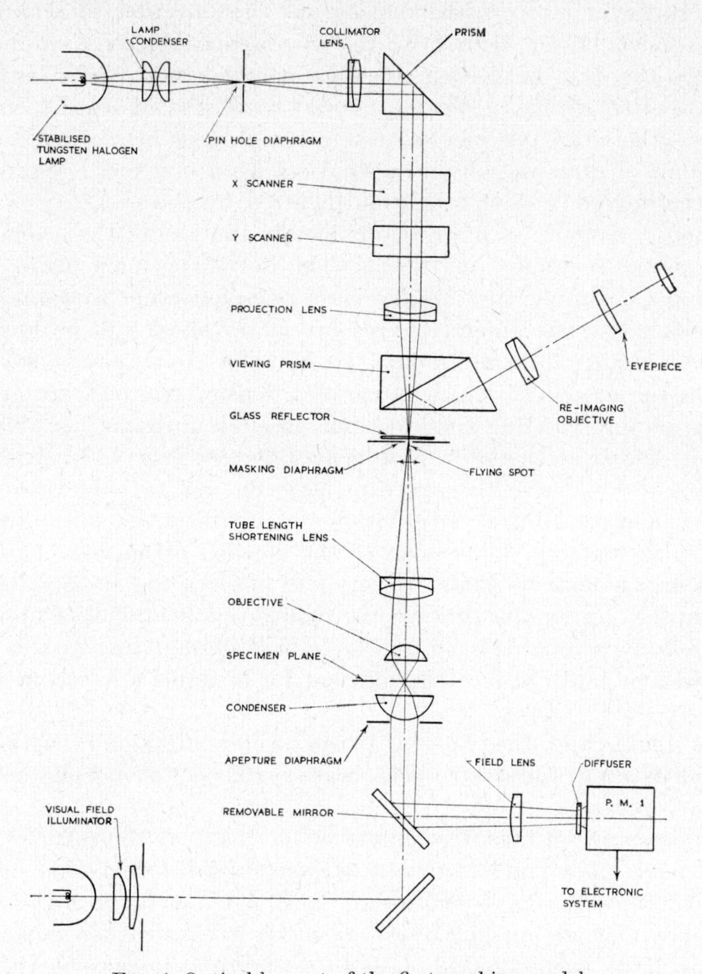

FIG. 4. Optical lay-out of the first working model.

At this early stage the star-collimator was fitted with a single, fixed "star" of a size calculated to achieve a geometrical image at the specimen measuring about 1 μm when a 40× microscope objective was employed. The subsequently standard set of five selectable apertures was not introduced until the final prototype design stage was reached, the purpose being to offer optimum trade-off between signal-to-noise conditions and lateral resolution. To attempt a rigorous treatment of this somewhat complex subject (e.g. Weinstein, 1955) would be beyond the scope of the present paper in which the main aim is to give a general account of the development of a complete instrument. However, it is worth pointing out that in order to achieve the full resolving power defined by the microscope objective's numerical aperture the size of the star aperture would have to be vanishingly small, so that virtually no light would reach the photo detector. In practice, therefore, one has to strike a compromise between the opposing claims of illumination and resolution. To a first order approximation, the degradation of resolution incurred by the necessity of using a scanning aperture of finite size is merely the size of the geometrical image of the scanning aperture as seen at the specimen plane. Thus, by using a scanning aperture having a size equivalent to 1 μm at the specimen plane the objective's resolution threshold will be increased by substantially the same amount, i.e. by 1 μm. An empirical investigation on the extent to which integrated density readings are affected by size of the scanning spot was not possible until the variable spot size was added at the later, prototype stage (see page 157). Meanwhile the aim was to have the aperture large enough to achieve a signal which would certainly be adequate for preliminary test purposes.

The glass reflector plate beneath the viewing prism was introduced at the very outset to enable the user to observe the spot (stationary or scanning) superimposed upon the microscope field while illuminated by the conventional transmitted light base illuminator. At this point the effective depth of focus is sufficient for a simple microscope cover-slip to be used.

This plate, and the viewing prism, were individually withdrawn manually when the instrument was switched to its photoelectric mode of operation.

The existence of an intermediate image plane at the reflector plate would have made the instrument excessively tall had we not inserted the positive tube-length shortening lens just above the microscope's nose-piece. The re-imaging objective in the inclined (then monocular) viewing head is required in order to relay the images at the glass reflector plate into the eye-piece. This re-imaging arrangement provides an erect image of the specimen.

The removable mirror beneath the substage condenser was manually traversed into position to direct the photometric beam into the photomultiplier. The latter has always been of the end-view pattern because of the possibility that the wire grid of side-view versions might cause moving shadows across the photocathode during scanning. The field-lens originally imaged the objective's pupil directly onto the photocathode but light reflected by the photocathode's mirror-like surface caused spurious photometric readings by being partially reflected back to the photocathode by air-glass surfaces in the optical system. This effect was completely cured by introducing the ground glass diffuser which also conferred the additional advantage of virtually neutralizing errors arising from slight lateral displacements of the pupil image. These errors, arising from unavoidable local variations in sensitivity over the surface of the photocathode, are largely integrated out by the ground-glass surface diffusing light from the pupil image over a relatively large area. Not all objectives have their pupils in quite the same position so these displacements cannot be completely avoided in practice.

The main purpose of the tests which were carried out with this first complete working model was to investigate the relative area of the microscope field which could be scanned with reasonable photometric accuracy. In retrospect it would seem that the need for photometric uniformity over the scanned field was over-emphasized, in so far as lack of uniformity can be compensated by the established practice of subtracting a clear-field integrated density reading from that obtained for the specimen. However, gross departures from uniformity distort the base line of oscilloscope-displayed transmission profile traces and curtail the available range of the optical density conversion system. Results indicated that the detected variation in transmission could be kept within 1% over an object field of 170 μm using a 40× flat-field objective, the main cause of attenuation at substantially larger fields being stop-vignetting within the higher-powered objectives. This did not appear a serious limitation in view of the fact that the specimen must in any case be brought within the boundary of some form of diaphragm to mask out the unwanted regions of the field. It just implied that this registration had to be achieved by moving the stage instead of the masking diaphragm.

V. The Transmission-Absorbance Conversion System

Having achieved what appeared to be a practicable optical system we concentrated upon the inverse logarithmic function which expresses the relationship between the light transmission, T, as measured by the photodetector and the required absorbance (optical density), O.D. $= \log_{10} 1/T$. Here we were faced with the difficulty that logarithmic amplifiers were in too early a stage of development to offer the reliability and reproducibility expected of a commercial instrument. A possible alternative which we discussed with Sira Institute was to simulate the function by a multiple diode-switching network (Yang, 1954). The resulting characteristic consists of a set of straight-line segments which can be approximately fitted to the ideal curve, each line being obtained by appropriate switching of the diodes controlling a network of resistors. This possibility was finally rejected since an excessive number of components would be required for accurate densitometry.

It was then realized that an exact analogue of the required optical density function exists in a capacitor discharging into a constant shunted load. This situation is illustrated by Figs. 5(a) and 5(b). The

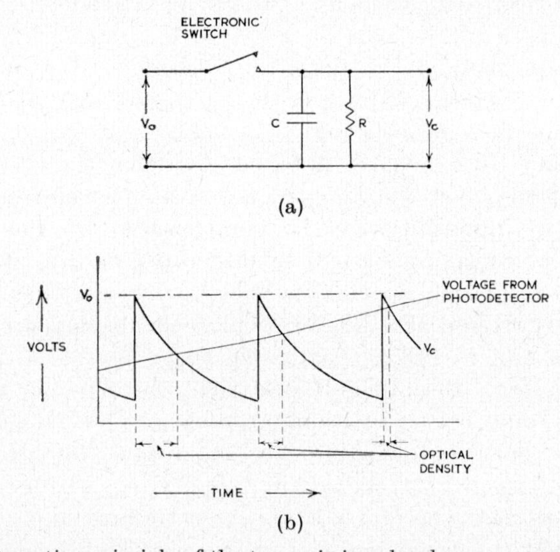

FIG. 5. Operating principle of the transmission-absorbance conversion system.

relevant mathematical expression is $t = 2{\cdot}3\ \mathrm{CR}\log_{10} V_0/V_c$, where t is the time interval between V_0 and V_c. Here, the voltage falls exponentially with respect to time from the instant when the voltage source is removed. This implies that optical density can be represented by the time taken for the capacitor to discharge to the same voltage as that being produced by the transmission-measuring photodetector. All one has to do is to ensure that the capacitor is initially charged up to a datum voltage which corresponds with the photodetector's voltage at 100% transmission. Since the two voltages coincide only at discrete, separated instants of time the same discreteness necessarily applies to the optical density measurements themselves. We therefore had to consider the optimum repetition rate for the recurrent re-charging of the capacitor. After prolonged consideration a rate of 10 KHz was decided upon as a reasonable compromise between photo-electric response to the highest optically resolvable spatial frequencies in the specimen and the practical electronic limits set by problems associated with fast pulse-rise times.

An analogue circuit, designed to operate at this frequency, was accordingly made and tried out by feeding it with the optical trans-mission signal from the photodetector shown in Fig. 4. The validity of the optical density conversion was easily established by making static, spot measurements through neutral density filters and plotting the resulting transmission factors against the outputs from the analogue circuit. This circuit was designed to deliver rectangular pulses whose leading edges are derived from the charging pulses applied to the condenser and which are terminated by a coincidence circuit which quenches the pulses at the instant when the exponentially decaying voltage across the load resistor R corresponds with the light-trans-mission voltage from the photoamplifier. This results in the generation of a 10 KHz rectangular wave having a width proportional to optical density.

Because our density conversion circuit is an exact analogue of the optical transmission to optical density relationship its range is theor-etically unlimited. However, signal-to-noise considerations and the effects of veiling glare in any optical system indicated the wisdom of confining the range to optical densities between 0·0 and 1·3.

Direct, instantaneous read-out of optical density was obtained by passing the rectangular wave-form signal through a moving coil meter which registered the average power in the rectangular pulses, the power being proportional to pulse duration. The required integration of optical density was performed by storing the density pulses in a con-denser of large capacity, the final voltage being displayed by a high impedance valve voltmeter connected across the condenser.

VI. Masking System

In the early version of the complete working model of the instrument, illustrated by Fig. 4, unwanted regions of the specimen were eliminated from the integration process by a selectable masking diaphragm at the intermediate image plane beneath the viewing prism. But experience soon revealed an unsatisfactory aspect of this simple and obvious instrumentation.

During the periods of partial interception of the scanning light-spot by the edge of the mask the resulting reduced light transmission was inevitably interpreted as optical density by the electronic system. This resulted in a spurious background density contribution which had to be subtracted from the integrated readings. While we thought this may be tolerable we did not like the fact that very low integrated densities could be considerably less than the spurious density contribution from the mask.

We were soon driven to the inescapable conclusion that the effect could be eliminated only by locating the field mask in a position where it would not intercept the main photometric beam. This was ultimately achieved by splitting off a small proportion of the scanning beam before it reaches the objective and focusing the resulting secondary raster through the field mask, as shown in Fig. 6. The effective aperture

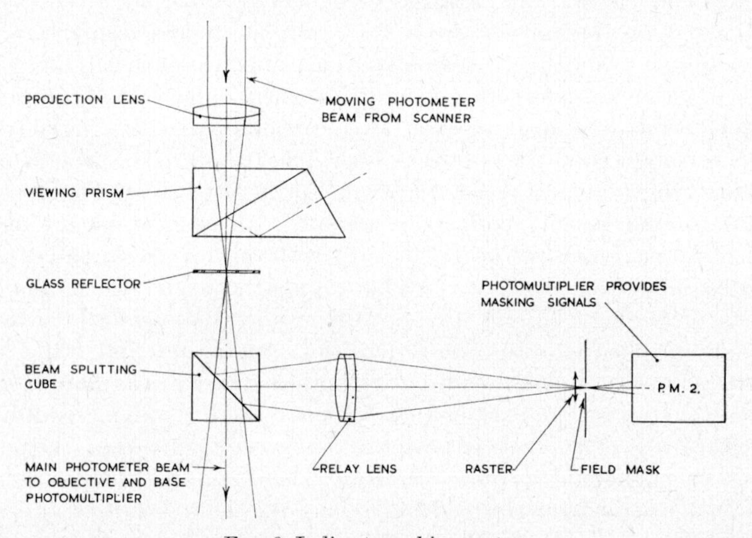

FIG. 6. Indirect masking system.

of the mask can be seen superimposed upon the specimen by replacing the secondary photodetector with a dimmer-controlled lamp. When the secondary photodetector is illuminated by the scanning spot the resulting voltage opens an "and" gate which allows the density pulses to reach the integrator. Thus, density pulses are accumulated only when the scanning spot is within the aperture of the mask. An initially unforeseen advantage of this indirect form of masking system derives from the resulting complete separation of the masking and density measuring functions. Thus, with direct systems, regions of the specimen which transmit less light than the pre-determined "off" trigger level of the masking circuitry are recorded as having an optical density of zero because the relevant photoelectric signals are instrumentally interpreted as masking-off signals. With our indirect system, on the other hand, the mutual independence of the same two functions permits exploitation of the full range of the density conversion system and even densities which exceed this range are recorded as having maximum measurable density (see Fig. 16), instead of being recorded at zero density as they are with direct systems. We prefer to set the masking trigger level at 50% because this corresponds to the situation when the scanning spot is half occluded by the edge of the mask, thus permitting the effective scanned area of the specimen to be determined by measuring the apparent size of the selected mask, e.g. by means of a stage micrometer. This masking system has proved so satisfactory and reliable that it has undergone no significant modification since its original inception.

VII. Area Measurement

When we considered that the density-integrating and field-masking aspects of the working model had reached a satisfactory stage of development, as shown by results obtained with artificial test objects consisting of geometrically-localized absorbing films evaporated onto microscope slides, we examined the possibility of providing additional read-out facilities. Of the various possibilities, it seemed that area might prove to be the most useful additional parameter and one which looked readily attainable merely by the addition of appropriate circuitry. All that was needed was a further gating circuit controlled by a variable threshold discriminator calibrated in terms of optical density. When the threshold discriminator is set at a particular density reading the associated gate passes only those density pulses which exceed this pre-set limit. These pulses are then shaped to be of fixed height and width before passing to an integrator for providing a final read-out which is proportional to the area of the masked region of the field in which the optical densities exceed the value set on the threshold discriminator. This enables the user to obtain the effective area of the masked region of the field by setting the threshold discriminator to zero density. The area of an isolated object inside the masked region can be obtained by setting the discriminator between zero and the lowest density presented by the object and the area of a more absorbing nucleus can be established by setting the discriminator between the maximum density within the cytoplasm and the minimum density within the nucleus. Using this procedure it is possible to build up density distribution histograms for single cells.

VIII. Monochromator System

In most applications of microdensitometry it is implicitly assumed that the Beer–Bouguer Law applies, i.e. that the absorbance measured at a given point of the specimen is directly proportional to the amount of light-absorbing material per unit area. For this to be true the light used should be monochromatic, or at least the range of wavelengths present should be small compared with the width of the peak of the absorption spectrum of the specimen. Imperfectly monochromatic light results in distortion of the absorption spectrum, the apparent absorbance being lower than the correct value at wavelengths near the absorption peak, and greater than the correct value at troughs of the spectrum. Further, at any given wavelength an admixture of light of a different wavelength may result in an error which varies with the absorbance, so that the apparent absorbance is not directly proportional to the true absorbance.

Some means of isolating and selecting monochromatic light is therefore essential in microdensitometry. Very little attention was given to this question until an advanced stage of development had been reached, it having been thought that adequate wavelength selection could be effected by the insertion of spectral filters or by the use of a single, graded spectrum interference filter. Some commercially available filters however pass light with a rather wide bandwidth, and interference filters tend to contain troublesome pin-hole defects which allow the passage of unfiltered white light. Accordingly, when Professor R. Barer first saw our working model in 1966 he strongly advised that a more conventional type of monochromator be provided. Fortunately, our optical system was especially adaptable to the inclusion of a simple monochromator, because it provides a region where the photometric beam is static and collimated.

Our choice of monochromator was strongly influenced by the desire to maximize the effective light transmission. It was thought that the light loss inherently incurred by gratings was too high a price to pay

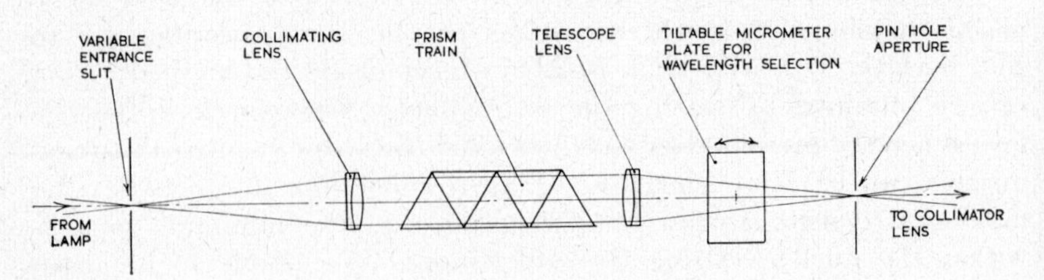

Fig. 7. Monochromator system.

for the relatively trivial advantage of spectral linearity. We therefore opted for a prismatic system, and had the good fortune to find an immediately available bought-out direct-vision dispersing prism-train. This offers the advantage of a direct, axial configuration in which the role of the conventional exit slit is played by the stationary photometric aperture (the moving image of which forms the flying spot in the plane of the specimen). Changing the diameter of the measuring spot therefore alters the band-width, further control over this being provided by a variable slit at the entrance aperture of the monochromator. This flexibility of control permits the user to obtain the best compromise between spectral band-width, spatial resolution and signal strength. An optical diagram of the monochromator system is shown in Fig. 7.

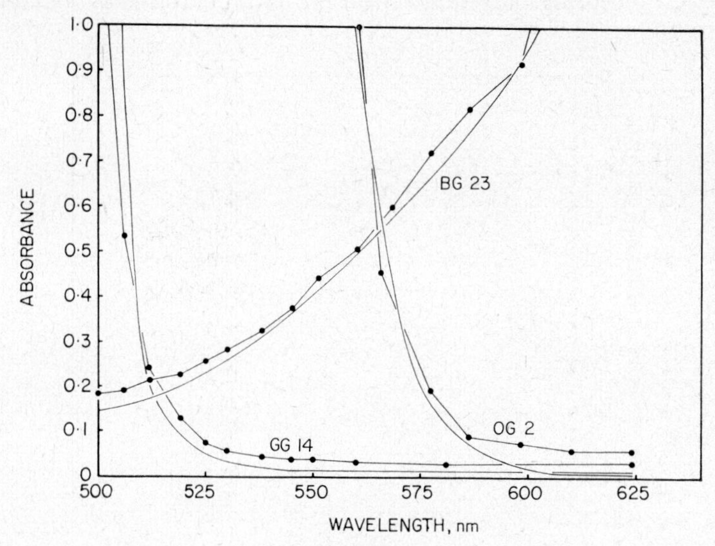

FIG. 8. Simple prism monochromator of the M85 tested against a high-quality diffraction grating monochromator (Bausch and Lomb Spectronic 505). Good agreement is seen in the spectra of three different types of glass.

The spectral resolution is relatively low with so simple a system: the band-width is of the order of 10 nm at the blue end of the spectrum if a moderately small measuring spot (exit slit) is used, but can be many times this value, e.g. at the red end of the spectrum with both monochromator apertures fully open. The monochromator performance however appears to be fully adequate for microdensitometry, good results being obtainable even with difficult test objects. Thus absorbance spectra of various coloured glass filters obtained using the M85 showed very satisfactory agreement with spectra produced by a high-quality conventional recording spectrophotometer (the Bausch and Lomb Spectronic 505) with a bandwidth of 0·5 nm (Fig. 8). In a more crucial series of experiments a variable pathlength spectrophotometer cell was filled with a test solution (e.g. of a biological stain, oxyhaemoglobin etc.) and placed in the optical path of the M85. Using a relatively small monochromator bandwidth the pathlength of the cell was found to be directly proportional to the absorbance at a given wavelength, i.e. Bouguer's Law was adhered to. Increasing the monochromator bandwidth altered the proportionality constant (altered the slope of the line) without materially affecting the linearity of the relationship (Fig. 9); microdensitometric results obtained using different monochromator bandwidths are therefore directly comparable, interconversion requiring only multiplication by an empirically established constant.

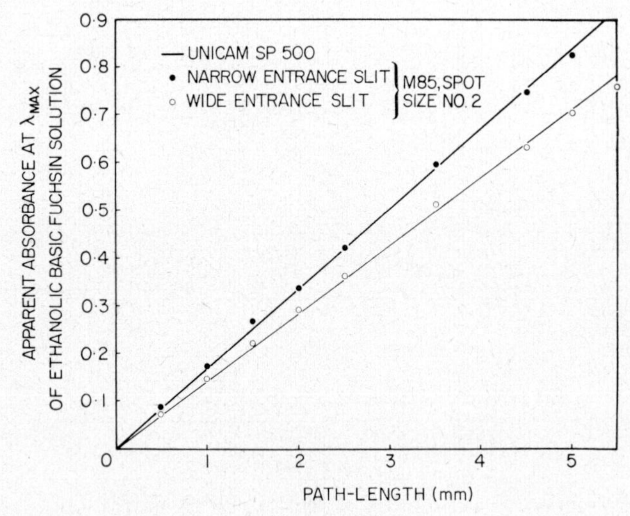

FIG. 9. Results of tests with a variable pathlength spectrophotometer cell showing adherence to Bouguer's Law.

IX. DESIGN FOR PRODUCTION

In 1967, when the relevant basic design parameters had been established on the working model, the project was transferred to the newly-formed Research and Development Department at York with the object of bringing the project forward to the production stage.

The general layout and appearance of the final instrument was worked out in collaboration with Messrs. London and Upjohn, Design

Consultants, taking into consideration such factors as the ergonomics of the positioning of controls and meters, mechanical stability, compactness, ease of assembly and ease of servicing. It was decided that all production microdensitometers should be capable of being readily converted to take the scanning interference system when this became available.

A modular system of construction was selected in order that the instrument could be easily dismantled for transportation and also so that servicing could be accomplished in the field by exchanging modules.

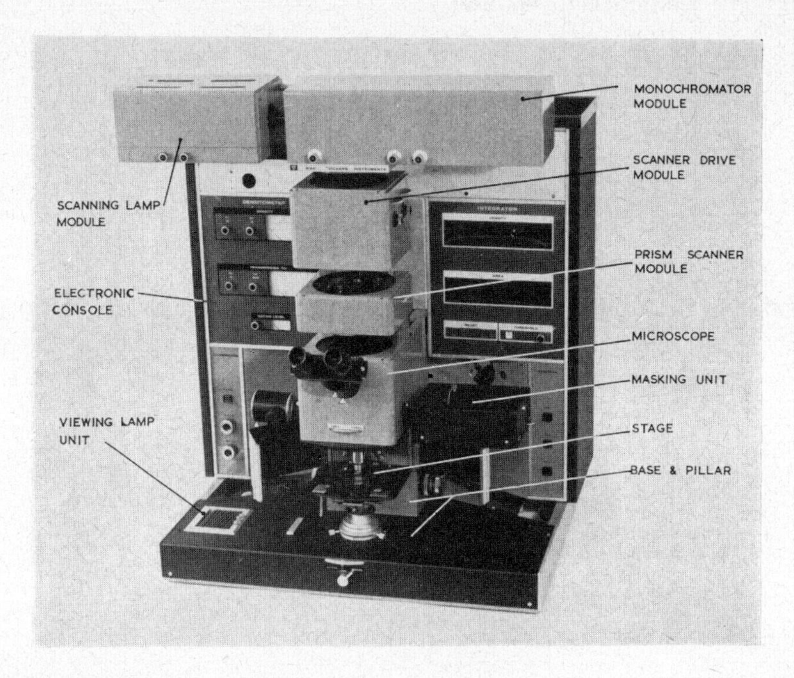

FIG. 10. Exploded view of M85 Scanning Microdensitometer illustrating the modular construction.

Figure 10 shows an exploded view of the M85 illustrating the modular method of its construction. The mechanical and optical system was subdivided into the following units:

Basic microscope block and base with masking unit.
Prismatic scanner unit.
Motor drive and gearbox.
Lamphouse and monochromator.

These sub-assemblies were coupled together by large diameter cone fittings providing a simple means of assembling and aligning the instrument together with good mechanical stability.

The electronics were housed in a console which was secured to the back of the main microscope block. Here again a modular system was employed by dividing the electronics into the following units:

Stabilized D.C. power supply for the scanning lamp.
Stabilized low tension power supply.
Stabilized E.H.T. supply (for photomultipliers).
Densitometer panel unit.
Integrator panel unit.

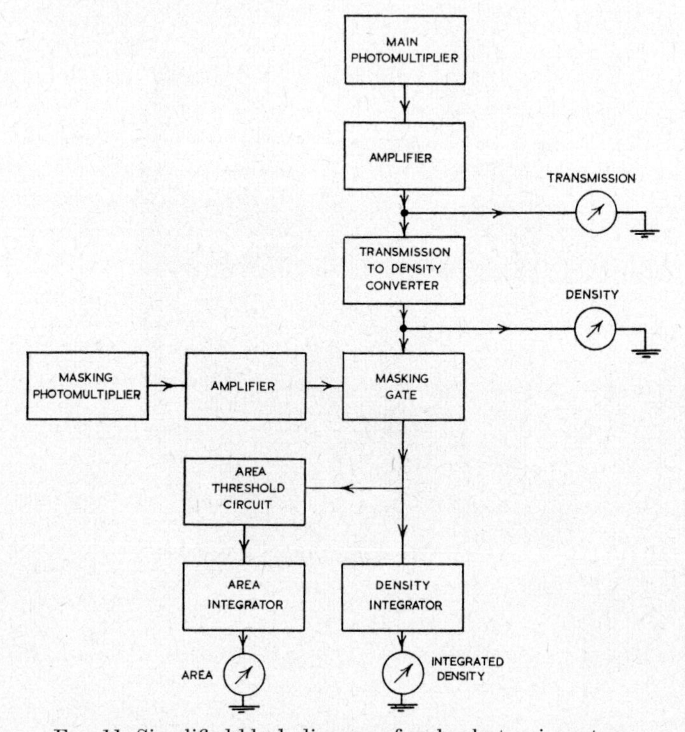

FIG. 11. Simplified block diagram of early electronic system.

The decision was taken to make three prototype instruments to this design so that testing, further development and marketing investigations could run concurrently when the prototypes were completed.

The first prototype instrument was completed ready for initial testing by the end of 1967.

Figure 11 shows a simplified block diagram of the electronic system at this stage.

X. Prototype Testing

Professor R. Barer had been consulted and his advice obtained during the early stages of development. He had kindly offered to have the prototype instrument tested by his staff in the Department of Human Biology and Anatomy at the University of Sheffield. The first prototype was installed at Sheffield for a short period early in 1968 for preliminary testing.

These tests immediately revealed some areas in which further investigations and development work were necessary.

Glare (stray light) in the optical system was estimated by measuring the apparent transmission of opaque particles in the microscope field. In the presence of glare, specimens have a lower apparent absorbance (higher apparent transmission) than they should have. The percentage error varies with the absorbance (Fig. 12), so that glare can cause serious systematic errors in the comparison of weakly and strongly absorbing specimens. In the first prototype an excessive amount of glare (more than 10%) was found. This was somewhat reduced by paying attention to various reflecting surfaces inside the instrument,

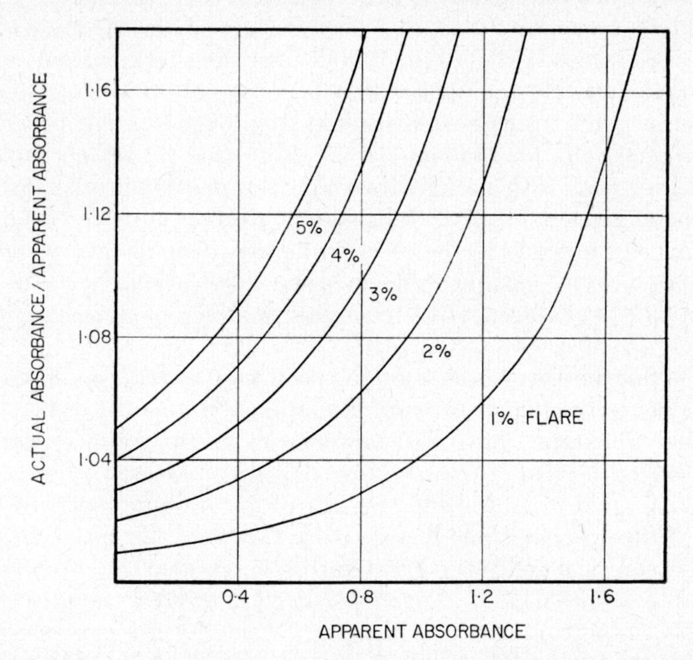

Fig. 12. Ratio of actual to apparent absorbance of a specimen in the presence of glare, plotted against the apparent absorbance. Reprinted from *J. Microscopy* (**92**, 1, 1970).

and was finally brought to a level acceptable for many types of investigation by the introduction of additional stops in the system, particularly an adjustable field stop between the microscope condenser and the main photomultiplier tube. Earlier work on glare (e.g. Naora, 1951, 1955; Lison, 1953) had shown that in conventional microscope systems, in which light passes through the specimen to be collected by the microscope objective, considerable reduction in glare results from limiting the field illuminated. It is interesting to note the identical effect on glare produced in M85 by a stop in a field plane below the microscope condenser, despite the fact that in this flying-spot system the specimen is illuminated during actual measurement by light passing *down* through the microscope objective. This may be regarded as an example of the general rule that in any optical system the direction of travel of light rays may be reversed without substantially altering the properties of the system; any principle which holds for conventional microscopy is therefore valid also in flying-spot microscopy.

A certain amount of residual glare seems however to be unavoidable if a reasonable area in the specimen plane is illuminated. Typically 2–4% glare is found with an illuminated area of (say) 40 μm diameter, and some means for correcting the resulting error is necessary for work of the highest accuracy. Now the effect of glare on absorbance measurements is analogous to that of photomultiplier tube dark-current, and the M85 was fortunately equipped with a dark-current off-setting device for adjusting the 0% transmission reading. It proved possible to compensate automatically for residual glare by adjusting the off-set current in such a way that with no light reaching the photomultiplier tube the instrument gives a *negative* transmission reading equal in magnitude (but opposite in sign) to the measured glare. This facility (Goldstein, 1970) has proved indispensable in some applications, e.g. the comparison of the Feulgen–DNA contents of different types of human leucocyte (Bedi and Goldstein, 1974).

In the first prototype the scanning spot was the reduced image of an adjustable rectangular aperture. Experimental investigation showed that the apparent integrated absorbance of a scanned specimen decreased as the size of the scanning spot was increased, and a similar effect was found when the effective size of the scanning aperture in the plane of the specimen was increased by throwing the specimen out of focus. Theoretical analysis (Goldstein, 1971) revealed that the apparent integrated absorbance of a given specimen is to a first approximation inversely proportional to the diameter (or side) of the measuring spot, provided this is of moderate size. With very large measuring spots the apparent integrated absorbance tends to a plateau value. By measuring a specimen with spots of two or more different sizes it is possible to estimate by extrapolation the integrated absorbance corresponding to a measuring spot of zero diameter. This correction for residual distribution error is not necessary in most routine applications, but can be of importance in high-accuracy work (see e.g. Bedi and Goldstein, 1974). To facilitate reproducible alteration of measuring spot size, the adjustable square aperture of the prototype was replaced in production instruments by a series of interchangeable circular apertures. It may be mentioned in passing that there appears in principle to be little to choose between square and circular measuring apertures, but the latter are somewhat easier to manufacture.

During the preliminary tests a considerable variability in integrated density readings was obtained when scanning the Feulgen-stained nuclei of frog red blood cells, which were representative of the type of biological object likely to be studied on the instrument in practice. A study of the repeated measurements made on a single nucleus showed that the readings followed a cyclic pattern with a basic period of eight readings. Investigations revealed that this variability in readings was due mainly to small differences between the eight scanning prisms in the slow rotor. Photographic examination of the raster pattern (Fig. 13a) further showed a spatial non-uniformity of the scan lines resulting from imprecise reduction gearing to the slow rotor. Development work was immediately put in hand to replace the slow rotor by a single prism driven by high precision gears. Execution of the raster was now initiated by pressing a button to cause a motor to drive the prism slowly through the appropriate angle. Successive scans now occurred in opposite directions, i.e. from top to bottom of the field and then from bottom to top of the field after reactivation of the motor. This had the advantage that a single frame could be initiated precisely at the moment desired, rather than having to wait for the preceding scan to be completed as was the case with the eight prisms mounted in the slow rotor.

When the single slow prism system was fitted to the prototype instrument extremely good reproducibility was achieved on repeated measurements of the same object, and the spatial uniformity of the raster was also much improved.

A further important development, resulting from the early testing of the instrument, was the provision of a meter permitting direct reading of the output of the secondary ("gating") photomultiplier tube and precise setting up of the masking system. As already described, integration of density pulses during a scan only occurs when light from the scanning spot falls on the secondary photomultiplier tube, and when correctly adjusted the integrator gate should open just as the centre of

(a)

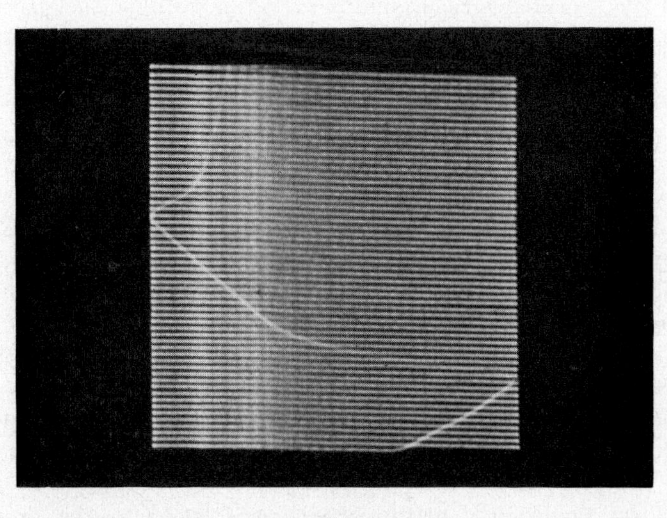

(b)

FIG. 13. (a) Curved-line raster of early prototype, showing spatial non-uniformity. (b) Square raster pattern obtained using the vibrating mirror scanner. The fly-back from the vertical scan can be seen as the flying spot returns to the top of the raster. The brighter area that appears in the left of this photograph is caused by the deliberate slowing down of the flying spot during the latter half of the horizontal fly-back period. This change in speed reduces overshoot and improves the linearity at the beginning of the horizontal scan.

Measurements are gated out during the fly-back periods so that they do not contribute to the integrated readings. In order to separate the lines for the purpose of illustration, a small spot was used with a large scan area. With more usual instrumental settings complete coverage of the specimen is obtained without gaps.

the flying spot crosses the edge of the field mask. In early prototypes this adjustment was difficult, so that sometimes the area measured was larger or smaller than the area apparently masked. This could result not only in objects being erroneously included in or excluded from a given measurement, but in faulty calibration of the instrument. Calibration, to give integrated absorbance measurements in absolute units, is best performed by scanning an empty field of radius R μm, set to a known absorbance d (e.g. by swinging a standard filter into the optical path). If the integrated absorbance of the scanned field, in arbitrary machine units, is e the calibration constant $K = \pi R^2 d/e$. Multiplication by this constant converts results in arbitrary units into absolute units with the dimensions of area; incorrect gating results in the area scanned, and hence the calibration constant, being erroneous.

These preliminary tests also showed that the integrators needed much greater sensitivity than that provided in order to obtain useful readings on small biological objects (e.g. human chromosomes). It also became apparent that several switched ranges of sensitivity should be available to cater for the wide range of objects which would be studied on the instrument.

At this particular time, integrated circuits were becoming readily available. The main photomultiplier amplifier had already been modified to utilize an integrated circuit operational amplifier with improved performance over the original discrete component, transistor amplifier. The reliability and simplicity of application of the new digital integrated circuits made it feasible to design suitable digital integrators to replace the original analogue integrators. A system using pulse-counting techniques was quickly developed for the densitometer. The 10 KHz rectangular wave-form, with pulse width proportional to optical density, was used to gate 1 MHz pulses, generated by an internal oscillator, into a six-decade digital counter to form the density integrator. In this way the number of pulses reaching the counter during a complete scan is proportional to the totalled duration of the density pulses and therefore proportional to the integrated density. The subdivision of each density pulse into a number of 1 MHz pulses resulted in an integrator with very high sensitivity. Only the five most significant digits of the six-decade counter are required in the final integrated density display.

The area integrator now took the form of a five-decade counter receiving a pulse from the area threshold circuit each time the density pulse exceeded a width determined by the setting of the threshold control.

With these new integrators the densitometer was capable of giving significant readings on very small, lightly stained objects. Large, dense objects could also be measured without any range changing facility due to the large capacity of the integrators.

Later tests at Sheffield on the improved prototypes established the excellent accuracy, resolution and reproducibility obtainable on the M85. The reliability of the instrument also looked very promising as during an intensive three months period of testing of this prototype no fault occurred. The precision (reproducibility) of measurements depends to a large extent on the nature of the specimen, being greater with specimens having a large surface area and neither too high nor too low a point absorbance. Typical figures obtained for the coefficient of variation (i.e. the standard deviation of a series of readings, expressed as a percentage of the mean) were 1·3, 2·1, 4·4 and 8·6 respectively for a Feulgen-stained frog erythrocyte nucleus, a single Feulgen-stained human chromosome in a metaphase plate, a single unstained human erythrocyte measured at the absorption peak of oxyhaemoglobin, and a single Gram-stained staphylococcus. Even with a relatively "difficult"

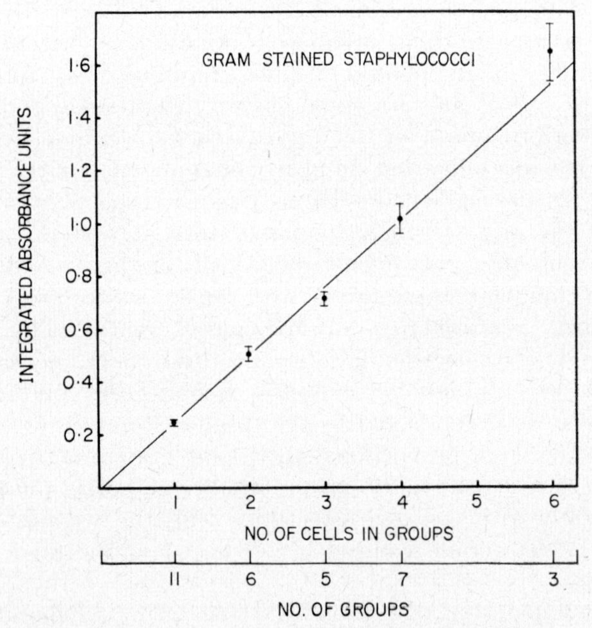

Fig. 14. Integrated absorbances of groups of Gram-stained Staphylococci. Each group contained between 1 and 6 bacteria. There were three groups containing six cells, seven groups containing four cells etc. The results show that the integrated absorbance is proportional to the number of cells in the group. This is a good test of temporal and spatial reproducibility.

object an estimate of the integrated absorbance can of course be obtained to any desired precision simply by increasing the number of replicate measurements. The reproducibility of measurements in different parts of the field is shown in Figs 14 and 15, which show that the integrated absorbance of a group of stained nuclei or bacteria is directly proportional to the number of specimens in the group.

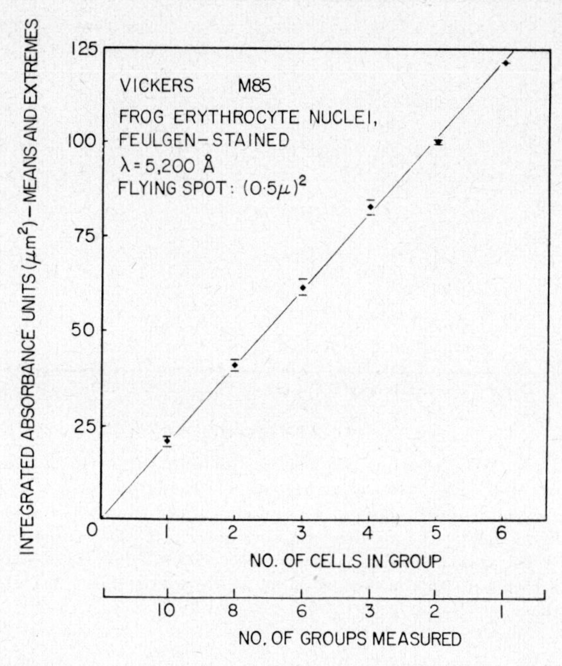

FIG. 15. Integrated absorbances of groups of Feulgen-stained frog red cell nuclei. See caption to Fig. 14. The integrated absorbance varies little between nuclei, making these useful test objects.

The accuracy of an instrument (i.e. how closely the results obtained coincide with the actual value of the parameter being measured) is more difficult to assess than the precision. In addition to investigations which have already been mentioned into glare, distribution error and monochromator slit-width error, tests using a series of neutral-density filters showed the photometric response and the transmission-absorbance converter of the M85 to give a linear response. Linearity was also assessed by a polarized-light method: with a fixed analyzer below the microscope condenser the apparent transmission and absorbance registered by the machine were recorded with alteration in the azimuth of a polarizer lying on the rotatable microscope stage (Fig. 16).

The main photomultiplier fitted to the prototype instruments was the E.M.I. Type 6094 which has an S11 photocathode. This tube has very little response above 650 nm and the usable spectral range of the instruments extended from 400 nm to 650 nm, the limiting factor at the short wavelengths being the low output in the blue end of the spectrum from the 50 W tungsten halogen lamp used as the scanning source. It

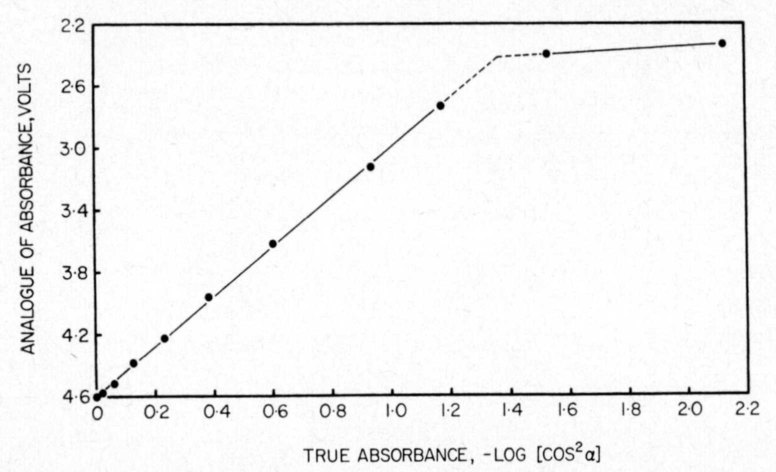

FIG. 16. Linearity of the photometric system (including the transmission-absorbance converter) of the M85. The "true absorbance" is calculated from the function $-\log (\cos^2 \alpha)$ where α is the azimuth angle of a polarizer on the rotatable microscope stage, a fixed analyzer being below the microscope condenser. On the ordinate is plotted the voltage output of the transmission-absorbance converter: the voltage is linearly related to the true absorbance up to the design value of absorbance of 1·3. Data by courtesy of Mr. K. S. Bedi.

was realized that some users would be interested in measurements in the range 650 nm to 700 nm (e.g. chlorophyll absorption bands). It was therefore decided to offer the EMI photomultiplier Type 9558 with S20 photocathode as an alternative to those users interested in measurements up to 700 nm. Figure 17 shows the relative spectral response characteristics using these different photomultipliers. At the extremes of the spectrum a large measuring aperture may have to be used to obtain sufficient signal for precise measurements. An example of such work is the measurement of the haemoglobin content of single, unstained erythrocytes (James and Goldstein, 1974) using the absorption peak of haemoglobin at *ca.* 415 nm (the so-called Soret band). Distribution error due to the large measuring aperture can then become a problem, but in the particular case cited it could be shown (by measuring erythrocytes with spots of two different diameters) that error from this cause was negligible.

The second prototype instrument was despatched to the USA for exhibition early in 1968. This instrument was carefully examined upon its return to check for any adverse results from the transportation. As a result of this study a comprehensive vibration and bump test was drawn up for including in the final test routine of all production M85's.

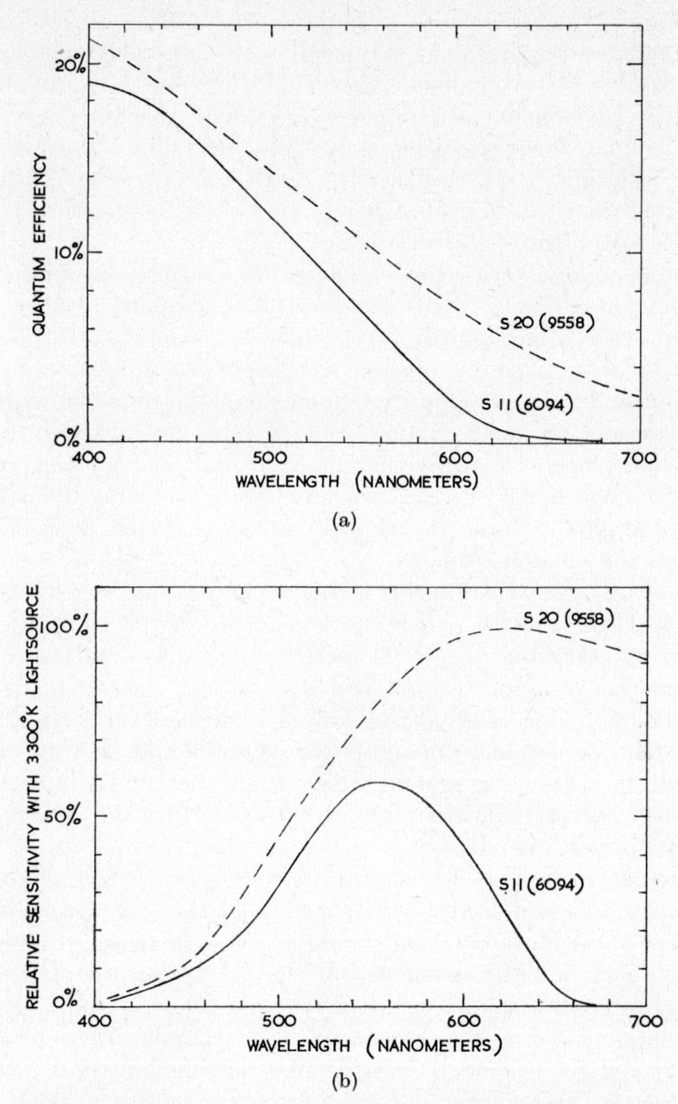

FIG. 17. Relative spectral response characteristics obtained using EMI photomultipliers types 6094 and 9558.

XI. Production and Further Development

The scanning microdensitometer was put into production early in 1969 following a demonstration of the instrument at a Royal Microscopical Society Meeting at Guys Hospital, London in January, 1969.

Delivery of production instruments commenced in January 1970, and good contact was established with the users in order to ensure that satisfactory performance was obtained. The close collaboration between the Research and Development team of Vickers Instruments and the staff of the Department of Human Biology and Anatomy at Sheffield during the final stages of development had gone a long way to ensuring that the resulting instrument met the market requirements. However, customer feedback of information was actively sought with a view to further development of the instrument.

The most common criticisms of the earlier instruments were centred on the microscope stage. This consisted of a standard rotating stage combined with a mechanical superstage. The controls of this stage could only be reached by opening the light excluding doors of the microscope and the precise positioning of small objects was a somewhat tedious process. A small gliding stage with a joystick control was specially developed to overcome this difficulty. The joystick control allows the stage to be operated with the light excluding doors closed and considerably reduces the time taken in changing over between objects on the same microscope slide.

Although a scan on these earlier instruments took only twenty seconds to complete, the routine and repetitive nature of the applications to which many users put the M85 created a demand for a faster scan. Also, since many users were interested in taking measurements from extremely small objects occupying only a very small proportion of the scanned field, it became apparent that there would be considerable advantage in a scanning system which would permit the area of the scan to be easily varied. Therefore, in October 1970, the development of a new scanner was initiated.

Electromagnetically deflected mirrors were selected as the basis for the new scanning system. It was desirable that the new scanner should be a direct physical replacement for the existing prismatic scanner and associated gear box. The geometry of the light beam dictated mirror diameters of greater than 1·5 cms, and to ensure good optical flatness the thickness of the mirrors had to be at least 2 mm. The high power capability and good inherent linearity of a moving coil system as used in loudspeakers showed more promise for the basic drive than did the galvanometer type of movement.

Initial tests were carried out on a prototype M85, the slow vertical scan being produced by a mirror deflected on a simple flexure pivot by a modified, small moving coil loudspeaker (Fig. 18). The loudspeaker coil was supplied with a ramp voltage waveform in order to produce the slow linear scan. Good reproducibility was immediately obtained on integrated density readings with this arrangement and photographs of the raster showed a high degree of uniformity of scan not achieved with the prism scanner (Fig. 13b).

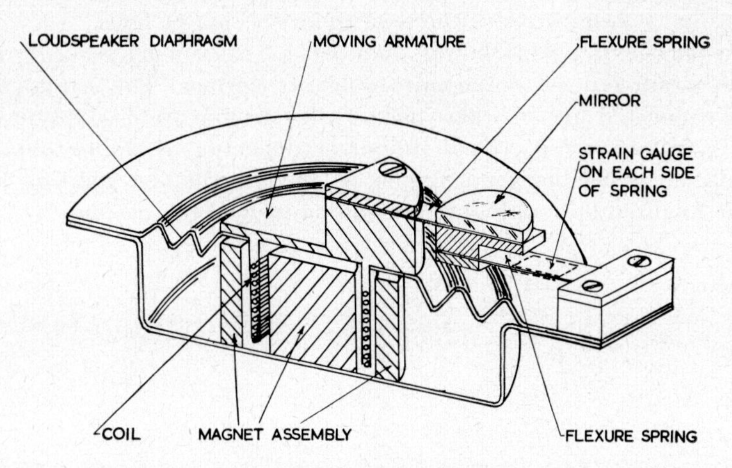

FIG. 18. Cut-away view of a "nodding" mirror scanner assembly based upon a conventional moving coil loudspeaker movement as used for a working model.

High-quality electromagnetic transducers were now ordered so that a complete prototype scanner unit could be constructed. While awaiting delivery of these transducers, development of the drive electronics continued, using a specially constructed electronic analogue simulator of the damped mass-spring system of the transducers. This simulator was easily constructed using integrated circuit operational amplifiers and it proved most valuable in optimizing the design of the fast scanner drive circuit.

It was soon apparent that overall negative feedback would be necessary to overcome overshooting and resonant oscillations at the ends of each line scan. Investigations carried out on the simulator showed that the overshooting and resonant oscillations could be effectively damped by feeding back a voltage proportional to the velocity of the moving coil system (i.e. velocity feedback). This voltage could be readily obtained in practice from the back e.m.f. developed across the moving coil.

The first prototype of the new scanner unit was tested at Sheffield. Extremely good short-term reproducibility of measurements was obtained but a long-term drift in scan amplitude was detected. Investigations showed that this drift was a result of a slow change in the resistance of the moving coils of the transducers due to the power dissipated in the coils. Further tests on the simple analogue simulator showed that displacement feedback would overcome the drift problem and also effectively increase the primary resonant frequency of the fast transducer system such that natural air damping would reduce overshoot and non-linearity of the scan to an acceptable level.

Displacement feedback was obtained eventually by securing resistance strain-gauges to the mirror flexure springs. The strain-gauges were connected into a bridge network fed from a stable d.c. source so that a voltage proportional to mirror deflection could be obtained. Displacement feedback was employed on both the fast and slow scans in the finalized design. Figure 19 shows a view of this scanner.

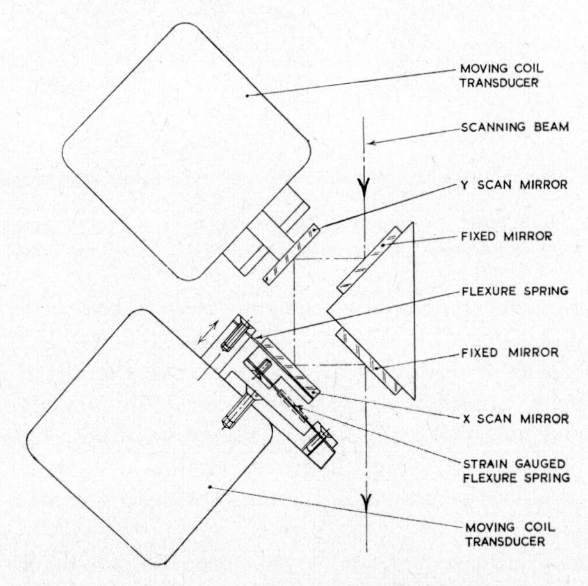

FIG. 19. Arrangement of final electromagnetic scanner unit.

There were many immediate benefits gained from the introduction of the new scanning system and several possibilities for further development became apparent now that the position of the scanning spot could be determined electronically. The scan was now rectilinear in form unlike the previous curved scan (Fig. 13b). The X and Y scan voltages

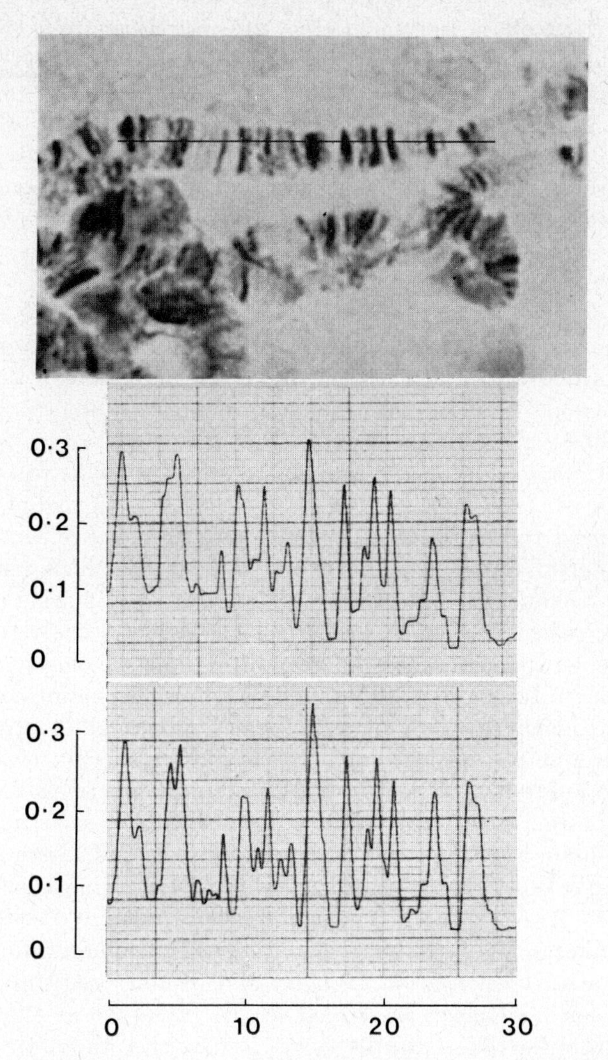

FIG. 20. Pen-recordings of line scans along part of chromosome 3R of *Drosophila melanogaster*. The photomicrograph illustrates the Feulgen-stained preparation scanned, the dark line indicating the position of the scan. The two recordings were made at wavelength 570 nm, with flying spots of 0·4 μm (upper recording) and 0·2 μm (lower recording) respectively. The improved resolution with the smaller spot can be seen. Markings on ordinate and abscissa are respectively absorbance and μm. Courtesy of Dr. I. J. Hartmann-Goldstein, University of Sheffield, Department of Genetics.

could be used in conjunction with an oscilloscope or pen recorder for obtaining synchronized displays of transmission or density profiles along a section of the specimen (Fig. 20). Furthermore a two-dimensional display of density distribution within a specimen could be obtained using a storage display unit.

The size of scan on the production units can be selected by switched controls to suit the size of specimen being measured. Standard instruments are set to give a complete scan in 5 seconds, but this can be varied on a pre-set control over the range 3 to 15 seconds.

XII. Conclusion

This case history of the development of a scanning microdensitometer has been primarily concerned with the technical aspects of the project. No reference has been made to the many difficult commercial problems and decisions which are inseparably associated with enterprises of this nature. Although a detailed discussion of such problems would, in any case, be out of place in the context of this article it seems but right and proper at least to record their existence and significance.

The project had to be planned and managed in such a way as to achieve a technically satisfactory instrument at a marketable price, allowing a reasonably short period for recovery of the research and development expenditure. Such expenditure must be closely controlled and a realistic target date for completion must be maintained. These commercial considerations made it seem impractical to carry out an exhaustive and possibly inconclusive literature survey, especially as there appeared to be sufficient in-house technical experience for the task in hand. In the event, this possibly questionable omission does not appear to have restricted the usefulness of the final instrument once certain minor shortcomings had been eliminated during the important phase of prototype evaluation at Sheffield University. It also turned out that a number of significant aspects of microdensitometer function and operation (e.g. compensation for glare and residual distribution error) had not been adequately described in the literature at the time the project was initiated, so that little was lost by the pragmatic approach adopted.

The performance of the M85 as finally produced owes much to the close cooperation between academic science and industry which existed during this project. In addition to the development of a new instrument, this cooperation led to the establishment of new approaches and techniques in the field of microdensitometry, and we feel that all parties concerned benefited from the collaboration.

Acknowledgements

The help of various people has already been mentioned in the text. In addition to these persons we would like to thank users of the M85 for their valuable comments and suggestions which have significantly contributed to the further development of the instrument.

Thanks are also due to the Science Research Council for providing a grant to the Department of Human Biology and Anatomy, Sheffield for the development of electronic methods of cytology.

References

Bedi, K. S. and Goldstein, D. J. (1974). *Nature, Lond.* **251**, 439–440.

Box, H. C. and Freund, H. G. (1959). *Rev. Scient. Instrum.* **30**, 28.

Deeley, E. M. (1955). *J. Scient. Instrum.* **32**, 263.

Dyson, J. (1964). *Nature, Lond.* **203**, 1300.

Engle, J. L. and Freed, J. J. (1968). *Rev. Scient. Instrum.* **39**, 307.

Goldstein, D. J. (1970). *J. Microsc.* **92**, 1–16.

Goldstein, D. J. (1971). *J. Microsc.* **93**, 15–42.

James, V. and Goldstein, D. J. (1974). *Br. J. Haemat.* **28**, 89–101.

Lison, L. (1953). *Science* **118**, 382–383.

Naora, H. (1951). *Science* **114**, 279–280.

Naora, H. (1955). *Expl. Cell Res.* **8**, 259–278.

Piller, H. (1973). *In* "Advances in Optical and Electron Microscopy" (R. Barer and V. E. Cosslett, eds.) Vol. 5, pp. 95–114. Academic Press, London and New York.

Weinstein, W. (1955). *J. opt. Soc. Am.* **45**, 1006–1008.

Yang, C. C. (1954). *Rev. Scient. Instrum.* **25**, 807.

Electron Microscopical Localization of Enzymes

PETER R. LEWIS

Physiological Laboratory, Cambridge, England

I. Introduction

Although biologists can learn much from purely morphological studies of tissues, knowledge of the chemical composition of individual structures is of far greater value. It is for that reason that histochemistry has become such an important part of microscopical science in the second half of this century. In a typical histochemical procedure for optical microscopy tissue sections are exposed to a chemical reaction which leads to the formation of a coloured end-product at sites with some specific reactivity. By techniques of this type it is now possible to produce highly specific staining of many of the chemical constituents that make up biological tissues. The very fact that the latest edition of Pearse's textbook of histochemistry was published in two volumes (Pearse, 1968, 1972) totalling over 1500 pages emphasizes the extent to which the subject has grown.

A particularly important branch of histochemistry is concerned with the localization of enzymes, the basic catalysts of living cells. To localize an enzyme, tissue sections are incubated in a medium containing a suitable substrate together with a substance which will form an insoluble precipitate with one of the products of enzyme action. Under optimum conditions the distribution of the precipitate accurately reflects the distribution of the enzyme. Modern enzyme histochemistry can be considered to date from the year 1939 when a procedure for alkaline phosphatase was proposed (in two independent publications by Gomori in America and Takamatsu in Japan). Before then there were only two or three methods which demonstrated enzyme activity of any kind and none of them was fully understood or very widely applicable. The technique for alkaline phosphatase, however, was specifically devised to demonstrate on tissue sections the distribution of a known enzyme with well-defined biochemical characteristics. Gomori went on to develop similar techniques for other enzymes and published the first book (Gomori, 1952) to deal with the subject in any depth. By 1960 techniques were available for some 40 to 50 enzymes at the optical microscope level; today the number is over a hundred and continuing to grow.

The study of enzyme distribution at the electron microscope level is a development essentially of the last 10 years. The rapid progress made in the previous 20 years at the optical microscope level meant that many of the problems were understood even if ways of solving them had not been found. Application to electron microscopy introduced new problems: better methods of ultrastructural fixation had to be found; better spatial resolution had to be obtained; the chemical end-products had to be rendered impervious to the necessary subsequent processing and had to be endowed with a sufficiently high density to give adequate contrast in the electron beam. Only for a very few enzymes has it been possible to solve these additional problems. None-the-less a wide range of biological applications is already possible and rapid advances are being made both in adding new enzymes and in improving existing techniques.

II. Some General Considerations

A consideration of the principles on which enzyme histochemistry is based is best introduced by discussion of the methods available for alkaline phosphatase. This was the first enzyme, historically, for which a histochemical technique was specifically devised, and the chemical reactions involved are relatively simple. In the original method, due to Gomori (1939), sections of fixed tissue were incubated in an aqueous medium containing the sodium salt of β-glycerophosphate and a high concentration of calcium ions buffered to an alkaline pH of about 9. Enzyme hydrolysis liberates free phosphate ions which are "captured" by the calcium ions to form insoluble calcium phosphate which is precipitated at or near the sites of enzyme activity. For routine optical microscopy the initial precipitate must be converted into a suitably coloured deposit. The standard procedure now adopted is to treat the sections with a cobalt salt and then with a sulphide solution to produce a stable black deposit of cobalt sulphide. The method has been somewhat modified for electron microscopy but the same basic principles are involved.

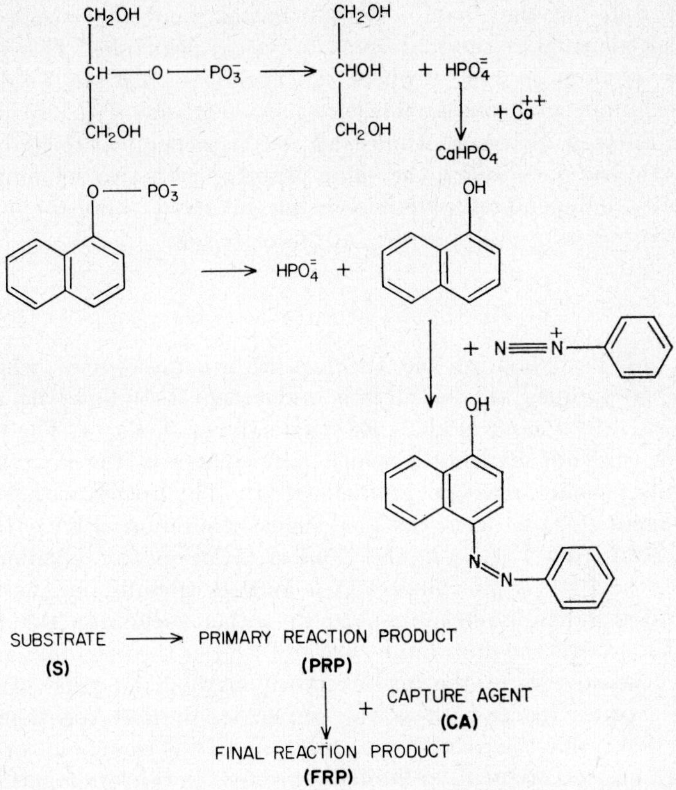

CH_2OH–$CH(O–PO_3^=)$–CH_2OH → CH_2OH–$CHOH$–CH_2OH + $HPO_4^=$; $HPO_4^=$ + Ca^{++} → $CaHPO_4$

α-naphthyl–O–$PO_3^=$ → $HPO_4^=$ + α-naphthol (OH); α-naphthol + $N{\equiv}\overset{+}{N}$–phenyl → 4-(phenylazo)-1-naphthol (OH, N=N–phenyl)

SUBSTRATE (S) → PRIMARY REACTION PRODUCT (PRP)

PRIMARY REACTION PRODUCT (PRP) + CAPTURE AGENT (CA) → FINAL REACTION PRODUCT (FRP)

FIG. 1. The chemical reactions in the two methods for alkaline phosphatase are displayed in a form to show the underlying similarities in the processes involved. Below is given an idealized scheme which is applicable to most histochemical techniques for hydrolytic enzymes; the abbreviations used in the text are shown in heavy type.

A second, quite different method for alkaline phosphatase was devised by Menten *et al.* (1944). Originally it was much less satisfactory but was soon improved to the point where it was the method of choice for optical microscopy (Pearse, 1954) although it has proved more difficult for electron microscopy. The technique consists of incubating fixed tissue sections in a medium containing α-naphthyl phosphate and a suitable diazonium salt. Enzyme hydrolysis releases free α-naphthol which rapidly couples with the diazonium salt at the alkaline pH used to form an insoluble, brilliantly-coloured azo-dye. For optical microscopy no further treatment is required, but for electron microscopy the azo-dye must be converted into a much denser deposit which will survive routine dehydration and embedding procedures.

Although the reactions involved in these two techniques are chemically so different they have a certain formal similarity which reflects the basic identity of the underlying physical principles. This identity is made obvious in Fig. 1 where the chemical reactions involved in each technique are compared with each other and with an idealized scheme referred to below. Almost all the histochemical techniques for hydrolytic enzymes follow the same pattern, with two sequential but essentially independent reactions being involved; first the enzyme-catalyzed reaction and then the capture reaction.

A. *The Theory of Enzyme Techniques*

For the discussion of the theoretical principles upon which the enzyme techniques are based it is convenient to follow the nomenclature used by Pearse (1960, 1968) as displayed in Fig. 1. The product from the enzymic hydrolysis which participates in the next stage is called the primary reaction product (PRP). This reacts with the capturing agent (CA) to form the final reaction product (FRP). Thus, as can be seen from Fig. 1, in the Gomori technique for alkaline phosphatase, the PRP is phosphate, CA is ionized calcium and the FRP is precipitated calcium phosphate. In the other technique the PRP is α-naphthol, CA is the diazonium salt, and FRP is the insoluble azo-dye.

Since the success of the histochemical method depends ultimately on the capture reaction, it is the properties of that reaction which usually determine the incubation conditions. Thus the choice of buffer, pH and temperature is often limited more by the capture reaction than by the properties of the enzyme. The degree of localization obtained is largely determined by CA. Specificity is usually determined by S but may occasionally be affected by CA if it produces significant enzyme inhibition. Many histochemical techniques are a compromise between conflicting chemical limitations, particular examples being techniques for ATPases (Section IV) and cholinesterases (Section VB).

A major factor limiting the degree of localization that can be hoped for is the distance which molecules of PRP can diffuse before they are captured and the distance which the FRP can diffuse before it precipitates onto some cellular structure where it remains firmly anchored.

Too little is known about the intimate physical environment existing in protoplasm for any accurate theoretical treatment to be possible. An idealized treatment has been published by Holt and O'Sullivan (1958), however, from which certain generally applicable conclusions can be drawn, and these can be extended on a qualitative basis by the application of simple reaction kinetics theory.

One extremely important conclusion concerns diffusion into the tissue of CA—this must be rapid enough to maintain an adequate concentration in spite of continual removal by reaction with PRP. Hence localization can be improved by increasing the diffusibility of CA into the tissue and by reducing the rate at which PRP is released. The latter effect is particularly important and accounts for the high proportion of good histochemical procedures in which the enzyme is constrained to work well away from its optimum pH and temperature. Adequate inward diffusion of CA is all-important, however, in many techniques; so sections or slices prepared for incubation should be as thin as possible. Some idea of the adequacy of diffusion can be gained from a comparison of the picture seen at the surface of the slice with that seen deeper within it. Whenever a cytochemical technique is used for the first time therefore ultrathin sections taken from different depths into the slice should be examined in the electron microscope and their appearance compared. A convenient way of doing so is to cut transverse sections across the thickness of the slice. Merely changing the fixation procedure can alter the rate of penetration: glutaraldehyde often gives poorer penetration than formaldehyde. The actual texture of the tissue may be important, a very open type of structure giving far better results than a very compact one. Most of the techniques will not work satisfactorily on unfixed tissue because the various lipid membranes impede diffusion unless their integrity has been destroyed by fixation.

For staining intensity to reflect enzyme activity the rate of production of PRP must be proportional to the concentration of enzyme at any site, but independent of substrate concentration. The reaction is then said to obey zero order kinetics. For optical microscopy this condition is normally thought to be obeyed (see Pearse, 1968) especially when thin sections are used, but it often will not hold for thick slices or small tissue blocks incubated for electron microscopy. Inadequate inward diffusion of substrate, as of CA, will reveal itself in differences of staining at different depths into the tissue and the same general comments apply. Diffusion difficulties, both of S and CA, are so common a feature of E.M. histochemistry that they must always be borne in mind. Material for incubation should always be as thin as possible, enzyme reactions kept as slow as practicable and adequate time allowed for solution changes. For most investigations accurate localization is more important than exact proportionality in staining intensity; adequate provision of CA rather than of S is therefore the more desirable. For that reason many E.M. techniques include a pre-incubation stage in which the tissue is treated with incubation medium complete except for the substrate.

Finally, in electron microscopy the physical nature of the FRP can be crucial. Ideally it should not be crystalline but should be deposited in an amorphous form. It should not be soluble in the dehydration solution and should not react with the embedding medium. Where one or other of these two criteria are not met it is necessary to convert the FRP into some substance which does meet them. Sometimes postfixation with osmium tetroxide is sufficient to produce an insoluble product; in other techniques a sulphide solution is used to convert the PRP into an insoluble heavy metal sulphide.

B. *The Aims of Enzyme Histochemistry*

There are basically four types of use to which E.M. histochemical techniques for enzymes may be put:

(a) to identify by its enzyme content a particular type of cell or cell process;

(b) to identify by its enzyme content a particular type of cell organelle or specialized region of a cell;

(c) to obtain accurate localization of the major concentration of an enzyme;

(d) to obtain an accurate quantitative assessment of the distribution of an enzyme.

These four categories are listed in order of increasing stringency required of the histochemical procedure.

An example of the first type is the use of cholinesterase techniques to stain selectively cholinergic neurones or nerve fibres in a heterogeneous population. Here, diffusion of reaction product even over many nm might not be disastrous and gross artefacts could be tolerated provided they were recognizable.

An example of the second type is the use of techniques for acid phosphatase to identify lysosomes and related structures. Some degree of diffusion can obviously be tolerated provided it does not obscure identification.

The third category includes the uses made of these techniques in the majority of papers so far published. Here diffusion artefacts have to be minimal but a quantitative relationship between amount of staining and enzyme concentration is not essential since attention is focused on the sites of maximal enzyme concentration. Inevitably the experimental procedure must be much more precisely defined than in the first two categories and control experiments become essential.

The fourth category has little relevance at the E.M. level at the present time since methods of quantitation are in their infancy but it is worth emphasizing that even at the optical microscope level very few histochemical techniques for enzymes meet the stringent requirements of this category. Here, however, lies the ultimate goal of many of the scientists who are developing and improving these techniques.

C. *Practical Methodology*

Although the individual details of the techniques will obviously vary with the particular problem being studied, there is a basic sequence common to all the methods for enzymes. After preliminary fixation and dissection the tissue must be prepared in a form suitable for the incubation stage, usually as thin slices or sections. After incubation some form of treatment is often necessary to convert the FRP into a stable end-product which will survive dehydration and embedding. Thus there are essentially four distinct stages in a typical technique:

(1) fixation and/or dissection of the tissue;

(2) preparation of the tissue in a form suitable for incubation;

(3) incubation of the tissue samples;

(4) further processing before dehydration and embedding.

1. *Problems of fixation*

A few enzymes do not withstand any form of fixation and special techniques have to be used for them. The majority of enzymes for which histochemical techniques exist will withstand some form of aldehyde fixation without too great a loss of activity, however, and the improvement in ultrastructural presentation is so great that most workers use fixed tissues routinely. For many investigations the use of a fixative before incubation is entirely valid, but some loss of enzyme activity is likely to occur and substrate specificity may occasionally be affected. Strict correlation with biochemical data may not always therefore be possible. Problems are particularly likely to arise when the tissue contains two or more enzymes demonstrable by the same histochemical technique since their sensitivity to fixation is unlikely to be identical. Similarly, what appears to be the same enzyme by biochemical criteria may show widely differing responses to fixation in different tissues and different animals. Species variations in enzymes is a very widespread phenomenon which has cropped up in a number of histochemical investigations. The classic E.M. fixative, osmium tetroxide, rapidly destroys the activity of almost all enzymes and potassium permanganate is little better. Furthermore, both these fixatives make most tissues so brittle that it is difficult to take samples through a complex incubation sequence without damaging them mechanically. It was therefore only with the introduction of aldehyde fixatives (Sabatini *et al.*, 1963) that routine E.M. enzyme histochemistry became a practical possibility. At the present time glutaraldehyde and formaldehyde, either separately or in combination, are the fixing agents most commonly used. Enzymes vary in their sensitivity to the two aldehydes. As a general rule formaldehyde, provided it is made by the depolymerization of paraformaldehyde, causes less loss of enzyme activity. Glutaraldehyde usually gives better ultrastructural preservation and is still preferred by many workers. Because of the improvement in tissue preservation that can be achieved, fixation by perfusion is often used, frequently followed by further fixation, at 4°C, of tissues after they have been dissected out. The tonicity and pH of the fixing solution are important, cacodylate and phosphate being the buffers most often used. Many problems concerning fixation remain to be solved, and for further information on these Stoward (1973) should be consulted, but for many enzymes fixation procedures are available that combine adequate retention of activity with moderate preservation of ultrastructure. It should be realized, however, that except in very special circumstances it is not possible to preserve the finest ultrastructural detail in material processed for enzyme demonstration.

2. *Problems of incubation*

For most enzyme techniques the tissue must be incubated in the form of thin slices or sections. Unless the tissue sample is thin enough the various reagents in the incubation medium will not diffuse in fast enough. In some techniques the substrate may be slow to diffuse and sites of very high activity will not be shown up in sufficient contrast because the enzyme there will be inadequately supplied with substrate. In other words, production of the PRP by the first reaction (Fig. 1) will not obey zero order kinetics at sites of high enzyme activity. In many techniques it is the capturing agent which is slow to diffuse. The second reaction will not then obey first order kinetics and some diffusion of the PRP will occur, especially around sites of high activity, before capture takes place and the FRP is precipitated. Hence sharpness of localization will be distinctly blurred. One method of minimizing this effect is to pre-incubate the slices in a medium lacking the substrate so that some of the capturing agent is able to diffuse into the tissue before the enzyme is exposed to the histochemical substrate. For some techniques the tissue needs to be sectioned as thin as 20–30 μm to obtain adequate penetration. Sections can be cut on a freezing microtome or in a cryostat, but ice-crystal formation can lead to undesirable artefacts. Various commercial devices are available for cutting tissues, either fixed or unfixed, without freezing; the best imparting a vibrational, sawing action to the cutting edge.

The precise conditions of incubation will naturally vary from one investigation to another and some of the special conditions appropriate for particular enzymes are mentioned later. Typically, however, tissue sections or slices are incubated in the complete incubation medium for a period of from 15–20 min up to 2 hours or more at a temperature between 0° and 37°C. A temperature of 4°C is most often used since diffusion artefacts are less serious at low temperatures because enzyme activity is depressed; tissue damage is also thought to be less serious at lower temperatures. Controls are an essential part of any enzyme study. For E.M. histochemistry the usual control is to incubate an adjacent section in parallel in a medium from which the substrate has been omitted. Another control is to heat a section to a temperature at which enzymes are destroyed or to incubate in the presence of a selective inhibitor of the enzyme being studied. Where a particular histochemical procedure can show up several enzymes it may be necessary to incubate in the presence of inhibitors for the unwanted ones.

With many techniques the FRP is sufficiently stable to withstand routine dehydration and embedding. With others some pre-treatment is necessary. In a number of the azo-dye techniques post-fixation with osmium tetroxide is essential in order to produce an insoluble end-product and in other techniques is desirable in order to give a sufficiently dense end-product. Treatment with a buffered sulphide solution is necessary in a number of techniques in order to convert the FRP into a heavy metal sulphide.

III. Methods for the Simple Phosphatases and Sulphatases

It is convenient to begin by discussing this group of enzymes because the histochemical techniques for their detection are relatively straightforward and chemically uncomplicated. All of them are hydrolytic enzymes which release either a phosphate or sulphate group by simple hydrolysis. Most of them have a wide specificity, so a variety of histochemical substrates can be used. Basically there are two groups of techniques available which are typified by those for alkaline phosphatase shown in Fig. 1. In the first technique developed for electron microscopy a heavy metal is used to capture the acidic radical. In the second type of technique the organic hydrolysis product is captured by a diazonium salt to form an insoluble azo-dye. It would be unreasonable to be very dogmatic about the relative merits of the two types of technique. The use of the heavy metal ones has certainly been more popular and has yielded many excellent published electron micrographs. The azo-dye techniques suffer from the disadvantage that diazonium salts are slow to diffuse into tissues and the final contrast in the electron microscope is seldom as good as in the heavy metal techniques. At the present time, the azo-dye methods are less satisfactory to use but they could have a great future since there is a wide potential for the synthesis of new substrates and diazonium salts. It is convenient to discuss the heavy metal techniques first.

A. *Acid Phosphatase and Sulphatases*

Although alkaline phosphatase was the first enzyme to be studied histochemically, it is more logical to begin a discussion of electron microscopic applications with acid phosphatase. This enzyme, or more correctly group of enzymes, will hydrolyze a wide variety of monophosphate esters and has a pH optimum well on the acid side of neutrality. Biochemically, the acid phosphatases were characterized early, liver, spleen and especially prostate being particularly rich sources among mammalian tissues. Histochemically, however, the situation was most unsatisfactory for some fifteen or twenty years after the first technique was published by Gomori (1941). In his technique he used sodium β-glycerophosphate as the substrate, acetate buffer at pH 5·0 and divalent lead ions as the capturing agent. He chose lead because its phosphate is sufficiently insoluble under acid conditions, and up to the present no better alternative heavy metal has been found. Various modifications have been made to the technique, particularly as a result of studies with the electron microscope (Holt and Hicks, 1961; Barka and Anderson, 1962). Acetate as a buffer tends to cause some damage to the ultrastructure and the most popular buffers are now tris-maleate and dimethylglutarate. The concentration of lead has also been slightly reduced to minimize the non-enzymic staining of various membrane structures and cell nuclei. The technique has proved more valuable to electron microscopy than to optical microscopy. The reasons are partly that the non-specific staining is more obtrusive in thick sections and partly that many of the structures stained are near the limits of optical resolution but are clearly delineated in the electron microscope. The technique gives very selective staining of lysosomes and of parts of the Golgi complex, as is illustrated in Fig. 2, and it is used routinely to stain up these structures in many cell types.

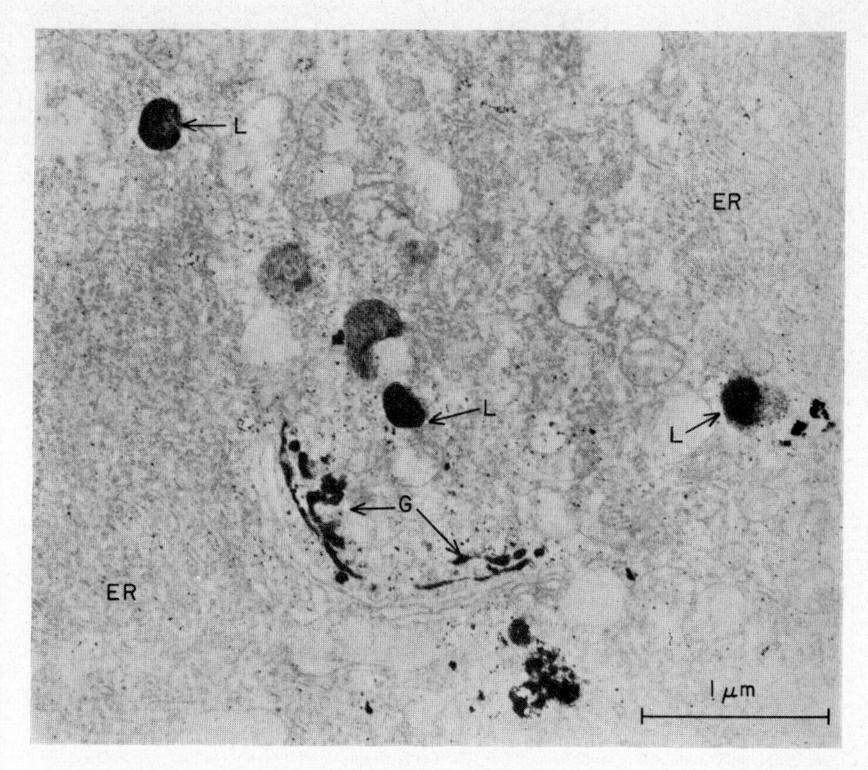

Fig. 2. An area of cytoplasm from a motor neurone in rat hypoglossal nucleus stained for acid phosphatase with β-glycerophosphate as substrate and lead ions as the capturing agent. No counter stain was used. The only structures intensely stained are the lysosomes (L) and the segment of the Golgi complex (G) thought to be concerned in their synthesis. Note that the rough endoplasmic reticulum (ER), large areas of which are present, is unstained: compare the picture for cholinesterase distribution shown in Fig. 8 (×25 000).

Another enzyme present in lysosomes is a sulphatase, perhaps better referred to as an acid arylsulphatase. In the histochemical technique for this enzyme the substrate usually used is p-nitrocatechol sulphate in an acetate buffer at a pH of 5·5 with either lead or barium ions as the capturing agent. In general the use of barium is to be preferred on both biochemical and electron microscopic grounds (Hopsu-Havu *et al.*, 1967). An example of lysosomal staining with this technique is shown in Fig. 3. This technique would appear to be the best for the selective staining of lysosomes, although, largely for historical reasons, the techniques for acid phosphatase have been more widely used.

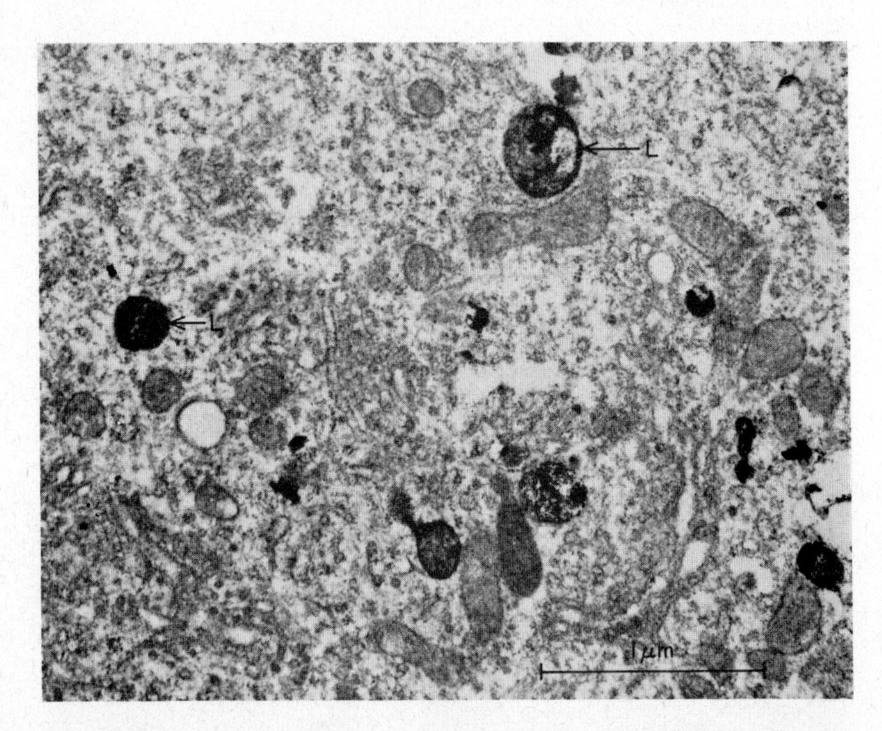

FIG. 3. An area of cytoplasm from a neurone in the mouse hypothalamus stained for arylsulphatase with barium ions as the capturing agent. In spite of the heavy counter-staining to show up the ultrastructure, staining of lysosomes (L) is quite evident. Here the incubation was deliberately kept short so that any internal structure within stained organelles would not be obscured, e.g. as in the lysosome at the top of the electron micrograph (× 30 000). (By courtesy of Mrs. Z. E. Y. Caffyn.)

Besides the lysosomal acid phosphatase there are a number of other phosphatases with pH optima in the acid or neutral region but with much narrower specificities. There are, for instance, several enzymes which will hydrolyze one or other of the various nucleotide mono- and diphosphates. Some of these enzymes can be demonstrated histo-chemically by simple modifications of the technique for acid phos-phatase. Although the functions of the various nucleotidases are largely unknown they are important histochemically because they can be used to produce selective staining of intracellular organelles. In many cell types highly selective staining of the Golgi apparatus can be obtained with thiamine pyrophosphate as the substrate at a pH of 7·2 with activation by manganous ions. This is perhaps the most reliable way of staining up the Golgi apparatus selectively. Goldfischer *et al.* (1971) have also shown that at a more alkaline pH staining of endoplasmic reticulum can be obtained in liver cells with some but not all of the nucleoside diphosphates.

Our understanding of lysosomes, the Golgi complex and the relationship between them owes much to the application of cytochemical methods at the E.M. level. The lysosome began as a biochemical concept: for extended accounts of the earlier work on lysosomes see de Duve (1959) and de Reuck and Cameron (1963). It was discovered that differential centrifugation of homogenates from tissues such as rat liver yielded a separate fraction, sedimenting between the mitochondrial and microsomal fractions, which contained a number of acid hydrolases. These enzymes become active only after treatments such as sonication, lipid extraction or exposure to hypotonic media. It was postulated that the cells possess membrane-bound particles containing a range of degradative enzymes that remain latent until the membrane is altered in some way. Acid phosphatase is one of the enzymes present and studies of its distribution were made with the optical microscope in a wide range of plant and animal tissues (for references, see Gahan, 1967), from which it became obvious that probably all cells possess acid phosphatase positive granules. Although lysosomes are within the resolving power of the optical microscope most of them are well below 1 μm in diameter and far more suited to study by electron microscopy. The techniques for acid phosphatase have been extensively used for E.M. studies (Etherton and Botham, 1970; Bowen, 1971) but in some respects arylsulphatase is a better marker for these organelles (Figs 2 and 3). The great advantage offered over the biochemical approach is the ability to examine lysosomes in individual cells rather than in whole tissues and to compare the enzyme content of differing forms of lysosome-like structures found in a single cell type. The yield of lysosomes in fractionation techniques is very low and much of our evidence for the origin and fate of lysosomes comes from E.M. histochemistry. Although there is still some disagreement in the literature, the cytochemical evidence favours the idea that lysosomes are formed in association with a part of the Golgi complex (Novikoff *et al.*, 1971; Decker, 1974) designated by the abbreviation GERL which stains intensely for acid phosphatase (Fig. 2). The life history of lysosomes is not fully known, but one of their functions is to combine with material taken in by the cell to form a so-called secondary lysosome in which digestion of the ingested material occurs. Lysosomes and the associated acid hydrolases are particularly prominent in degenerating neurones (Decker, 1974) and are usually implicated in the processes of cell death.

B. *Alkaline Phosphatase*

The method for alkaline phosphatase introduced and later improved by Gomori (1939, 1952) has been widely applied at the optical level. Calcium ions are used as the capturing agent (see Fig. 1) and it soon became obvious that calcium phosphate is not sufficiently insoluble to give impeccable localization. Prolonged incubation produces some blurring of the picture with the optical microscope and cell nuclei become stained in the neighbourhood of strong cytoplasmic enzyme activity (Pearse, 1968, should be consulted for a detailed discussion of these problems). Reale and Luciano (1967) used the technique for electron microscopy with a slightly higher calcium concentration in the incubation medium, but their only major modification was to interpose a wash in dilute lead nitrate solution after incubation to convert the FRP into the more insoluble and more dense lead phosphate. Their results were better than might be expected; although diffusion artefacts are obvious at high magnifications they do not obtrude too seriously at low ones (Fig. 4). Nonetheless the techniques cannot be regarded as entirely satisfactory. Somewhat earlier Mizutani and Barrnett (1965) had investigated the possibility of using other cations as the capturing agent in place of calcium. Their best results were obtained with cadmium but unfortunately the precipitate of cadmium phosphate is not entirely stable to routine processing for electron microscopy. Several workers have proposed techniques in which lead is used as the capturing agent, but a major technical difficulty is contamination with lead carbonate formed from carbon dioxide adsorbed by the alkaline medium. The use of citrate as a chelating agent reduces the risk of contamination and a reliable method incorporating citrate was published by Mayahara *et al.* (1967). Resolution is still not good but an impressive degree of contrast can be obtained at low magnifications in tissues such as intestine or kidney where the enzyme is concentrated at the brush border (Fig. 4).

A related enzyme for which a similar histochemical technique has been developed is glucose-6-phosphatase. It has a much narrower substrate specificity and will hydrolyze only a small group of sugar phosphates.

Its pH optimum is near neutrality and in all the histochemical procedures suggested a pH in the range 6·5–7·0 is used with lead ions as the capturing agent. There are conflicting reports in the literature concerning its stability to fixatives; certainly the liver enzyme is very

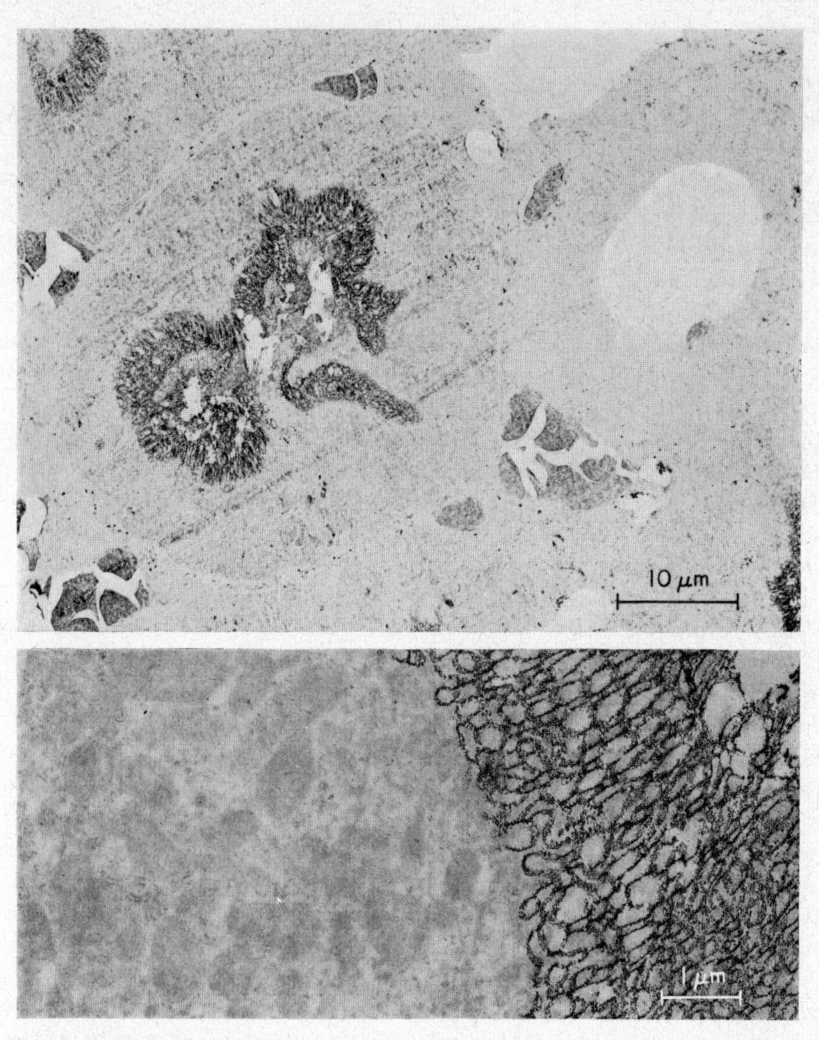

FIG. 4. Staining of proximal tubules in rat kidney for alkaline phosphatase. In the top electron micrograph at very low power (×1620) the lumenal border of the proximal tubule on the left is intensely stained whereas that of the distal tubule on the right is completely unstained. The lower micrograph (×11 000) emphasizes the lack of staining in the general cytoplasm, with the intense staining along the membranes of the microvilli making up the so-called brush border of the tubular cells.

sensitive to fixation (Sabatini *et al.*, 1963; Drochmans *et al.*, 1972). It is probable that there are a number of other specific phosphatases which could be studied histochemically if their activities were not swamped by the very non-specific alkaline phosphatase. The recent introduction of new potent specific inhibitors of alkaline phosphatase (Borgers, 1973) may well overcome the difficulty.

The very wide specificity of alkaline phosphatase has enabled the development of several histochemical techniques which make use of very unphysiological substrates. The phosphate group is retained but the other part of the substrate molecule is tailored to give the required histochemical properties. The majority of such techniques for the phosphatases depend upon coupling with diazonium salts and the simplest example of such a technique is shown in Fig. 1. These azo-dye techniques have the particular attraction for optical microscopy that the FRP is intensely coloured. They have several disadvantages, however, for electron microscopy. Most azo-dyes are not much denser than ordinary E.M. embedding media and therefore give poor intrinsic contrast in the electron microscope; also, many of them are too soluble in lipid solvents to survive routine embedding. A third disadvantage is the poor penetrating power of diazonium salts which virtually precludes the use of unfixed tissue and necessitates the cutting of very thin slices for incubation if diffusion artefacts are to be avoided. Nonetheless the technique has great potential value because of the extensive chemical variations which can be played on the structure of both the substrate and the capturing species. Most of the variations tried have been aimed at producing a denser, more insoluble end-product. Basically two approaches have been used. The more direct is to synthesize a suitable diazonium salt which has a heavy metal already incorporated into it. Tice and Barrnett (1965) were the first to explore this approach by synthesizing several phthalocyanin derivatives with copper or lead in them together with one or more amino groups that could be diazotized. Livingston *et al.* (1969) explored the use of some of the lead series further but abandoned them in favour of triphenyl lead derivatives, with one of which good localization of lysosomal acid phosphatase was obtained by Beadle *et al.* (1971). The use of the diazotate of p-mercury-acetoxyaniline was explored by Smith and Fishman (1969). The full procedure which they had to use was rather complex and the results did not compare favourably with other techniques. Recently, rather better results have been obtained by Sasaki and Fishman (1973).

The second, and up to the present more fruitful, approach is to choose a variant of substrate and/or coupling reagent that produces an azo-dye which is osmiophilic. Usually this means the introduction of groups which are able to reduce osmium tetroxide, and leads to a highly dense, insoluble end-product. This approach has been used to study various esterases as well as phosphatases but in view of their basic similarity it is more convenient to concentrate discussion of all

TABLE I

A list of some of the azo-dye procedures which have been proposed for hydrolytic enzymes at the E.M. level

Substrate ester	Capture agent diazotate of	References
α-Naphthyl acetate	pararosaniline	Lehrer and Ornstein (1959)
Naphthol ASTR phosphate	pararosaniline	Bowen (1968)
Naphthol ASBI glucuronide	p-nitroaniline	Bowen (1971)
Indoxyl esters	pararosaniline	Hayashi *et al.* (1968); Holt and Hicks (1966)
Indoxyl acetate	BAXD	Kawashima and Murata (1969)
2-Naphthylthiol acetate	BBN	Bergman *et al.* (1967)
2-Thiononanoylbenzanilide	BBN	Seligman *et al.* (1966)
Naphthol AS esters	lead phthalocyanins	Tice and Barrnett (1965)
Naphthol AS esters	lead triphenyl-p-aminophenethyl	Beadle *et al.* (1971)
Naphthol AS esters	p-Hg-acetoxyaniline	Smith and Fishman (1969)
Peptide esters of 4-Methoxy-β-naphthylamine	o-dianisidine	Rutenburg *et al.* (1969)
	4-aminophthalhydrazide	Seligman *et al.* (1970)

Notes
Naphthol AS is 2-hydroxy-3-naphthoic acid anilide
Naphthol ASTR and ASBI are derivatives with extra substituents
BAXD is N,N′-bis(p-aminophenyl)-1,3-xlylenediamine
BBN is 4′-amino-2′,5′-diethoxybenzanilide

the azo-dye techniques into this section. Only a few highlights will be picked out here; a list of various procedures which have been proposed is given in Table I. Perhaps the first major advance at the E.M. level was the introduction by Lehrer and Ornstein (1959) of hexazonium pararosaniline as a coupling agent. (Pararosaniline is the simplest member of the triphenylmethane dyes and carries three amino groups all of which are rapidly diazotized by nitrous acid.) The azo-dye that it produces with simple naphtholic substrates withstands routine dehydration and embedding but does not give good contrast in the electron microscope. With a substrate that is itself osmiophilic, however, excellent contrast can be obtained and such a substrate was successfully introduced by Holt and Hicks (1966). Holt (1952) had earlier used indoxyl acetate as a substrate in a technique where free indoxyl liberated by esterase activity was oxidized to the blue dye indigo. The original technique is not suitable for electron microscopy but free indoxyl readily reacts with diazonium salts and the resulting azo-dye is very strongly osmiophilic because of the presence of the indoxyl radical. Results obtained with this technique are referred to later in the discussion of motor end-plate histochemistry, but it should be emphasized here that, because a wide range of indoxyl derivatives are available, the technique is potentially very valuable and some recent developments are particularly promising (Davis *et al.*, 1972).

A somewhat different approach has been pursued over a number of years by Seligman and his collaborators. This is to start with the standard phenolic esters and diazonium salts and to introduce into them chemical groups which will confer special reactivity on the azo-dye, such as osmiophilia or the ability to chelate a heavy metal. (They have used the same approach to find tetrazolium salts suitable for the E.M. demonstration of dehydrogenase activity.) One of the early examples of this approach was the use of aromatic thio-esters. The free thiol compound reacts rapidly with diazonium salts to produce a yellow diazothiolether which will react with osmium tetroxide, although only if warmed to 37°–45°C (Hanker *et al.*, 1964). The results obtained with this and related methods (Bergman *et al.*, 1967) do not compare very favourably with those from methods in which the capturing agent is a heavy metal and they have not been widely used. For some enzymes, however, only an azo-dye technique is available. Thus the only histochemical method for aminopeptidases depends upon the use of amino acid esters of β-naphthylamine as substrates. Rutenberg *et al.* (1969) used a method in which the azo-dye formed with the β-naphthylamine was first chelated with cupric ions and then reacted with thiocarbohydrazide to form an osmiophilic complex. A better method is that introduced by Seligman *et al.* (1970) in which freshly diazotized 4-aminophthalhydrazide was used as the capturing agent to form a strongly osmiophilic azo-dye. With this method they obtained excellent staining of endoplasmic reticulum in renal tubular cells and pancreatic acinar cells.

IV. Enzymes that Hydrolyze ATP

Of the hydrolases so far discussed only a minority have functions which are known or suspected at the present time. The lysosomal hydrolases are mostly degradative and those in the Golgi apparatus are concerned with a variety of functions still not fully understood. Alkaline phosphatase probably fulfils several different functions depending upon its situation; where it is concentrated on the cell membrane, as at the brush border of the intestine or proximal kidney tubules, it is probably concerned with the transport of solutes such as glucose. A group of enzymes of particular interest to physiologists consists of those which use adenosine triphosphate (ATP) as their substrate. Known collectively as ATPases these enzymes have a wide range of functions and it is important to realize that they are a family of enzymes of very differing properties with a variety of responses to histochemical methods. The very fact that these enzymes share a common name has been instrumental in compounding much of the confusion which has arisen in the histochemical literature since minor changes in technique from one laboratory to another can lead to the demonstration of a quite different enzyme, similar only in that it can hydrolyze ATP. Some of the early histochemical work is therefore suspect and difficult to interpret, but the situation has markedly improved in the last 3 or 4 years with the development of more specific methods for individual ATPases.

The great importance of this group of enzymes to the biologist arises from the fact that ATP is the key substance used by animal tissues for the provision of energy. Whenever thermodynamic work has to be fed into a system to drive it in the direction required, the rate limiting process is usually tightly coupled to some other reaction which can yield the necessary free energy. In many of the processes so far studied the energy yielding reaction is the hydrolysis of ATP. Its polyphosphate bonds can be thought of as possessing a large amount of stored energy which is released when hydrolysis occurs by the addition of water. As anyone who has ever added phosphorus pentoxide or acetic anhydride to water will know, the amount of energy released by such a process is not inconsiderable. Under the appropriate conditions the spontaneous hydrolysis of ATP can be very slow; in the living cell it is probably a reasonably stable molecule in the chemical sense and undergoes hydrolysis only when that is coupled to an energy-requiring process. Two types of hydrolysis can occur, as is illustrated in Fig. 5. In the commonest type the terminal anhydride bond is split to release inorganic phosphate, sometimes as a two stage process in which some active intermediate is first phosphorylated and then hydrolyzed. In the less common type ATP is split at the first anhydride bond with the release of inorganic pyrophosphate. It should be mentioned that ATP also acts as a phosphorylating agent; the first step in the metabolism of glucose, for instance, is its conversion to glucose-1-phosphate with the release of ADP. Histochemical techniques are available, however, only for those processes in which inorganic phosphate or pyrophosphate is formed.

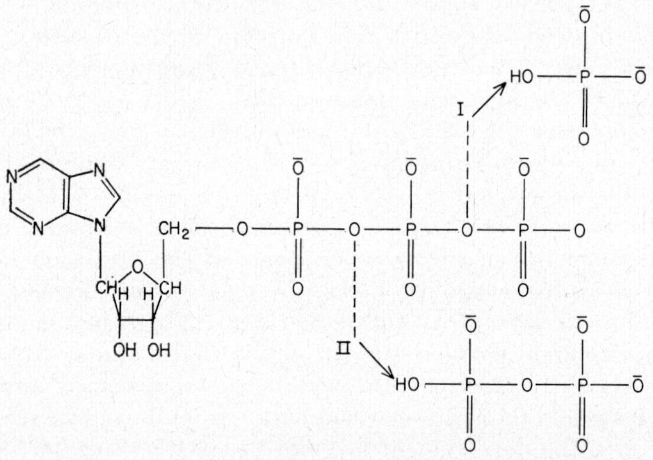

FIG. 5. Diagram to illustrate the two types of hydrolysis of ATP. Type I is the commonest studied histochemically in which the terminal bond is split to yield a free phosphate group either directly or via breakdown of a phosphorylated intermediate. Type II, in which the first anhydride linkage is broken with the release of inorganic pyrophosphate, is represented histochemically by the method for adenyl cyclase illustrated in Fig. 7.

Histochemical localization of the individual ATPases is discussed in detail later, but it should be emphasized here that several different enzymes or enzyme systems are known to exist with markedly differing responses to the cationic composition of the incubation medium. A biochemically well documented enzyme is the myosin ATPase of muscle which is activated by calcium ions. Mitochondria contain a well defined ATPase activated by magnesium ions. Of particular physiological interest is a membrane ATPase activated by monovalent cations such as sodium and potassium, present in tissues known to have powerful ionic pumps. A fourth type of ATPase activity demonstrable histochemically is adenyl cyclase in which pyrophosphate is produced. Alkaline phosphatase will also hydrolyze ATP under alkaline conditions, a fact which has not always been allowed for in histochemical studies.

One of the great difficulties in the developing of histochemical techniques for ATPases has been the discovery of suitable capture conditions for the liberated inorganic phosphate. In the Gomori-type techniques for acid and alkaline phosphatases calcium and lead ions have been used successfully but neither of these cations are ideal for the ATPases. Under many conditions calcium phosphate is not sufficiently insoluble to give good enzyme localization and free Ca^{++} ions have a marked effect on the activity of some ATPases. For electron microscopy lead has enormous advantages since lead phosphate is highly insoluble over a wide range of conditions and forms a very dense precipitate giving high contrast in the microscope. But the use of lead has many disadvantages some of which have been realized only in recent years.

Naidoo and Pratt (1951) were the first to produce a method for ATPases with lead as the capturing divalent cation. They used their method for an optical microscope study of various p oblems concerning the nervous system. A similar approach was used by Wachstein and Meisel (1957) who showed that it was possible to demonstrate ATPase activity in bile canaliculi. This paper attracted a great deal of attention, both at the time and in subsequent years; so much so that many people have been encouraged to apply the technique at the electron microscope level in situations where it can give misleading results. Pratt (1954) had earlier shown in a parallel biochemical study that high concentrations of lead can inhibit ATPase activity and markedly reduce the effectiveness of magnesium activation. Wachstein and Meisel (1957) made no reference to these results and used a high concentration of lead nitrate (approximately 3·6 mM) in their procedure which has been copied by many subsequent workers. Several early attempts to use the lead technique for electron microscopy were not very successful (Essner *et al.*, 1958; Ashworth *et al.*, 1963) but they were before the general introduction of aldehyde fixatives. Tice and Smith (1965) obtained staining for myosin ATPase in insect flight muscle that was reasonably well localized along the sarcomeres at least at low magnifications. But good localization at high magnifications has only come with the realization of the particular problems associated with the technique.

One of the problems has already been mentioned: that lead ions can cause enzyme inhibition and can reduce the normal activating effect of some other divalent cations. Some degree of enzyme inhibition by the capturing agent is not uncommon in enzyme techniques, but the situation here is complicated by the fact that the various ATPases differ in their sensitivity to lead and that these differences may vary from species to species. In general the membrane ATPases are most sensitive but can still be shown up where their activity is particularly strong (Fig. 6). A much more serious problem arises from the fact that ATP will form a relatively stable chelate complex with lead ions which can undergo a slow spontaneous hydrolysis to yield a precipitate of lead phosphate even in the absence of any ATPase activity. The extent

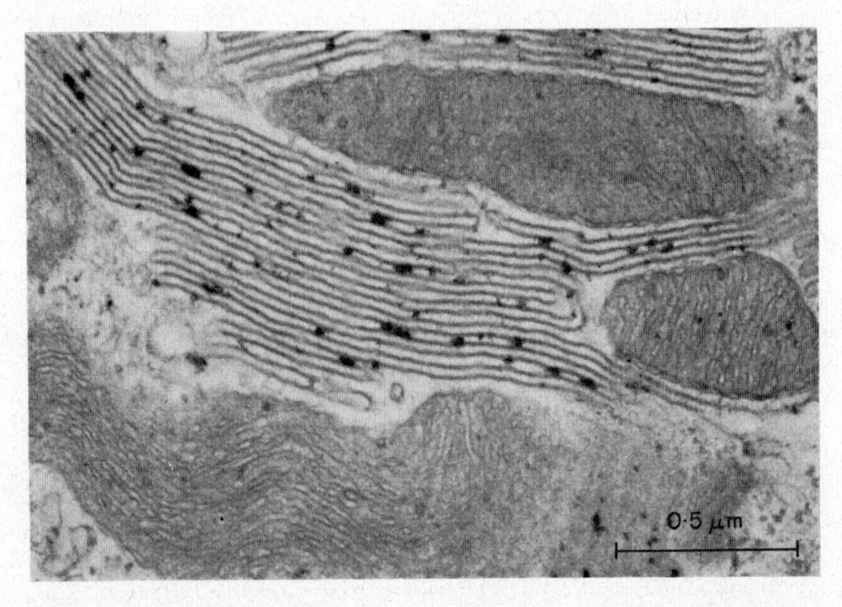

FIG. 6. An example of the distribution of staining obtained with the lead technique of Wachstein and Meisel (1957) for ATPase. The electron micrograph shows an area of a cell from the rectal papillae of the blowfly. Note the absence of staining in the mitochondria but the discrete deposits of histochemical reaction product between the stacked lamellae where active pumping of ions is believed to take place (×50 000). (By courtesy of Dr. B. L. Gupta.)

to which the existence of this chelate complex and the risk of its hydrolysis might invalidate the histochemical findings has been the subject of much debate in the literature. The progress of this debate can be traced back by anyone interested from the final papers by the principal protagonists (Rosenthal, *et al.*, 1970; Novikoff, 1970). The general conclusions to be drawn would appear to be that some of the earlier work must be suspect and that even with optimally designed incubation media the pattern of staining obtained should be interpreted with the greatest caution. Nonetheless biologically useful results can and have been obtained where the problem is a suitable one.

Many of the attempts to demonstrate the various ATPases have been unsatisfactory. A major difficulty is the sensitivity of most of these enzymes to fixatives and to lead ions. Tice and Smith (1965) appeared to obtain good localization of myosin ATPases, at least at low power, in insect flight muscle but they used adipaldehyde (a very mild fixative) and a high concentration of calcium ions which might have reduced the degree of inhibition by the lead. At higher magnifications the localization appears less satisfactory and may be influenced by the presence of structures with a specific affinity for lead ions. Gillis and Page (1967) using glycerinated mammalian muscle fibres produced strong evidence that regions of specific affinity could mimic enzyme distribution. It is perhaps fair to say that histochemical studies of myosin ATPase distribution are consistent with present theories of muscle contraction but have not helped in their further elucidation. Mitochondrial ATPase has been the subject of several investigations with somewhat varied techniques and more than somewhat varied results. It is a Mg-activated enzyme which is moderately inhibited by lead and is rather sensitive to fixation. There is no doubt about its existence inside mitochondria but there is dispute over its precise location in relation to the membrane system of mitochondria. Perhaps the best degree of localization has been obtained by Ogawa and Mayahara (1969) who used a modification of their alkaline phosphatase technique with citrate to keep down the free lead ions. Overall, however, the histochemical results have not added anything significant to our knowledge from other disciplines. The Na, K-activated, membrane ATPase is even more sensitive to both lead ions and fixatives, so much so that many attempts to demonstrate it by the standard method have been unsuccessful. Where the enzyme is present in a very concentrated situation sufficient activity may exist to give obvious staining (see Fig. 6), but the distribution of the staining has to be interpreted with caution.

A quite different approach to the histochemical localization of membrane ATPase has been explored by Ernst (1972). There is now extensive biochemical and biophysical evidence that the sodium transport system operates in two stages. The first stage requires ATP and magnesium and is activated by sodium ions, ADP being released and a phosphorylated intermediate formed. In the second stage this intermediate is split by a phosphatase which is potassium activated. Many of the preparations used for biophysical studies possess potent K-dependent phosphatase activity towards various acyl and aryl phosphates. The present available evidence supports the view that this phosphatase activity is a manifestation of the second stage of the membrane ATPase system. Accordingly, Ernst (1972) investigated ways of demonstrating this phosphatase activity. With p-nitrophenyl phosphate as substrate and strontium ions as the capturing agent at a pH of 8·5–9·0 he was able to obtain excellent localization in the avian salt gland. The staining was concentrated on the inside surface of the basal and lateral membranes of the secretory cells. Such a distribution is not inconsistent with present theories as to how the salt gland produces a hypertonic secretion. The technique has also been applied successfully to rat kidney (Ernst, 1973) and cornea (Leuenberger and Novikoff, 1974). It seems certain that the histochemical procedure will provide a very valuable addition to methods of studying ionic transport mechanisms.

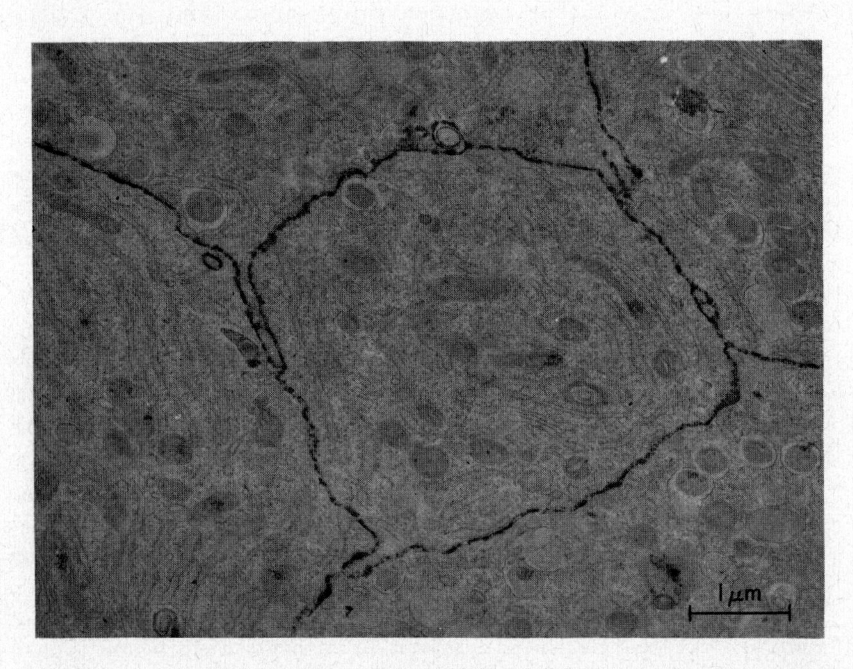

FIG. 7. An example of histochemical staining for adenyl cyclase. The electron micrograph shows a group of β-cells in a rat pancreatic islet with histochemical reaction product concentrated at the outer surface of the plasma membranes (× 13 500). (Reproduced by permission from Howell and Whitfield, 1972.)

Another enzyme which hydrolyzes ATP is adenyl cyclase. It belongs to the second category (Fig. 5) in which the first polyphosphate link is broken to release inorganic pyrophosphate. The other product formed is adenosine 3′, 5′-cyclic monophosphate (cyclic AMP), a substance which is now known to play an important role in the control of many cell processes, especially in the control of cell secretions. Reik *et al.* (1970) published a method for the histochemical localization of this enzyme in rat liver with lead as the capturing agent for the released pyrophosphate. Howell and Whitfield (1972) have improved the specificity of the method by using adenyl-imidodiphosphate in which the oxygen linking the second and third phosphorous atoms is replaced by an —NH— group. This substance is not hydrolyzed by the ordinary ATPases but is an effective substrate for adenyl cyclase. They obtained excellent localization of staining to the plasma membranes of the islet cells in rat pancreas (Fig. 7). Owing to the widespread importance of cyclic AMP, this histochemical technique promises to be especially valuable.

V. Methods for the Esterases

The esterases, or more correctly the carboxylic ester hydrolases, are enzymes which are able to hydrolyze esters of an alcohol or phenol with a carboxylic acid. In other words they catalyze the general reaction:

$$R_1 . O . CO . R_2 + H_2O \rightleftharpoons R_1 . OH + R_2 . CO_2H$$

The group comprises a number of biochemically distinct enzymes with differing specificities towards substrates and inhibitors. In general, each enzyme will hydrolyze a range of structurally-related esters and there is often considerable overlap in specificity between individual enzymes. Inevitably, therefore, the terminology is in a very confused state, with biochemists and histochemists using different nomenclatures. The problem is discussed at some length by Pearse (1972), but for the present purpose it is sufficient to divide the enzymes for which histochemical techniques exist into three groups: lipases, non-specific esterases and cholinesterases. Some subdivisions within the second and third groups will be discussed later.

The esterases occupy a very special position in the development of histochemistry at both the optical and electron microscopical levels. Several histochemical techniques involving new principles were first developed for enzymes in this group. In general, carboxylic esters are much easier to synthesize than the corresponding phosphate esters and hence many exotic esterase substrates have been tried out. A list of those that have achieved any sort of success is given in Table II, which illustrates the dominant position of the esterases in histochemical research. A contributory factor to this dominance is the great interest in the enzyme acetylcholinesterase, which is physiologically very important and has an unusually discrete type of localization. Many of the studies of esterase histochemistry have been prompted by interest in this particular enzyme, and have in return contributed greatly to our knowledge of its distribution and function.

TABLE II

Some of the methods introduced for localization of carboxylic esterases at the E.M. level

Substrate	Capture agent	References
α-naphthyl acetate	diazonium salt	Lehrer and Ornstein (1959)
Naphthol AS-D acetate	diazonium salt	Bowen (1968)
2-Naphthyl thioacetate	diazonium salt	Bergman *et al.* (1967)
2-Thiolacetoxybenzanilide	diazonium salt	
Indoxyl esters	diazonium salt	Holt and Hicks (1966)
Thiolacetic acid	Pb^{++}	Barrnett and Palade (1959)
Acetyl disulphide	Au^{+}	Koelle *et al.* (1968)
Thiocholine esters	Cu^{++}	Lewis and Shute (1964)
Thiocholine esters	Au^{+}	Davis and Koelle (1967)
Thiocholine esters	ferricyanide	Karnovsky and Roots (1964)
Selenocholine esters	ferricyanide	Kokko *et al.* (1969)
p-Nitrophenyl esters	Au^{+}	Vatter *et al.* (1968)
8-Hydroxyquinoline esters	Bi^{+++}	Deimling and Madreiter (1972)

A. *A survey of Available Methods*

The first two satisfactory esterase techniques were both devised by Gomori (1945; 1948) and depend on closely related chemical principles. In the technique for lipase, the substrate used is a long chain fatty acid ester of a high molecular weight polyglycol or polymannitol. In that devised for cholinesterase the substrate is a fatty acid ester of choline. Enzyme action releases the fatty acid, usually stearic or oleic acid in the first technique and myristic or lauric acid in the second, which can be precipitated by a wide range of heavy metal cations, those chosen by Gomori being respectively calcium and cobaltous ions. Both techniques suffer from the serious disadvantage that their histochemical specificity is not known with certainty. Although their range of specificity is much narrower than the coupling azo-dye techniques discussed below it has never been defined biochemically. They both suffer from the further disadvantage that in their original form the final reaction product is too coarse for ultrastructural studies and neither has been satisfactorily adapted for electron microscopy.

In 1949 two techniques for esterases were introduced which have both been adapted for electron microscopy. One of these (Koelle and Friedenwald, 1949) is specific for the cholinesterase group of enzymes and is discussed in detail later in this section. The other (Nachlas and Seligman, 1949) was an adaptation of the coupling azo-dye technique for alkaline phosphatase already discussed (Section IIIB). The first substrate used was β-naphthol acetate but it was soon replaced by the α-naphthol ester to give a faster capture reaction with diazonium salts, and ultimately a wide range of substituted naphthol derivatives were utilized (Pearse, 1972). The range of enzymes that can hydrolyze the acetate esters is very wide and probably includes most of the enzymes hydrolyzed by the other esterase techniques. It is therefore an excellent method of studying the broad distribution of esterases and has been widely used at the level of the optical microscope. It is also valuable where only one esterase enzyme is known to be present. The range of specificity can be narrowed by the use of esters of long chain carboxylic esters.

Another esterase technique was introduced only a year or so later (Barrnett and Seligman, 1951; Holt, 1952) which depended upon a quite different principle from those already considered. The substrate used is indoxyl acetate and enzyme action releases the free indoxyl radical which is spontaneously oxidized in the presence of molecular oxygen to the highly insoluble and brightly coloured dye indigo. In one sense therefore no specific capture agent is needed. The rate of spontaneous oxidation is too slow, however, for good localization even at the level of the optical microscope and attempts to speed up the reaction with a catalyst such as the ferricyanide ion have not proved entirely satisfactory. Subsequently (Davis and Ornstein, 1959) it was realized that free indoxyl would couple very rapidly with diazonium salts in the same way as naphthols do. And it is as substrates for coupling azo-dye techniques that indoxyl esters are now mostly used.

As already mentioned in Section III, a variety of procedures suitable for electron microscopy have been developed from these coupling azo-dye techniques with naphtholic and indoxyl substrates. The particular procedures for esterases were included for completeness in Table I together with those for the phosphatases. Consideration here will be restricted to the techniques which are relevant to the study of the non-specific esterases and lipase; application of azo-dye techniques to the study of the cholinesterases is dealt with later in the discussion of motor end-plate histochemistry.

As already mentioned in Section III, a major problem with the azo-dye techniques is to produce a sufficiently dense reaction product. The various ways of achieving this for the phosphatases can all be applied to the esterases and were often first tried out on the cholinesterases. With ordinary naphtholic esters as substrates it is necessary to use a diazonium salt that incorporates a heavy metal such as lead (Beadle *et al.*, 1971) or mercury (Smith and Fishman, 1969) in order to obtain adequate contrast in the electron microscope. For the non-specific esterases in general a far better solution is to use the indoxyl substrates since these produce richly osmiophilic azo-dyes with a wide variety of diazonium salts. Thio-esters of β-naphthol also produce osmiophilic azo-dyes but have found little application beyond the group in Seligman's laboratory who pioneered their use (Hanker *et al.*, 1964). They do provide, however, the best procedure so far published (Seligman *et al.*, 1966) for the demonstration of lipase at the electron microscope level, but there has been rather little physiological interest in the distribution of this enzyme over the last decade. Potentially at least the coupling azo-dye techniques have a great future in E.M. histochemistry. Because both the substrate and the capturing agent are aromatic molecules almost infinite possibilities exist for the synthesis of new reagents and some progress has already been made in producing compounds dove-tailed to specific cytochemical requirements (Davis *et al.*, 1972).

Chronologically the next technique to be introduced for esterases was that by Crevier and Belanger (1955). They based their technique on the observation by Wilson (1951) that the thiolacetic acid ($CH_3 . CO . SH$) molecule reacts with the esteratic site of acetylcholinesterase. An exchange reaction with water is thereby catalyzed with the release of H_2S. In the histochemical procedure plumbous ions are used to produce a black precipitate of lead sulphide. It was soon realized that many enzymes give a positive result but the technique has a special importance in the development of E.M. histochemistry. It was introduced in an almost unmodified form into electron microscopy by Barrnett and Palade (1959) and was used shortly after by Zacks and Blumberg (1961) and by Barrnett (1962) to study the ultrastructural distribution of cholinesterase at the motor end-plate. Together with the technique for acid phosphatase (Holt and Hicks, 1961) it gave the first clear indications of the degree of localization and contrast that might be obtained for enzymes with the electron microscope. Because of its lack of inherent specificity the technique has never rivalled the thiocholine technique in popularity and the pungent, unpleasant smell of the substrate must have further restricted its use.

The original technique gave very variable results which were ultimately traced to variations in the purity of the commercial substrate used. Koelle and his collaborators carried out an extended series of investigations into the technique. It was discovered that the true histochemical substrate was not the thiolacetic acid itself but an impurity, acetyl disulphide ($CH_3 . CO . S . S . CO . CH_3$), which forms spontaneously by oxidation (Koelle and Horn, 1968). Nevertheless acetyl disulphide is a satisfactory substrate with lead as the capturing agent only if a large excess of thiolacetic acid is present, the reason appearing to be that the lead must be present as an uncharged complex with the thiolacetic acid to prevent inactivation of the primary enzyme hydrolysis step. These complexities have been fully investigated only in the case of true cholinesterase and it is uncertain how far they apply to other enzymes. To that extent the histochemical technique with lead still leaves much to be desired. The use of other heavy metals as capturing agents has been investigated and a method with gold has been published by Koelle *et al.* (1968). The starting material is specially purified thiolacetic acid and the gold is converted into a complex with the thiolacetic acid. This complex acts as both substrate and capturing agent in the incubation procedure which has produced excellent pictures of enzyme distribution at motor end-plates. Other tissues, however, appear to have yielded less satisfactory results and the value of the method for non-specific esterases has yet to be assessed.

Another technique for esterases that makes use of gold as a capturing agent has been published by Vatter *et al.* (1968). They used acetate and butyrate esters of p-nitrothiophenol as substrates and sodium aurous thiosulphate in the incubation medium to provide the capturing agent. As a result of enzyme activity, aurous p-nitrothiophenolate precipitates out as a fine, non-crystalline deposit which gives excellent contrast in the electron microscope. The range of specificity of this technique has still to be defined, but in addition to some esterases it may reveal other hydrolytic enzymes including peptidases (O'Hare *et al.*, 1971). The same histochemical principle could probably be extended to other enzymes by the use of suitable substrates.

More recently a new technique designed specifically for certain non-specific esterases has been published by Deimling and Madreiter (1972). The substrate used is 8-acetoxyquinoline and the capturing agent is trivalent bismuth kept in solution as the complex with tartrate. The FRP is a chelate of bismuth with 8-hydroxyquinoline and is converted to brown bismuth sulphide which gives good contrast in the electron microscope. It is too early to know how valuable this method and any subsequent modifications may prove to be, but it has the merit of introducing a new heavy metal into enzyme histochemistry.

B. *Methods Specifically for Cholinesterases*

The cholinesterases are a well defined subgroup of the esterases with, as their name implies, the ability to hydrolyze esters of choline, particularly acetylcholine. Biochemically, two distinct enzymes with differing ranges of specificity have been identified: acetylcholinesterase (AChE), which will hydrolyze a range of acetate esters in addition to acetylcholine but not choline esters of long chain fatty acids; and pseudocholinesterase (PsChE), which will hydrolyze a wide range of choline esters including acetylcholine. Both are present in normal mammalian blood, AChE associated with the red blood cells and PsChE with the plasma. PsChE is widespread in tissues but its function is still uncertain. AChE is also widespread but is especially associated with the nervous system where its prime function is to hydrolyze acetylcholine that has been released as a neurotransmitter by a particular class of neurones. This class of neurones, known as cholinergic, includes all the motor supply to skeletal muscle, much of the autonomic nervous system and a significant proportion of nerve cells in parts of the brain. Hence the great interest in methods of localizing AChE. To be of use, however, a high degree of specificity is essential except where other esterases are known to be absent. The only group of techniques which satisfy that criterion are those based on the method introduced by Koelle and Friedenwald (1949), in which acetylthiocholine is used as substrate. The histochemical substrate therefore differs from the physiological one only in the replacement of an oxygen by a sulphur atom. Specificity is only marginally affected and the substrate is hydrolyzed quite as fast as acetylcholine itself. Hydrolysis does, however, release thiocholine instead of choline and the presence of the free sulphydryl group leads to enhanced reactivity which can be utilized in several ways for histochemical purposes. In the original method the thiocholine was captured with cupric ions; and several other heavy metals can be used instead. The sulphydryl group has appreciable reducing power; and this property too has been utilized in the design of techniques for electron microscopy.

An important refinement available with the thiocholine techniques is a further improvement in enzyme specificity by the use of alternative substrates and selective inhibitors. Both cholinesterases will hydrolyze acetylthiocholine but in many species only PsChE will hydrolyze butyrylthiocholine at a rate adequate to show up with the histochemical technique. A range of inhibitors are available; so that it is possible to use either acetyl- or proprionylthiocholine with an appropriate inhibitor to demonstrate either AChE or PsChE alone.

The method has been extensively used for optical microscopy to study a wide range of biological problems. Because it was introduced before the use of fixatives had become generally accepted in enzyme histochemistry the original technique has been modified in a number of ways. In fact, the cholinesterases in most situations survive aldehyde fixation extremely well, and some degree of fixation is usually necessary

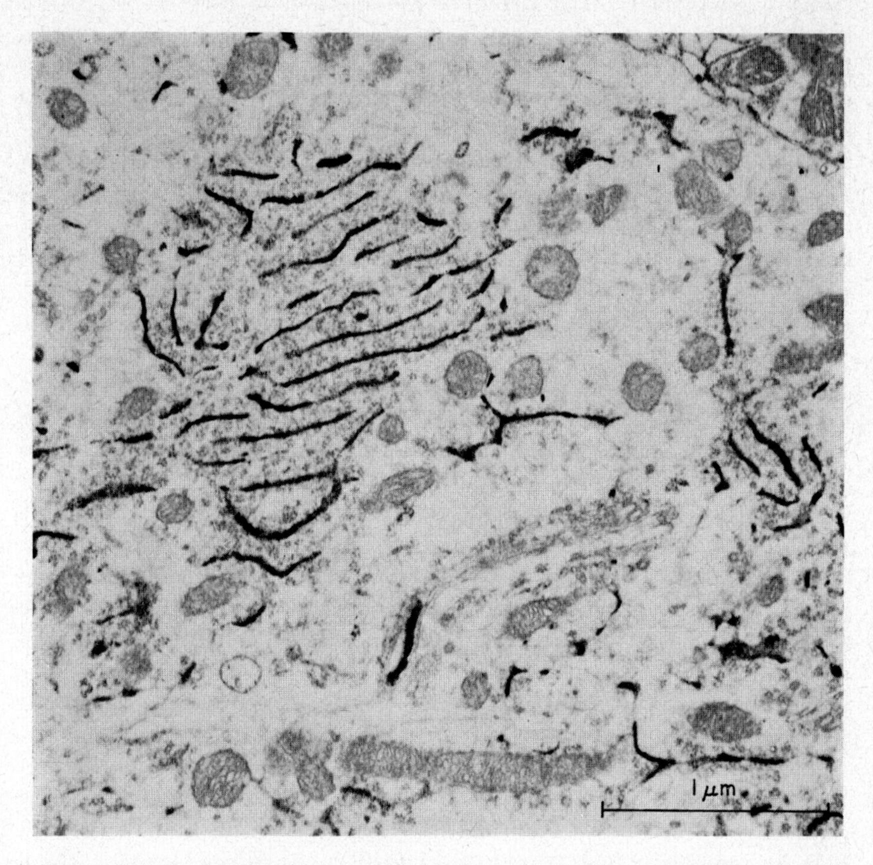

FIG. 8. Staining for true cholinesterase by the method of Lewis and Shute (1969) in the cytoplasm of a motor neurone in rat spinal cord. The rough endoplasmic reticulum is intensely stained but mitochondria show staining due only to the lead citrate counterstain. The plasma membrane (at the top right hand corner) is also unstained as is most of the Golgi complex (×30 000).

to ensure adequate diffusion of the histochemical reagents across cell membranes. With cupric ions the released thiocholine forms a white, almost amorphous precipitate which is normally converted after incubation into highly insoluble brown copper sulphide by treatment with a buffered sulphide solution. Since cupric ions tend to inhibit the enzyme activity they are normally buffered to a tolerable level by addition of an appropriate concentration of glycine which forms a reversible chelate complex with them. Formation of the chelate complex is pH dependent: as the pH is reduced the equilibrium concentration of free cupric ions increases, and the sensitivity of the method can be thereby controlled. At a pH of 6 the method is very sensitive but localization is often poor. As the pH is reduced sensitivity falls and localization becomes more accurate; reducing the ratio of glycine to copper (from 5:1 in the original technique down to about 5:3 in some modifications) has a somewhat similar effect. At a pH of 5 with a low ratio of glycine to copper the method is particularly valuable for studying the precise location of sites of high enzyme activity, as for instance in the mammalian motor neurone (see Fig. 8). Many minor variants of the method have therefore been devised for the study of particular problems both at optical and at electron-microscopic levels. Only the six major variants currently used in electron microscopy will be discussed here.

The first of the modifications designed specifically for electron microscopy was published in 1964 by Karnovsky and Roots. This modification was extended by the introduction in 1969 of the selenium analogue of acetylcholine as a substrate by Kokko *et al.* The other four modifications were all published in full in 1966 and 1967 although preliminary reports had appeared earlier (Lewis and Shute, 1964; Joo *et al.*, 1965; Tennyson *et al.*, 1965). There was therefore a brief period when several groups of researchers were looking for ways of adapting the thiocholine technique for cytochemical studies; the question of priorities is largely academic and what is much more important is an appreciation of the relative merits of the different techniques when applied to particular problems. An extensive assessment of their relative merits would be out of place in the present review, but a few comments on the subject are necessary here in order to justify the selection of papers for special mention.

It is convenient to discuss the six techniques in the order in which they diverge from the original technique introduced for optical microscopy. The method of Brzin *et al.* (1966) differs only very slightly. The concentration of the copper-glycinate is somewhat higher but the ratio is kept the same (1:5) and so is the pH. After incubation the material is post-fixed with potassium permanganate but could be treated with a sulphide solution instead to produce a reaction product stable to embedding. The technique is very sensitive because such a high pH is used and will detect low concentrations of enzyme such as occur in developing tissues. Kasa and Csillik (1966) use the somewhat lower pH of 5·4 and add a low concentration of lead ions in an attempt to obtain greater contrast in the electron microscope: their method has not been extensively used. Lewis and Shute (1966, 1969) use a still more acid pH, usually 5·0–5·5, and a lower ratio of glycine to copper. Although the technique is thereby made less sensitive, incubation times can be increased because the degree of localization is much improved by the rise in concentration of free cupric ions in equilibrium with the glycine. Several examples are shown in Figs 8, 10 and 11.

The variant introduced for electron microscopy by Karnovsky and Roots (1964) uses a different principle for the capture reaction. Copper is still used in the incubation medium but with citrate instead of glycine to control the concentration of free cupric ions. Also included are ferricyanide ions which are reduced by the enzymatically-released thiocholine to ferrocyanide. The FRP is an orange-brown precipitate of the highly insoluble cupric ferrocyanide known as Hatchett's brown. The technique has been widely used because of its intrinsic simplicity, but the quality of localization has been very variable. The main problem appears to be the poor diffusibility of the ferricyanide ion; certainly the technique works best on tissues with an open type of structure where diffusion barriers are minimal. A further problem is that the FRP sometimes has a microcrystalline structure. In summary, this modification is easy and quick to use but often gives poor cytochemical localization. A further variant of this technique, which may prove valuable in the future even though it does not solve the localization problems, has been proposed by Kokko *et al.* (1969) who synthesized and used the selenium analogue of acetylcholine as a substrate.

The last thiocholine technique, introduced at the E.M. level by Davis and Koelle (1967), makes use of monovalent gold as the capturing agent. A wide range of incubation media was tested before a satisfactory procedure could be devised, and a very high concentration of phosphate ions (approximately 4·0 M) is essential in order to obtain satisfactory precipitation of a FRP assumed to be gold-thiocholine phosphate. The FRP is converted into gold sulphide before embedding. Under optimum conditions this technique has produced the clearest-looking micrographs of AChE distribution at mammalian motor end-plates, but has not proved so satisfactory on other tissues.

C. *Some Biological Applications*

The literature on the distribution of AChE at the E.M. level is expanding rapidly. The number of papers in which one or other of the thiocholine modifications has been used is exceeded only by those in which methods for ATPases and peroxidases have been employed. The biological value of many of the ATPase studies is very much in doubt because of the serious reservations that still exist with regard to lead techniques based upon the original method of Wachstein and Meisel (1957). Most of the peroxidase studies are concerned essentially with the use of exogenous horseradish peroxidase in the form of a tracer, as discussed in Section VII. It is probably therefore reasonable to assert that AChE is the enzyme whose ultrastructural localization has proved the most biologically rewarding of study. Certainly, its value in a wide range of neurophysiological problems is unquestionable

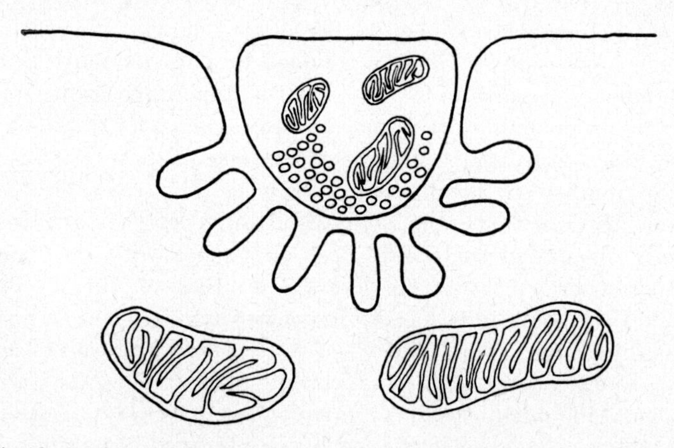

FIG. 9. An idealized representation of a cross-section through a synaptic region forming part of a motor end-plate. The diagram illustrates the morphological relationship of the nerve ending, with its complement of ACh-containing microvesicles, to the muscle membrane with its deep infoldings that form the subsynaptic clefts. These run in from the walls of the "gutter" in which the nerve process lies and comprise a major part of the inter-synaptic space into which the ACh is released on nerve stimulation.

and merits this separate discussion of some of the biological applications. The mammalian motor end-plate has figured prominently in the development of esterase methods and forms the logical starting point for this discussion. Of much greater potential significance, however, is the application of these techniques to neurological problems, particularly those involving the brain. This section therefore ends with some examples of problems being studied in the central nervous system.

Cytochemical studies of enzyme distribution at the myoneural junction have been doubly valuable, both to neurophysiology and to E.M. histochemistry. Given the ultrastructural morphology observed in the electron microscope, electrophysiological studies place specific

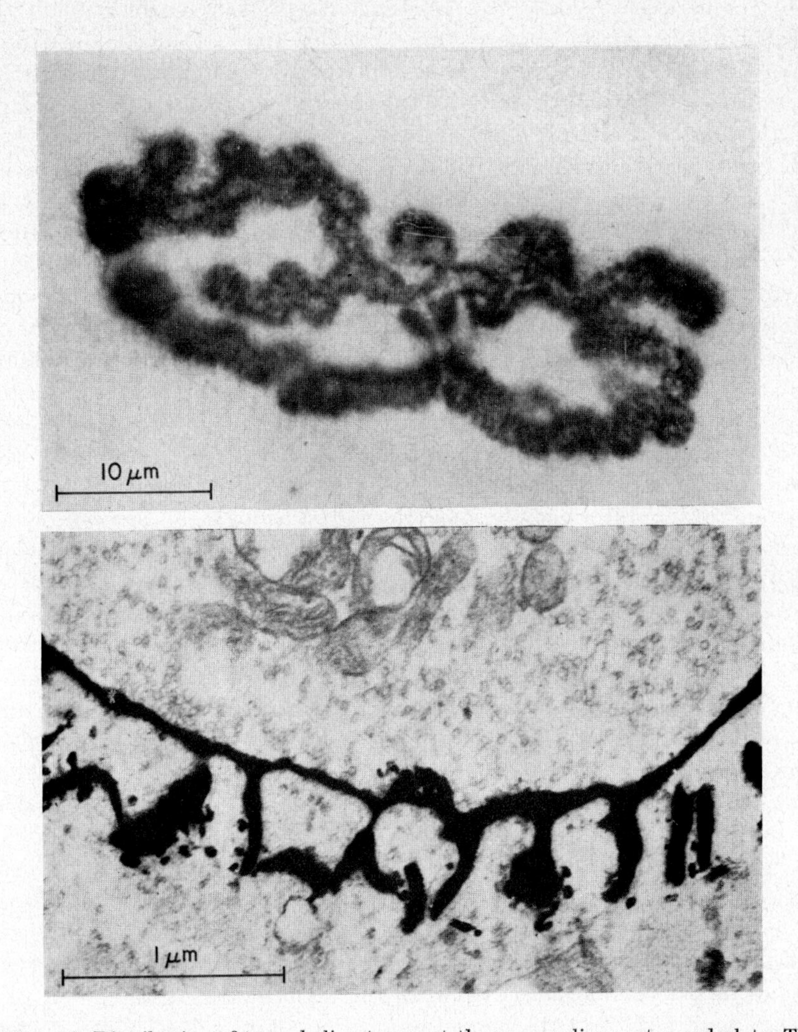

FIG. 10. Distribution of true cholinesterase at the mammalian motor end-plate. The optical micrograph (×2100) shows a plan view of a complete motor end-plate on a rabbit muscle fibre. The nerve can just be seen entering from the top left of the picture. The electron micrograph (×30 000) shows a cross-sectional view of part of a synaptic gutter in a rat diaphragm muscle fibre. Comparison with Fig. 9 shows that staining is concentrated in the synaptic space and the subsynaptic clefts. There is no staining of the mitochondria or the synaptic vesicles. (Optical micrograph by courtesy of Dr. R. I. Woods; electron micrograph reproduced from Lewis and Shute, 1966.)

restrictions on where AChE might be found. The force of these restrictions is best appreciated by reference to Fig. 9 which depicts an idealized junction in cross-section. Various lines of biochemical and physiological evidence place much of the acetylcholine (ACh) in the microvesicles but some free in the nerve cytoplasm. Fully active AChE cannot therefore exist associated with the microvesicles or in any other situation where it has free access to the cytoplasm. Following nerve stimulation ACh is released into the synaptic space: it must in fact be within at least 1 μm which is the distance an ACh ion could move in free solution in the time available. The presence or absence of AChE within the myoplasm would appear to be irrelevant since micro-injection of ACh directly into the muscle cell appears to have no physiological effect. Similarly its presence is largely unnecessary outside the zone of contact between nerve and muscle. Originally the histochemical evidence was conflicting but as the techniques have been refined so the picture has become clearer. Even at the level of the optical microscope the restriction of staining to the general area of myo-neural contact was obvious (Fig. 10) although comparison of the various techniques indicated that more than one esterase was present. At the E.M. level the general esterase techniques stain a number of structures, but the specific thiocholine method produces staining restricted to the nerve membrane and the region of the synaptic cleft (Fig. 10). In particular, no staining is observed of microvesicles, mitochondria or any other structures actually within the nerve process, and the staining of the muscle membrane stops abruptly at the point where contact with the nerve is lost (Lewis and Shute, 1966); PsChE but not AChE is sometimes associated with the teleoglial cells that cover the myoneural junction. Published electron micrographs of the thiocholine technique do not allow of any firm conclusion about relative localization of AChE within the width of the synaptic space. Better resolution is claimed for some of the less specific cytochemical methods but there is still some disagreement over the extent to which AChE is membrane bound rather than situated between the membranes.

Although enzyme cytochemistry may not have added much to our functional knowledge of the normal mammalian motor end-plate, it does offer a valuable method of studying myoneural junctions in other species and under conditions where electrophysiological techniques would be very difficult to use. Cytochemical studies of AChE distribution are becoming particularly important in other parts of the nervous system, both peripheral and central, where it is now accepted that ACh can act as a neurotransmitter. As at the motor end-plate it is to be expected that AChE will be associated with sites of ACh release, and evidence to this effect is fast accumulating. Although there are exceptions to any such generalization, it provides a valuable working hypothesis for tackling problems difficult to study by other means. It

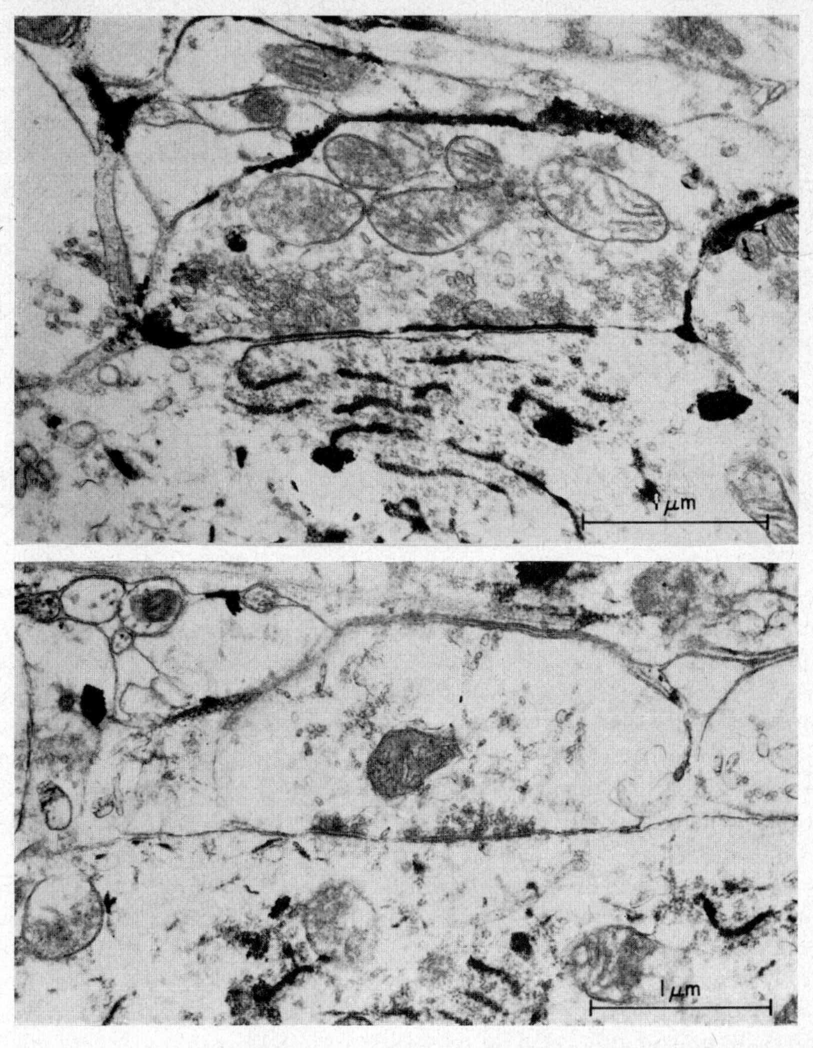

FIG. 11. The distribution of true cholinesterase around two large axo-somatic endings on the same rat ventral horn cell. In the upper electron micrograph almost the whole of the membrane of the presynaptic process is stained, but the mitochondria and microvesicles are unstained. In the lower electron micrograph none of the synaptic components are stained. Note the staining of the rough endoplasmic reticulum in the cell cytoplasm. It is a reasonable assumption that the upper synapse is cholinergic and the lower one is not (×25 000). (Reproduced from Lewis and Shute, 1966.)

has long been known from studies with the optical microscope that nerve cells which are known to be cholinergic contain AChE not only at their terminations (as at the motor end-plate) but also along the length of their axons and in their cell bodies. In their axons, staining is concentrated at the outer membrane, which enables cholinergic fibres to be identified in the electron microscope (Lewis *et al.*, 1971). In the cell bodies the rough endoplasmic reticulum is intensely stained (Fig. 8), the reaction product being concentrated in the intralaminar space. Perhaps the most valuable application of the cytochemical technique is in the study of synapses in the central nervous system. At a cholinergic synapse the membranes of the presynaptic process are normally intensely stained whereas non-cholinergic synapses are unstained. Fig. 11 shows examples of two synapses on the same motor neurone in rat spinal cord and illustrates the ability to distinguish nerve endings which have different neurotransmitters.

VI. Enzymes that Catalyze Oxidation-Reduction Reactions

The principles on which techniques for these enzymes are based differ somewhat from those for hydrolytic enzymes, largely because the type of reaction being catalyzed is different. In hydrolytic reactions water is added to the substrate which is thereby split into two products, one of which is captured in a second reaction to form the insoluble end-product. In oxidation-reduction reactions two different substrates are involved, one being oxidized or reduced at the expense of the other. In the histochemical techniques one of the reactants is normally the physiological substrate for the enzyme to be detected and the other is some substance which gives rise to an insoluble product on oxidation or reduction. The second reactant is usually an unphysiological substance which may affect both the sensitivity and specificity of the overall histochemical reaction.

The three types of reactants most commonly used are (a) aromatic amines, (b) ferricyanide, (c) tetrazolium salts. An aromatic amine was first used more than 50 years ago to detect the activity of what we now know as the enzyme cytochrome oxidase (for early references see Lison, 1936). The dramatic advance, however, was the introduction of 3:6-diaminobenzidine (DAB) by Graham and Karnovsky (1966) since this polyamine is oxidized by a number of enzymes to a highly insoluble polymer which has the great advantage for electron microscopy of being osmiophilic; a postulated sequence of reactions for the oxidation of DAB is given in Fig. 12. Ferricyanide has long been used by biochemists to study the oxidative processes occurring in mitochondria (for references see review by Palmer and Hall, 1972) where a number of enzymes are present that can catalyze its reduction to ferrocyanide (Dixon, 1971). In the presence of cupric ions the ferrocyanide is rapidly captured to produce an insoluble precipitate of Hatchett's brown as in the cholinesterase technique of Karnovsky and Roots (1964) already discussed. The tetrazolium salts are water-soluble compounds and are readily reduced by a number of dehydrogenases to the corresponding formazans which are insoluble in water and deeply-coloured. They were developed for optical microscopy and technical problems have arisen in the attempt to apply them to electron microscopy. The best results have been obtained with specially developed tetrazolium salts that give osmiophilic formazans (Seligman *et al.*, 1971).

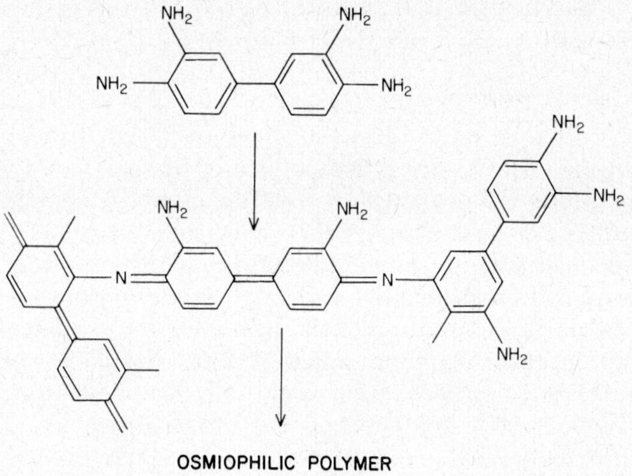

NH_2 NH_2 NH_2 NH_2

NH_2 NH_2 N N NH_2 NH_2 NH_2

OSMIOPHILIC POLYMER

Fig. 12. The type of reaction postulated to occur in the oxidation of 3,3′-diaminobenzidine (DAB) to produce an insoluble polymer that will react with osmium tetroxide.

For the purpose of histochemical discussion it is convenient to divide the enzymes which catalyze oxidation-reduction reactions into two main groups. In the first group molecular oxygen or its equivalent is one of the reactants; such enzymes are usually termed oxidases. The second group includes a variety of enzymes which do not normally interact with oxygen; many of the reactions catalyzed are more logically thought of as the abstraction of hydrogen rather than the addition of oxygen and the enzymes concerned are then called dehydrogenases. Most of the best histochemical techniques for oxidases depend upon DAB or related aromatic polyamines. The tetrazolium salts used for electron microscopy are mostly used to detect individual dehydrogenases although often a system of several enzymes or co-enzymes may be involved in the overall histochemical reaction. Ferricyanide occupies an intermediate, and sometimes ambiguous, position. It principally reacts with a wide range of flavoproteins (Dixon, 1971) that lie between the oxidases and the dehydrogenases in the oxidation-reduction economy of the cell. Because the hydrogens, or their equivalent, abstracted from a substrate molecule by a dehydrogenase can often be handed on to a flavoprotein, ferricyanide can be used in a number of histochemical methods for dehydrogenases. For the very same reason, however, the accuracy of localization and specificity of these methods can be open to doubt. Potentially, ferricyanide could be a valuable hydrogen-acceptor for a range of histochemical studies but strict biochemical controls will be essential. DAB has already proved its worth in the field of E.M. histochemistry and the authenticity of the pictures it produces is not really in doubt.

1. *The use of diaminobenzidine*

The first full paper on the histochemical use of DAB was published by Graham and Karnovsky (1966) who used it to follow the fate of injected molecules of horseradish peroxidase in mouse kidney. The use of horseradish peroxidase and other exogenous enzymes as tracer molecules is dealt with in the next section, but the success of DAB in this application soon stimulated its use to study various endogenous oxidative enzymes. Application of the technique for peroxidase is very simple both in theory and in practice. The enzyme is robust and survives routine aldehyde fixation very well. The incubation medium consists of 0·05–0·20% w/v DAB (as the tetrahydrochloride) in a tris buffer at a pH between 7 and 8. A small quantity of hydrogen peroxide (usually sufficient to give a final concentration of 0·01% w/v or less) is added either immediately before use or after the DAB has been allowed 15–20 min to penetrate the tissue. Oxidation of the DAB proceeds rapidly in the presence of hydrogen peroxide to form an amorphous polymer according to a scheme of the type shown in Fig. 12. The polymer is insoluble in ordinary dehydrating and embedding solvents; it is also highly osmiophilic so that postfixation with osmium tetroxide produces excellent contrast in the electron beam. The technique works equally well in plant and animal tissues with little modification necessary for most studies (Poux, 1969; Strum and Karnovsky, 1970; Novikoff *et al.*, 1971). Incubation for an hour or less is quite sufficient to give adequate contrast and the staining is highly specific (Fig. 13).

A number of other haem-proteins have weak peroxidatic activity which can be shown up if the concentration of the reagents is increased and the pH adjusted appropriately. (Haemoglobin and myoglobin can be shown up when present in high concentration, but their activity does not interfere in peroxidase studies.) Another enzyme which can be shown up with suitable modification of the technique is catalase. This enzyme, which catalyzes the breakdown of hydrogen peroxide into

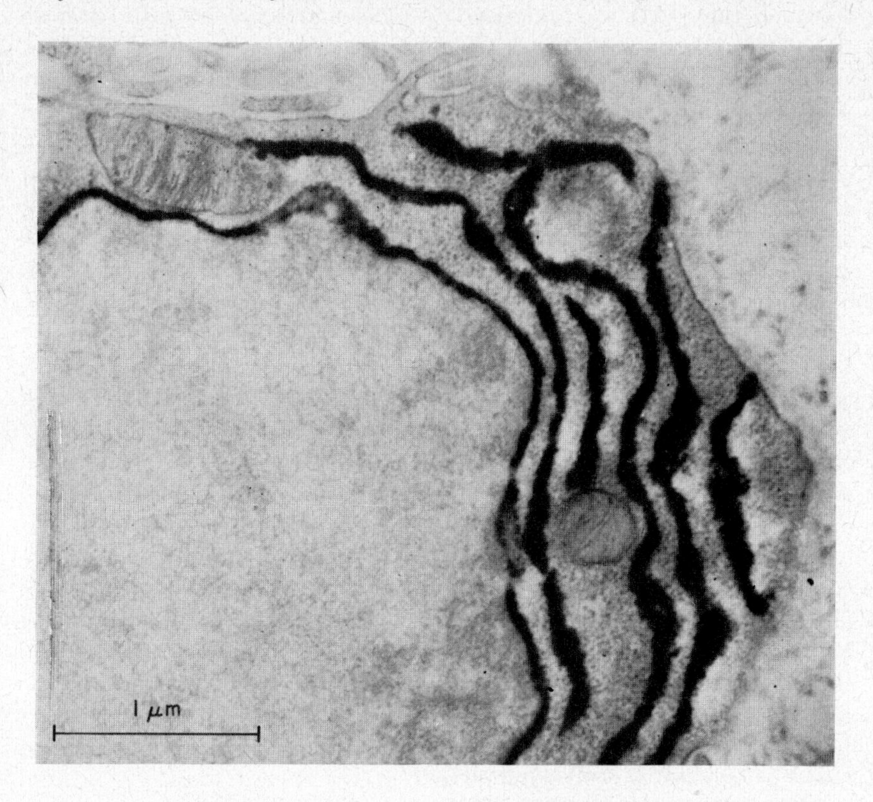

Fig. 13. The localization of peroxidase by the DAB method. In the field shown, from the lachrymal gland of the rat, reaction product is concentrated in the intra-cisternal space of the endoplasmic reticulum and the nuclear envelope. The mitochondria are unstained (×28 500). (By courtesy of Dr. R. L. Tapp.)

water and molecular oxygen, when in its native state has no effect on DAB. After fixation and certain other treatments, however, it will catalyze a rapid oxidation of DAB by hydrogen peroxide; as a histochemical procedure it works best if the concentration of reagents is high and the pH is raised to 8·5–9·0 (Fahimi, 1969). The staining produced is definitely due to catalase because it is inhibited completely by the specific inhibitor, 3-amino-1, 2, 4-triazole.

Mitochondria will also catalyze the oxidation of DAB, but the precise biochemical system responsible is still in doubt. Under some conditions cytochrome oxidase is responsible. This is the enzyme that catalyzes the final step in the respiratory chain, or looked at the other way round catalyzes the initial attack on the oxygen molecule. Seligman *et al.* (1968) set out to devise a histochemical technique for this enzyme in which the DAB was oxidized by molecular oxygen (with catalase added to prevent a build up of any hydrogen peroxide and so eliminating any possible interference from peroxidase activity). Excellent staining of mitochondria was obtained provided only brief fixation was used, and the staining was inhibited by cyanide. At a pH of 6·0 with a low concentration of hydrogen peroxide present (and no added catalase) Novikoff and Goldfischer (1969) obtained even better staining of mitochondria and under these conditions it is by no means certain what enzyme system is involved. As a specific stain for mitochondria, however, it has the advantage of simplicity over other published methods.

The use of DAB has had a profound effect on several lines of cytological investigation. One of these is the study of peroxisomes, which are microbodies present in many if not all cell types. As with many cytochemical investigations a marriage of biochemistry and electron microscopy has been necessary in order to elucidate the role of these intracellular organelles and of particular importance is the use of DAB to demonstrate the peroxidatic activity of the catalase which they contain. In many mammalian tissues peroxisomes contain a terminal flavin oxidation system coupled to catalase which breaks down the highly reactive hydrogen peroxide produced before it can escape into the general cell cytoplasm. In plant tissues the oxidative system contained in the peroxisomes is an essential link in the formation of carbohydrate by photosynthesis. The present state of our knowledge of these organelles is well summarized in the record of a symposium published in the November, 1973, number of *Journal of Histochemistry and Cytochemistry*. The study of endogenous peroxidase in, for instance, thyroid cells (Strum and Karnovsky, 1970) has been helped, but the greatest consequences of the introduction of DAB have been in the study of exogenous peroxidase used either as a molecular tracer or as a marker in immuno-cytochemistry. These uses of DAB are more appropriately discussed in the next section.

2. *The use of tetrazolium salts*

Tetrazolium salts have been used in histochemistry for many years (for references, see Pearse, 1972) to study a wide range of dehydrogenase enzymes. A number of problems were encountered, some of which were resolved with the introduction by Tsou *et al.* (1956) of Nitro-BT which contains two tetrazolium rings linked by a biphenylene group. The structural formula of Nitro-BT is illustrated in Fig. 14 (X = H) which

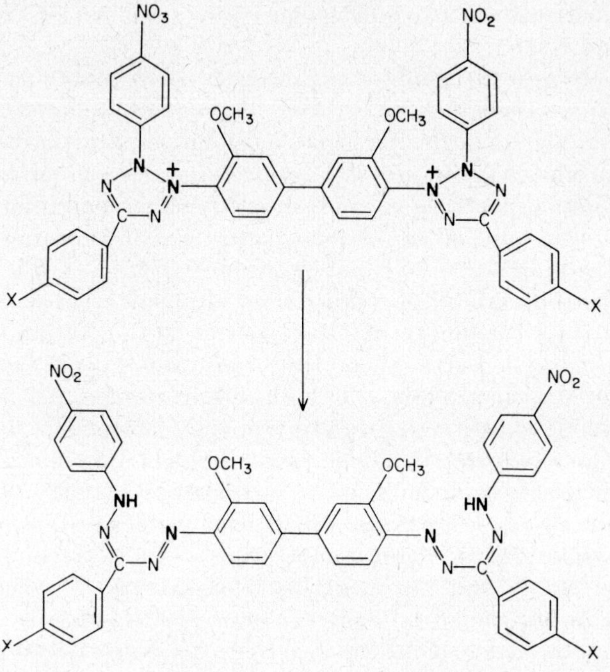

FIG. 14. The reaction by which a soluble tetrazolium salt is converted into an insoluble diformazan. The atoms and bonds involved in the change are shown in heavy type. The tetrazolium salt used routinely for optical microscopy, NBT, is unsubstituted at the position marked X in the above formulae, i.e. X = H. As explained in the text, various tetrazolium salts have been synthesized for electron microscopy in which the extra substituents are, in chronological order of their introduction: $-X = -NO_3$; $-X = -C{:}S-NH_2$; $(-C_6H_4-X) = (-CH = CH-C_6H_5)$.

also shows the changes that occur in the tetrazolium ring when two hydrogens are accepted from the dehydrogenase to produce the formazan. Since its introduction Nitro-BT has becomes the tetrazolium salt of choice for optical microscopy and in the latest edition of his textbook Pearse (1972) recommends its use in incubation media for over twenty dehydrogenases. It is sufficiently stable chemically but has a redox potential which allows it to accept hydrogens from a wide range of enzymes; its diformazan is relatively insoluble in lipids (an important property for accurate intracellular localization) and is intensely coloured. For electron microscopy, however, the diformazan suffers from two serious deficiencies: it is liable to be affected by some embedding procedures and it is not dense enough to give good contrast in the electron beam. It has been used successfully (Anderson and Personne, 1970) but the tetranitro derivative (Fig. 14; X $= -NO_3$) is definitely more satisfactory and has been preferred by a number of workers (e.g. Fahimi and Karnovsky, 1966).

In an attempt to produce an electron-microscopic technique for dehydrogenase Seligman and his collaborators adopted the same approach as they had for the azo-dye techniques. Starting with Nitro-BT as the model substance they synthesized a series of tetrazolium salts containing substituents which should react with osmium tetroxide. Their first successful substance (Seligman *et al.*, 1967) contained a thiocarbamyl group ($-C{:}S-NH_2$) in position X in Fig. 14. The formazan containing this group is osmiophilic although the recommended treatment, with osmium tetroxide vapour at 60°C, is rather too drastic for some tissues. Furthermore, problems arose over the chemical stability of the tetrazolium salt, both during storage and in solution. Other possible substituents were investigated (Seligman *et al.*, 1971) in which an unsaturated double bond supplied the osmophilia. The most satisfactory compound has a styryl group ($-CH = CH-C_6H_5$ in place of $-C_6H_4-X$) which would react with an aqueous osmium tetroxide solution at room temperature. This tetrazolium salt is chemically stable and gives excellent localization of succinic dehydrogenase in mitochondria (Seligman *et al.*, 1971); since it appears to have much the same biological reactivity as Nitro-BT, it should be applicable to many dehydrogenases. Other tetrazolium salts have been synthesized (Jones, 1969; Kalina *et al.*, 1972) which can be used for electron microscopy but they suffer from certain disadvantages compared with Di-styryl-Nitro-BT. So far, not many dehydrogenases have been studied at the electron microscope level, but the histochemical procedures for doing so are now available.

3. *The reduction of ferricyanide*

The third histochemical reactant used in the study of oxidation-reduction enzymes is ferricyanide. Like DAB it was first used at the electron microscope level, two very similar techniques appearing within a few months of each other (Ogawa *et al.*, 1968; Kerpel-Fronius and Hajos, 1968). Before then it had long been used by biochemists for quantitative studies especially of some of the intermediate enzymes along the mitochondrial respiratory chain. In the biochemical sense, ferricyanide fills the gap between the histochemical use of DAB at the molecular oxygen end of the chain and of tetrazolium salts at the dehydrogenase end. In practice there are several unresolved problems concerning the specificity of ferricyanide when used histochemically.

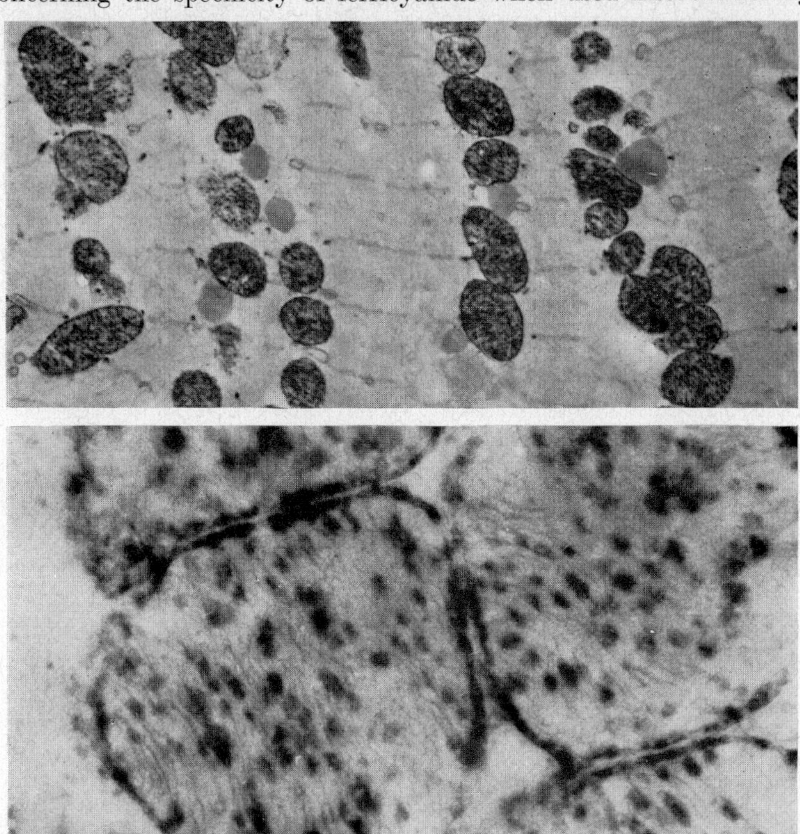

FIG. 15. Staining of mitochondria in mouse heart muscle by the ferricyanide technique for dehydrogenases. The low power picture (×6000) shows how well the method for succinate dehydrogenase picks out the mitochondria. The high power picture (×60 000) shows the somewhat diffuse localization obtained with β-hydroxybutyrate as the substrate. (By courtesy of Mr. B. A. Weavers.)

It does, however, give excellent staining of mitochondria, as is illustrated in Fig. 15, although the precise localization of staining at high power within the mitochondria depends upon the incubation conditions in a way that is still not fully understood (Kalina *et al.*, 1971). There are three complications which arise in the histochemical use of ferricyanide to study mitochondrial oxidation. The first is purely biochemical; ferricyanide can be reduced at two sites at least along the oxidative chain. The second is due to the poor diffusibility of the highly charged ferricyanide ion both in tissue generally and across the mitochondrial and plasma membranes in particular. The third is a specifically histochemical problem; conditions in the incubation medium have to be such that rapid precipitation of ferrocyanide ion occurs. In both the published techniques cupric ions are used to capture the ferrocyanide ions, and either citrate (Ogawa *et al.*, 1968) or tartrate (Kerpel-Fronius and Hajos, 1968) is used as a chelating agent to keep the concentration of cupric ions at a suitable level. The two techniques produce significant differences of staining which can only be due to the differences in the incubation media but for reasons that are not fully understood (Kerpel-Fronius and Hajos, 1970).

Both techniques have been used principally to study the distribution of succinic dehydrogenase which is a particularly important enzyme in mitochondrial respiration. With a short period of incubation, staining is confined to the inner membranes that make up the internal cristae of typical mitochondria. With longer incubation some spread of staining is observed but the cristae are the structures most strongly stained. The picture seen in heart muscle is illustrated in Fig. 15; the method is an excellent specific stain for mitochondria in most tissues but is perhaps slightly less reproducible than the method with DAB. The accuracy with which enzyme localization is revealed by the histochemical procedure can be checked biochemically. Techniques are available for separating the different structural components of mitochondria (Palmer and Hall, 1972) and determining the enzymes associated with them. Succinic dehydrogenase, like cytochrome oxidase, is found to be a constituent part of the inner cristae membrane; so here at least the histochemical technique gives accurate localization at the electron microscope level. Various other dehydrogenases can be studied by the use of appropriate substrates. With β-hydroxybutyrate, staining is associated more with the matrix within the cristae, again in agreement with biochemical data, but localization is less precise (Fig. 15). The great potential advantage of the histochemical techniques is in showing up differences between individual mitochondria within the same tissue, but research on such problems is only just beginning.

VII. The Use of Enzymes as Tracers

Histochemical methods for enzymes are so sensitive at the electron microscopic level that it is potentially possible to detect the presence of a single molecule of some enzymes. This great sensitivity has prompted the use of exogenous enzymes (that is, enzymes added from outside the tissue) in a number of ways. In essence, the enzyme molecules are used as tracers, either alone or in combination with other molecules, such as antibodies, which have specialized functions. Several of the enzymes mentioned in previous sections could be used for tracer studies, but for various technical reasons very few of them have been tried and most of the published work has been with just one enzyme, horseradish peroxidase (HRP).

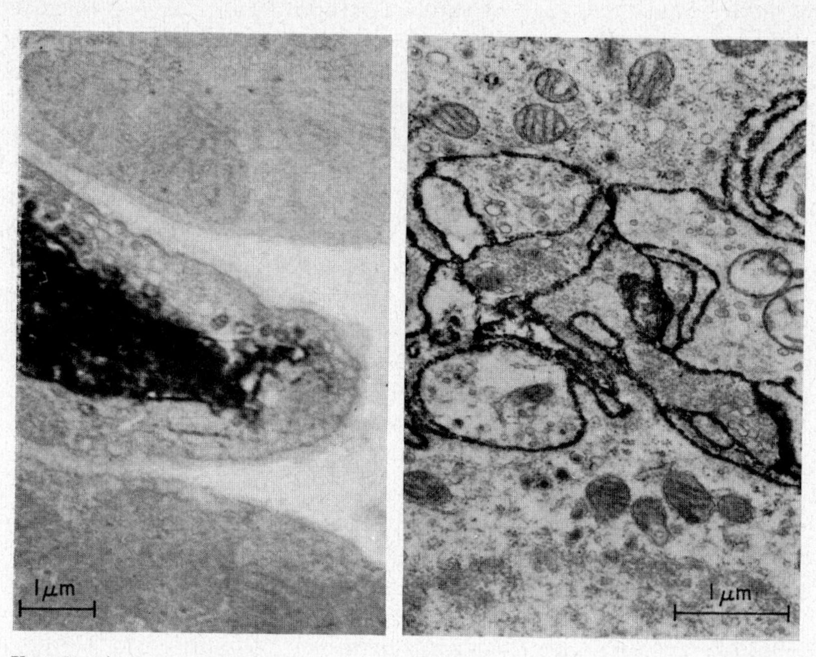

FIG. 16. An example of the use of horseradish peroxidase (HRP) to study tissue permeability in the anaesthetized rat. The left-hand electron micrograph shows a blood vessel in brown fat fixed 10 minutes after intravenous injection with HRP; no leakage of enzyme molecules out of the blood vessel has occurred (×10 000). The right-hand micrograph shows an area of carotid body fixed 2 min after a similar injection; HRP has passed out of the blood vessels and has penetrated deep between all the cells (×15 000). (By courtesy of Dr. R. I. Woods.)

The use of HRP as a successful *in vivo* molecular tracer at the E.M. level was introduced by Graham and Karnovsky (1966). They used HRP to study protein uptake in the proximal tubules of mouse kidney. With a molecular weight of about 40 000 HRP is very suitable for investigating many problems of permeability within tissues, such as the mechanisms involved in transcapillary exchange (Karnovsky, 1968). Its use to study diffusion between cells beyond the capillary wall is illustrated in Fig. 16. Other proteins with peroxidatic activity have been used with molecular weights ranging from about 160 000 in the case of myeloperoxidase down to 1900 in the case of microperoxidase. By injecting a peroxidase solution and then fixing tissues after measured time intervals estimates of diffusion rates can be obtained. HRP can also be used to study pinocytosis and related mechanisms by which cells take up large molecules from the extracellular medium. There are several reasons why HRP is particularly favoured for these studies: it is easy to obtain in a pure form; it is well tolerated *in vivo*; its enzyme activity survives aldehyde fixation extremely well; it is easily detected by the DAB technique; and endogenous peroxidase activity is relatively uncommon in mammalian tissues. Few other enzymes available in soluble form would satisfy these criteria to the same extent.

The use of HRP as an enzyme marker has proved particularly fruitful in the field of immuno-histochemistry. In the original technique of Nakane and Pierce (1967) HRP was chemically coupled to molecules of specific antibodies by treating a mixture with the reagent p,p′-difluoro-m,m′-dinitro-diphenyl sulphone. A complex protein molecule is thereby produced with at one end a specific antigenic site which will attach itself to a particular chemical grouping and at the other end a site with unimpaired peroxidase activity. A number of other coupling agents have been proposed, of which glutaraldehyde (Avrameas and Ternynck, 1969) seems likely to prove the most useful. The bridge between the antigenic and enzymic sites can be formed entirely by immunochemical techniques. Sections are treated with a total of four reagents in sequence: first, an immunoglobulin specific for the particular antigen under investigation; second, an antiserum against gamma globulins of the species used to prepare the first antibody; third, an immunoglobulin (from the same species as the first) specific for HRP; fourth, a preparation of HRP itself. The logic of this procedure is probably better appreciated visually, by reference to the pictorial scheme displayed in Fig. 17. This type of approach is likely to become popular for routine studies since the last three preparations are available commercially—only the initial specific antibody has to be specially prepared, and antibody preparations specific for some antigens are also available commercially. Nonetheless, immunochemical techniques suffer from many disadvantages when compared with routine histochemical staining methods. Because proteins are involved, purity of reagents is much more of a problem. Even more serious is the poor diffusion of globulins into tissues and, in particular, actually into cells. In practice, the study of specific antigens on the outside surfaces of cells presents few problems; study of intracellular antigens is a much more specialized field of research, but one which may rapidly become easier.

Apart from the various technical problems, which are likely to be gradually overcome, the main factor limiting the application of these techniques is the restricted range of substances which will act as effective antigens. Many intracellular enzymes are effective; so immunocytochemical techniques offer a way of studying the distribution of enzymes for which no direct histochemical procedure exists. The major problem is obtaining a pure antibody preparation, and that can be difficult even if it is possible to purify the enzyme. In theory, at least, it should ultimately become possible to study a wide range of enzymes by these techniques.

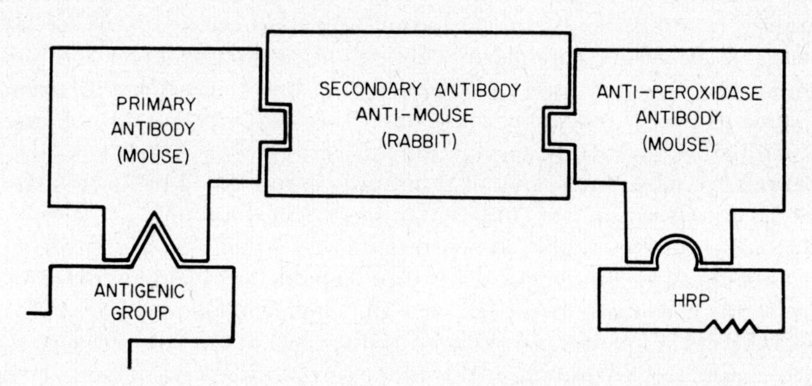

FIG. 17. A schematic diagram to show how horseradish peroxidase (HRP) can be coupled to specific chemical sites by application of appropriate immunoglobulins in the correct sequence. The species from which each immunoglobin was obtained is given in parenthesis.

VIII. Future Developments

Enzyme histochemistry at the electron microscopic level spans little more than a decade and only in the second half of that period have many of the techniques been improved to the point where they are easy to use by the general biologist and not just by the specialized histochemist. One of the future developments which can be confidently predicted, therefore, is the increasing use of these techniques as adjuncts to research in a wide range of subjects. A much more liberal attitude to the use of these techniques is already becoming evident among workers in a number of more rigid scientific disciplines. It has to be admitted, however, that in many investigations information from

E.M. histochemistry about enzyme localization has so far done little more than confirm conclusions reached from biochemical or physiological studies. The distribution of enzymes between the various compartments of mitochondria, for instance, has been largely worked out from fractionation studies (Palmer and Hall, 1972) and the main contribution of histochemistry has been in comparing enzyme distribution in mitochondria in different metabolic states or in different cells of the same tissue. With existing techniques there are ample opportunities for the use of E.M. histochemistry in solving problems falling in the first two categories mentioned in Section II: that is where the primary aim is to identify and study a particular type of cell or organelle. Several examples with universal applications have been quoted in this review: the use of acid hydrolases to study lysosomes, of succinic dehydrogenase and cytochrome oxidase to study mitochondria, and of catalase to study peroxisomes. Various enzymes are available for the study of the rough endoplasmic reticulum depending upon the particular cell type, glucose-6-phosphatase in liver, acetylcholinesterase in some neurones and so on; these and other enzymes can also be used to distinguish and study specific cell types.

Much the most likely area of rapid progress in the immediate future lies in the third of the four categories: the obtaining of accurate localization of the major concentrations of a specified enzyme. Some histochemical techniques are already capable of yielding excellent results on a routine basis (e.g. the thiocholine technique for AChE, the DAB technique for peroxidase) and several others are satisfactory with many tissues provided the proper precautions are taken. It is difficult to predict which of these techniques will prove valuable; much depends upon the biological importance of the problems to which they can be applied. Illustrations of studies with the two techniques mentioned above have been already given (Figs 8, 10, 11 and 13). Localization of adenyl cyclase (Fig. 7) is particularly important because of its involvement in the process of secretion by many cells, but modifications of the technique are likely to be necessary for specific applications because of the difficulties encountered with lead ions as the capturing agent. The same problem has arisen in the study of transport ATPases but here the technique with strontium ions introduced by Ernst (1972) offers a powerful alternative to the original lead technique. The role of the transport ATPases in the pumping of sodium and other ions by cells is of the greatest topical interest in biology at present and is certain to provide a fruitful field for histochemical study.

Improvements in histochemical techniques are to be expected in a number of directions. Existing techniques are sure to be improved by further research by both histochemists and biochemists. Considerable improvement can be expected from the discovery of heavy metals other than lead as capturing agents. The replacement of lead by barium in the aryl sulphatase technique produced a considerable improvement (Hopsu-Havu *et al.*, 1967) and so did the use of strontium for ATPase studies. The introduction of bismuth as a capturing agent by Deimling and Madreiter (1972) may well be extended to other enzyme techniques. There is wide scope for improvement in the azo-dye techniques by the development of new substrates but particularly by the synthesis of new diazonium salts. An interesting innovation is the synthesis by Davis *et al.* (1972) of substrates with an ester group at one end of the molecule and a diazonium group at the other so that on hydrolysis the substrate acts as its own capturing agent to produce a highly polymeric azo-dye. Their most promising substrate was the 3-acetoxy-indolediazonium ion which forms a strongly osmiophilic polymer on hydrolysis.

With a few notable exceptions, histochemical techniques are not as specific as biochemical ones. Inhibitors are routinely used in histochemical studies of the carboxylic esters and improvements in the specificity of other techniques are to be expected as new inhibitors for various enzymes become available. The existence of isoenzymes, which are different forms of an enzyme catalyzing the same reaction, has become an important concept in understanding the way cell processes are controlled. Methods of distinguishing isoenzymes histochemically (of lactic dehydrogenase, for instance) are being developed at the level of optical microscopy and are certain to be extended to electron microscopy.

Various recent advances in instrumentation are certain to have their repercussions on enzyme histochemistry. Techniques which selectively stain up particular cell organelles such as the Golgi apparatus or endoplasmic reticulum are eminently suitable for studies with the high voltage electron microscope. The high contrast which they can provide also makes them suitable for stereological studies and for methods of automated particle counting. Improvements in techniques for cutting ultrathin frozen sections of biological tissue without prior embedding (Appleton, 1974) may enable more accurate localization to be obtained by incubating the sections after they have been mounted on grids.

The most dramatic advances in enzyme histochemistry are likely to result from recent improvements in techniques of X-ray microanalysis at the E.M. level. In the first instance it should be possible to improve the quality of histochemical techniques. One of the perennial problems has been to adapt quantitative biochemical information to histochemistry because of the constraints imposed by having to use slices of fixed tissue for the incubation stage. With X-ray microanalysis it will be possible to measure directly on ultrathin sections the quantitative effects on the FRP of varying the incubation conditions. It will then be possible to devise conditions which are optimal for the histochemical demonstration of an enzyme in a particular tissue or organelle. It will also become much easier to test out entirely new techniques or modifications of old techniques with different heavy metals as capturing agents. A comparison of two techniques for acid phosphates, for instance, has been made by Ryder and Bowen (1974) with the aid of X-ray microanalysis, and the stoichiometry of the FRP in the ferricyanide technique for succinic dehydrogenase has been investigated by Weavers (1974). It should prove possible to use X-ray microanalysis as a method of quantifying enzyme techniques, but a number of difficulties have still to be overcome. Where comparative figures between similar structures in the same ultrathin section are required, at least semi-quantitative estimates of enzyme activity can be made already (Lewis, 1973). Where comparisons have to be made between different ultrathin sections the problem of measuring section thickness arises to introduce further uncertainty. The determination of absolute concentrations of enzymes will be much more difficult because of problems of standardizing the incubation stage but the use of X-microanalysis will help to improve standardization itself.

Enzyme histochemistry is entering a period of rapid expansion, not so much in the range of techniques as in the range of biological applications. The rate of introduction of new enzyme techniques at both optical and electron microscope levels has definitely slackened in the last 3 or 4 years, but during the same period the histochemical approach has become much more scientifically acceptable to biologists in general. It is perhaps this wider acceptance that may prove to be the most important factor in fashioning future developments.

Acknowledgements

I wish to thank my colleagues who have generously provided illustrations from their research; they are mentioned by name in the legends to the individual figures. I am grateful to the Histochemical Society, Inc. for permission to reproduce Fig. 7, and to the Company of Biologists for permission to reproduce Figs 10b and 11. I am grateful to Miss Fiona Hake for drawing Figs 9 and 17 and to Miss Coral Way for the typing. I wish particularly to thank Mrs. Anita Shelley for skilled technical assistance with some of the histochemical methods and for help with the bibliography.

References

Anderson, W. A. and Personne, P. (1970). *J. Histochem. Cytochem.* **18**, 783–793.
Appleton, T. C. (1974). *J. Microscopy* **100**, 49–74.
Ashworth, C. T., Luibel, F. J. and Steward, S. C. (1963). *J. Cell. Biol.* **17**, 1–18.
Avrameas, S. and Ternynck, T. (1969). *Immunochem.* **6**, 53–66.
Barka, T. and Anderson, P. J. (1962). *J. Histochem. Cytochem.* **10**, 741–753.
Barrnett, R. J. (1962). *J. Cell. Biol.* **12**, 247–262.
Barrnett, R. J. and Palade, G. E. (1959). *J. biophys. biochem. Cytol.* **6**, 163–169.
Barrnett, R. J. and Seligman, A. M. (1951). *Science* **114**, 579–582.
Beadle, D. J., Livingston, D. C. and Read, S. (1971). *Histochemie* **28**, 243–249.
Bergman, R. A., Ueno, H., Morizono, Y., Hanker, J. S. and Seligman, A. M. (1967). *Histochemie* **11**, 1–12.
Borgers, M. (1973). *J. Histochem. Cytochem.* **21**, 812–824.
Bowen, I. D. (1968). *J. Roy. microsc. Soc.* **88**, Pt. 2, 279–289.
Bowen, I. D. (1971). *J. Microscopy* **94**, Pt. 1, 25–38.
Brzin, M., Tennyson, V. M. and Duffy, P. E. (1966). *J. Cell. Biol.* **31**, 215–242.
Crevier, M. and Belanger, L. F. (1955). *Science* **122**, 556.
Davis, B. J. and Ornstein, L. (1959). *J. Histochem. Cytochem.* **7**, 297–298.
Davis, D. A., Wasserkrug, H. L., Heyman, I. A., Padmanabhan, K. C., Seligman, G. A., Plapinger, R. E. and Seligman, A. M. (1972). *J. Histochem. Cytochem.* **20**, 161–172.
Davis, R. and Koelle, G. B. (1967). *J. Cell. Biol.* **34**, 157–171.
de Duve, C. (1959). *In* "Subcellular Particles" (T. Hayashi, ed.) pp. 128–159. Ronald Press, New York.
de Reuck, A. V. S. and Cameron, M. P. (Eds.) (1963). "Lysosomes", Ciba Foundation Symposium. Churchill, London.
Decker, R. S. (1974). *J. Cell Biol.* **61**, 599–612.
Deimling, O. V. and Madreiter, H. (1972). *Histochemie* **29**, 83–96.
Dixon, M. (1971). *Biochim. biophys. Acta* **226**, 269–284.
Drochmans, P., Penasse, W. and Menard, D. (1972). *In* "Proceedings of the Fifth European Congress on Electron Microscopy", pp. 276–277. Institute of Physics, London.
Ernst, S. A. (1972). *J. Histochem. Cytochem.* **20**, 23–38.
Ernst, S. A. (1973). *J. Cell Biol.* **59**, A93.
Essner, E., Novikoff, A. B. and Masek, B. (1958). *J. biophys. biochem. Cytol.* **4**, 711–716.
Etherton, J. E. and Botham, C. M. (1970). *Histochem. J.* **2**, 507–519.
Fahimi, H. D. (1969). *J. Cell Biol.* **43**, 275–288.
Fahimi, H. D. and Karnovsky, M. J. (1966). *J. Cell Biol.* **29**, 113–128.
Gahan, P. B. (1967). *Intern. Rev. Cytol.* **21**, 1–63.
Gillis, J. M. and Page, S. G. (1967). *J. Cell Sci.* **2**, 113–118.
Goldfischer, S., Essner, E. and Schiller, B. (1971). *J. Histochem. Cytochem.* **19**, 349–360.
Gomori, G. (1939). *Proc. Soc. exp. Biol. N.Y.* **42**, 23–26.
Gomori, G. (1941). *Arch. Path.* **32**, 189–199.

Gomori, G. (1945). *Proc. Soc. exp. Biol. N.Y.* **58**, 362–364.

Gomori, G. (1948). *Proc. Soc. exp. Biol. N.Y.* **68**, 354–358.

Gomori, G. (1952). "Microscopic Histochemistry". University of Chicago Press, Chicago.

Graham, R. C. and Karnovsky, M. J. (1966). *J. Histochem. Cytochem.* **14**, 291–302.

Hanker, J. S., Seaman, A. R., Weiss, L. P., Ueno, H., Bergman, R. A. and Seligman, A. M. (1964). *Science* **146**, 1039–1043.

Hayashi, M., Shirahama, T. and Cohen, A. S. (1968). *J. Cell. Biol* **36**, 289–297.

Holt, S. J. (1952). *Nature, Lond.* **169**, 271–273.

Holt, S. J. and Hicks, R. M. (1961). *J. biophys. biochem. Cytol.* **11**, 47–66.

Holt, S. J. and Hicks, R. M. (1966). *J. Cell Biol.* **29**, 361–365.

Holt, S. J. and O'Sullivan, D. G. (1958). *Proc. Roy. Soc. B* **148**, 465–480.

Hopsu-Havu, V. K., Arstila, A. U., Helminen, H. K., Kalimo, H. O. and Glenner, G. G. (1967). *Histochemie* **8**, 54–64.

Howell, S. L. and Whitfield, M. (1972). *J. Histochem. Cytochem.* **20**, 873–879.

Jones, G. R. N. (1969). *Histochemie* **18**, 164–167.

Joo, F., Savay, G. and Csillik, B. (1965). *Acta histochem.* **22**, 40–45.

Kalina, M., Plapinger, R. E., Hoshino, Y. and Seligman, A. M. (1972). *J. Histochem. Cytochem.* **20**, 685–695.

Kalina, M., Weavers, B. and Pearse, A. G. E. (1971). *J. Histochem. Cytochem.* **19**, 124–130.

Karnovsky, M. J. (1968). *J. Gen. Physiol.* **52**, 64–95S.

Karnovsky, M. J. and Roots, L. (1964). *J. Histochem. Cytochem.* **12**, 219–221.

Kasa, P. and Csillik, B. (1966). *J. Neurochem.* **13**, 1345–1349.

Kawashima, T. and Murata, F. (1969). *Acta histochem. cytochem.* **2**, 54–60.

Kerpel-Fronius, S. and Hajos, F. (1968). *Histochemie* **14**, 343–351.

Kerpel-Fronius, S. and Hajos, F. (1970). *J. Histochem. Cytochem.* **18**, 219–221.

Koelle, G. B. and Friedenwald, J. S. (1949). *Proc. Soc. exp. Biol. N.Y.* **70**, 617–622.

Koelle, G. B. and Horn, R. S. (1968). *J. Histochem. Cytochem.* **16**, 743–753.

Koelle, G. B., Davis, R. and Devlin, M. (1968). *J. Histochem. Cytochem.* **16**, 754–764.

Kokko, A., Mautner, H. G. and Barrnett, R. J. (1969). *J. Histochem. Cytochem.* **17**, 625–640.

Lehrer, G. M. and Ornstein, L. (1959). *J. biophys. biochem. Cytol.* **6**, 399–406.

Leuenberger, P. M. and Novikoff, A. B. (1974). *J. Cell Biol.* **60**, 721–731.

Lewis, P. R. (1973). *J. Physiol.* **233**, 12–13P.

Lewis, P. R. and Shute, C. C. D. (1964). *J. Physiol.* **175**, 5–7P.

Lewis, P. R. and Shute, C. C. D. (1966). *J. Cell. Sci* **1**, 381–390.

Lewis, P. R. and Shute, C. C. D. (1969). *J. Microscopy* **89**, 181–193.

Lewis, P. R., Flumerfelt, B. A. and Shute, C. C. D. (1971). *J. Anat.* **110**, 2, 203–213.

Lison, L. (1936). "Histochemie Animale". Villars, Paris.

Livingston, D. C., Coombs, M. M., Franks, L. M., Maggi, V. and Gahan, P. B. (1969). *Histochemie* **18**, 48–60.

Mayahara, H., Hirano, H., Saito, T. and Ogawa, K. (1967). *Histochemie* **11**, 88–96.

Menten, M. L., Junge, J. and Green, M. H. (1944). *J. biol. Chem.* **153**, 471–477.

Mizutani, A. and Barrnett, R. J. (1965). *Nature, Lond.* **206**, 1001–1003.

Nachlas, M. M. and Seligman, A. M. (1949). *J. natn. Cancer Inst.* **9**, 415–425.

Naidoo, D. and Pratt, O. E. (1951). *J. Neurol. Neurosurg. Psychiat.* **14**, 287–294.
Nakane, P. K. and Pierce, G. B. Jr. (1967). *J. Cell. Biol* **33**, 307–318.
Novikoff, A. B. (1970). *J. Histochem. Cytochem.* **18**, 916–917.
Novikoff, A. B. and Goldfischer, S. (1969). *J. Histochem. Cytochem.* **17**, 675–680.
Novikoff, A. B., Beard, M. E., Albala, A., Shied, B., Quintana, N. and Biempica, L. (1971). *J. de Microscopie* **12**, 381–404.
Ogawa, K. and Mayahara, H. (1969). *J. Histochem. Cytochem.* **17**, 487–490.
Ogawa, K., Saito, T. and Mayahara, H. (1968). *J. Histochem. Cytochem.* **16**, 49–57.
O'Hare, K. H., Reiss, O. K. and Vatter, A. E. (1971). *J. Histochem. Cytochem.* **19**, 97–115.
Palmer, J. M. and Hall, D. O. (1972). *Prog. Biophys. molec. Biol.* **24**, 125–176.
Pearse, A. G. E. (1954). *Intern. Rev. Cytol.* **3**, 329–358.
Pearse, A. G. E. (1960). "Histochemistry: Theoretical and Applied", 2nd edn. Churchill, London.
Pearse, A. G. E. (1968). "Histochemistry: Theoretical and Applied", 3rd edn. Vol. I. Churchill, London.
Pearse, A. G. E. (1972). "Histochemistry: Theoretical and Applied", 3rd edn. Vol. II. Churchill, London.
Poux, N. (1969). *J. de Microscopie* **8**, 855–866.
Pratt, O. E. (1954). *Biochem. biophys. Acta* **14**, 380–389.
Reale, E. and Luciano, L. (1967). *Histochemie* **8**, 302–314.
Reik, L., Petzold, G. L., Higgins, J. A., Greengard, P. and Barrnett, R. J. (1970). *Science* **168**, 382–384.
Rosenthal, A. S., Moses, H. L. and Ganote, C. E. (1970). *J. Histochem. Cytochem.* **18**, 915.
Rutenburg, A. M., Kim, H., et al. (1969). *J. Histochem. Cytochem.* **17**, 517–526.
Ryder, T. A. and Bowen, I. D. (1974). *J. Microscopy* **101**, 143–151.
Sabatini, D. D., Bensch, K. and Barrnett, R. J. (1963). *J. Cell. Biol.* **17**, 19–58.
Sasaki, M. and Fishman, W. H. (1973). *J. Histochem. Cytochem.* **21**, 653–660.
Seligman, A. M., Karnovsky, M. J., Wasserkrug, H. L. and Hanker, J. S. (1968). *J. Cell Biol.* **38**, 1–14.
Seligman, A. M., Nir, I. and Plapinger, R. E. (1971). *J. Histochem. Cytochem.* **19**, 273–285.
Seligman, A. M., Ueno, H., Morizono, Y., Wasserkrug, H. L., Katzoff, L. and Hanker, J. S. (1967). *J. Histochem. Cytochem.* **15**, 1–13.
Seligman, A. M., Wasserkrug, H. L., et al. (1970). *J. Histochem. Cytochem.* **18**, 542–551.
Seligman, M. L., Ueno, H., Hanker, J. S., Kramer, S., Wasserkrug, H. and Seligman, A. M. (1966). *Expl. molec. Path.* (*Suppl.* **3**), 21–35.
Smith, R. E. and Fishman, W. H. (1969). *J. Histochem. Cytochem.* **17**, 1–22.
Stoward, P. J. (Ed.) (1973). "Fixation in Histochemistry", pp. 216. Chapman and Hall, London.
Strum, J. M. and Karnovsky, M. J. (1970). *J. Cell. Biol.* **44**, 655–666.
Tennyson, V. M., Brzin, M. and Duffy, P. E. (1965). *J. Cell. Biol.* **27**, 105A.
Tice, L. W. and Barrnett, R. J. (1965). *J. Cell Biol.* **25**, 23–41.
Tice, L. W. and Smith, D. S. (1965). *J. Cell. Biol.* **25**, Part II, 121–135.
Tsou, K. C., Cheng, C. S., et al. (1956). *J. Am. Chem. Soc.* **78**, 6139–6144.
Vatter, A. E., Reiss, O. K. and Newman, J. K., et al. (1968). *J. Cell Biol.* **38**, 80–98.

Wachstein, M. and Meisel, E. (1957). *Amer. J. Clin. Path.* **27**, 13–27.
Weavers, B. A. (1974). *Histochem. J.* **6**, 133–145.
Wilson, I. B. (1951). *Biochim. biophys. Acta* **7**, 520–525.
Zacks, S. I. and Blumberg, J. M. (1961). *J. Histochem. Cytochem.* **9**, 317–324.

Recent Advances in the Application of Negative Staining Techniques to the Study of Virus Particles Examined in the Electron Microscope

R. W. HORNE

Department of Ultrastructural Studies, John Innes Institute, Norwich, England

I. INTRODUCTION

THE possibility of surrounding small spherical viruses with an opaque material was observed by Hall (1955) and later applied to preparations of tobacco mosaic virus by Huxley (1956). The method of using neutralized phosphotungstic acid to surround bacteriophage and other viruses was described by Brenner and Horne (1959) as the "negative staining" technique. These authors found that by changing the pH of phosphotungstic acid (PTA) to near neutrality there was little interaction between the electron dense stain solution and bacteriophage particles which were considerably disrupted at the low acid pH of the normal PTA material in solution. Moreover, they recorded electron micrographs which provided a high degree of preservation and contrast capable of resolving considerable detail in the phages. It was perhaps fortuitous that the experiments were performed on one of the most sophisticated virus groups which possessed a series of complex protein structural features and consequently showed the immediate usefulness of the method to full advantage (Brenner *et al.*, 1959).

Subsequent applications of the original negative staining method using potassium phosphotungstate (KPT) as the electron opaque material followed rapidly and resulted in the morphological analysis of a number of animal, plant and bacterial viruses (see Horne and Wildy, 1961, 1963; Bradley and Kay, 1960; Bradley, 1962 and others). The results obtained from the virus experiments stimulated interest in developing or modifying the negative staining technique for studying the morphology of other biological structures including membranes, large protein macromolecules, bacteria and subcellular components. The latter experiments led to the conclusion that the basic method of Brenner and Horne (1959), although successful for the more stable animal viruses, had disruptive effects on certain other viruses when mixed with potassium phosphotungstate (KPT) or sodium phosphotungstate (NaPT). Further studies were carried out to extend the range of heavy metal salts which could be used in place of the original negative staining material and these were coupled with a more critical approach to the electron microscopy of negatively stained specimens. A number of highly satisfactory electron dense negative stains are now available for use with specific viruses and biological particles which will be discussed in more detail below. However, the question of the precise mechanism of negative staining is still a matter for some discussion and the simple interpretation of merely surrounding the specimen with a suitable stain is no longer acceptable. For the reasons described below relating to the conditions required for negative staining and interpretation of the image it has been suggested that the technique could better be described as "contrast embedding" (R. C. Williams, private communication).

In several publications the negative staining technique has been described as "negative" contrast which may be confused with the current procedures for contrast reversal under certain electron optical conditions where negative contrast results in the final image. For this reason the use of the term "negative" contrast should be discouraged.

One of the more interesting developments associated with negative staining is the application of optical diffraction techniques and photographic averaging to the analysis of the images recorded on the electron micrographs. These methods have important implications in determining the sizes, shapes and structural features of periodic detail recorded from negatively stained specimens of virus particles.

There is a voluminous literature relating to the viral structures studied with the aid of negative staining methods and for this reason the present article will of necessity be selective in order to present some of the more recent findings.

II. Experimental Procedures and Negative Stains

A. *Methods for Preparing Isolated Virus Particles*

The main purpose of negative staining is to surround or embed the biological object in a suitable electron dense material which will provide high contrast and good preservation. In the simple case a small volume of the negative stain is mixed with an equal volume of the biological material in suspension and deposited directly on to a specimen support film with the aid of a fine Pasteur pipette. The specimen is then allowed to dry at room temperature and examined in the electron microscope. In order to achieve ideal images several requirements have to be met and these include the correct concentration of stain, concentration of biological material, minimal interaction between the stain and specimen and an even distribution of the stain and specimen over the supporting film surface. Since there are many variables relating to the above requirements, there are at present no strict quantitative procedures available for preparing the majority of specimens and much is a matter of trial and error. However, it is possible to indicate some general guide lines which may be of value in the initial preparative stages.

It is generally accepted that the smaller the physical dimensions and the lower molecular weight of the viral components the more difficult it is to achieve satisfactory conditions for negative staining. In the case of large specimens such as virus crystals together with subcellular components, similar problems have also been encountered in establishing the optimum state for contrast and preservation. In addition, there is the problem of the charge effects associated with the biological specimens and the surface of the carbon support films used in electron microscopy.

B. *Requirements for Spreading Viral Material and Negative Stain*

The first problem in attempting to prepare viruses in the presence of negative stain is that of achieving an equal distribution of stain and viral material. An incorrect ratio of virus to stain can result in two situations, as illustrated in Fig. 1. In the case of too much virus or highly concentrated suspensions, large amorphous or electron transparent areas are observed in the presence of small, widely separated regions of electron dense material. Moreover, the relative lack of negative stain to preserve the viral components may result in the collapse or partial disruption of the particles. At low concentrations of virus mixed with negative stain, large areas of thick electron dense

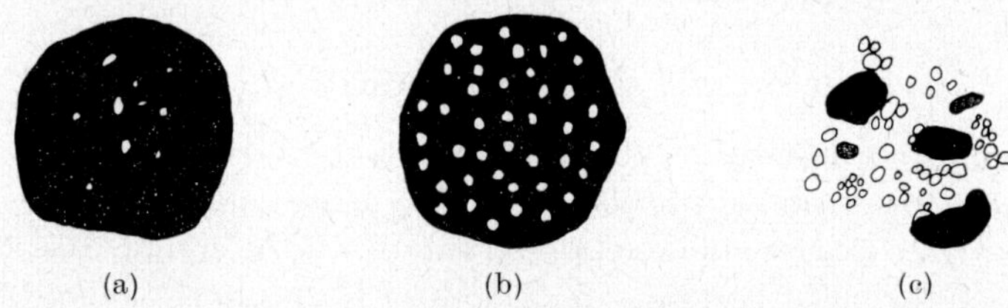

FIG. 1. The diagram shows the distribution of negative stain and virus particles deposited on the electron microscope specimen supporting film. (a) Large electron dense area in the presence of too much stain. (b) Ideal spreading conditions following correct mixture of stain and virus material. (c) The appearance of virus material at low concentrations of negative stain showing separation of electron dense material from the particles. (From R. W. Horne. Negative staining techniques. "Practical Methods in Electron Microscopy", North-Holland. In press.)

material are seen and frequently mask or even obscure the virus particles. Under the latter conditions the true size and shape of the particles may be considerably underestimated. In practice, the virus suspension is mixed with a given volume of negative stain and the volume of stain increased or decreased until acceptable spreading is observed. However, it should be stressed that the application of negative staining procedures to purified viruses is rarely successful at low concentrations of virus (i.e. below 10^5 particles/ml), although demonstration of the presence of virus in disrupted animal cells or taken directly from infected plant tissues has been achieved (see Parsons, 1963; Almeida and Howatson, 1963; Hitchborn and Hills, 1965). The spreading conditions in these experiments resulted from the presence of fragments in the cell cytoplasm disrupted by the negative stain.

The most direct method for mounting the virus material and negative stain is with the aid of a fine Pasteur pipette as illustrated in Fig. 2(a). As mentioned earlier the virus suspension is mixed with the solution of negative stain which will provide a reasonable balance of material for spreading purposes. Once the droplets are deposited on to the specimen support films the surplus liquid is removed by touching a pointed piece of filter paper to the rim of the liquid area (Fig. 2(b)). Ideally, the amount of stain and virus left to dry on the film surface should be sufficient to form a monolayer of the particles, but this is a matter of trial and error and for the reasons mentioned above the spreading characteristics may vary considerably from one area to another on the support film.

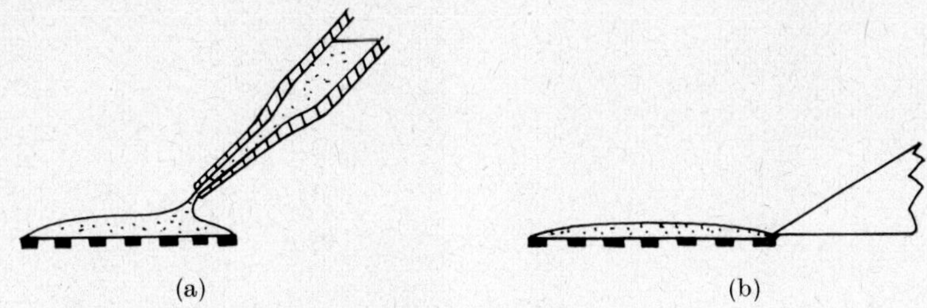

FIG. 2. The mixture of stain and virus material is deposited using a Pasteur pipette as shown at (a) and slowly drained with the aid of a pointed filter paper (b).

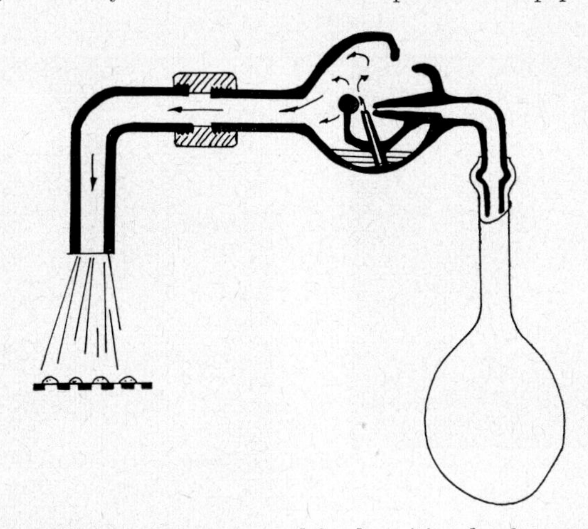

FIG. 3. Diagram of a glass spray gun used for depositing droplet patterns of negative stain and viruses onto specimen supports. (From R. W. Horne. Negative staining techniques. "Practical Methods in Electron Microscopy." North-Holland. In press.)

The second approach is essentially similar to the method of applying droplets to the grid, but the virus suspension and negative stain are deposited separately. This technique avoids mixing the stain and virus beforehand as the virus suspension may subsequently be required for other analytical or biological purposes, particularly where only very small volumes of virus suspension are available.

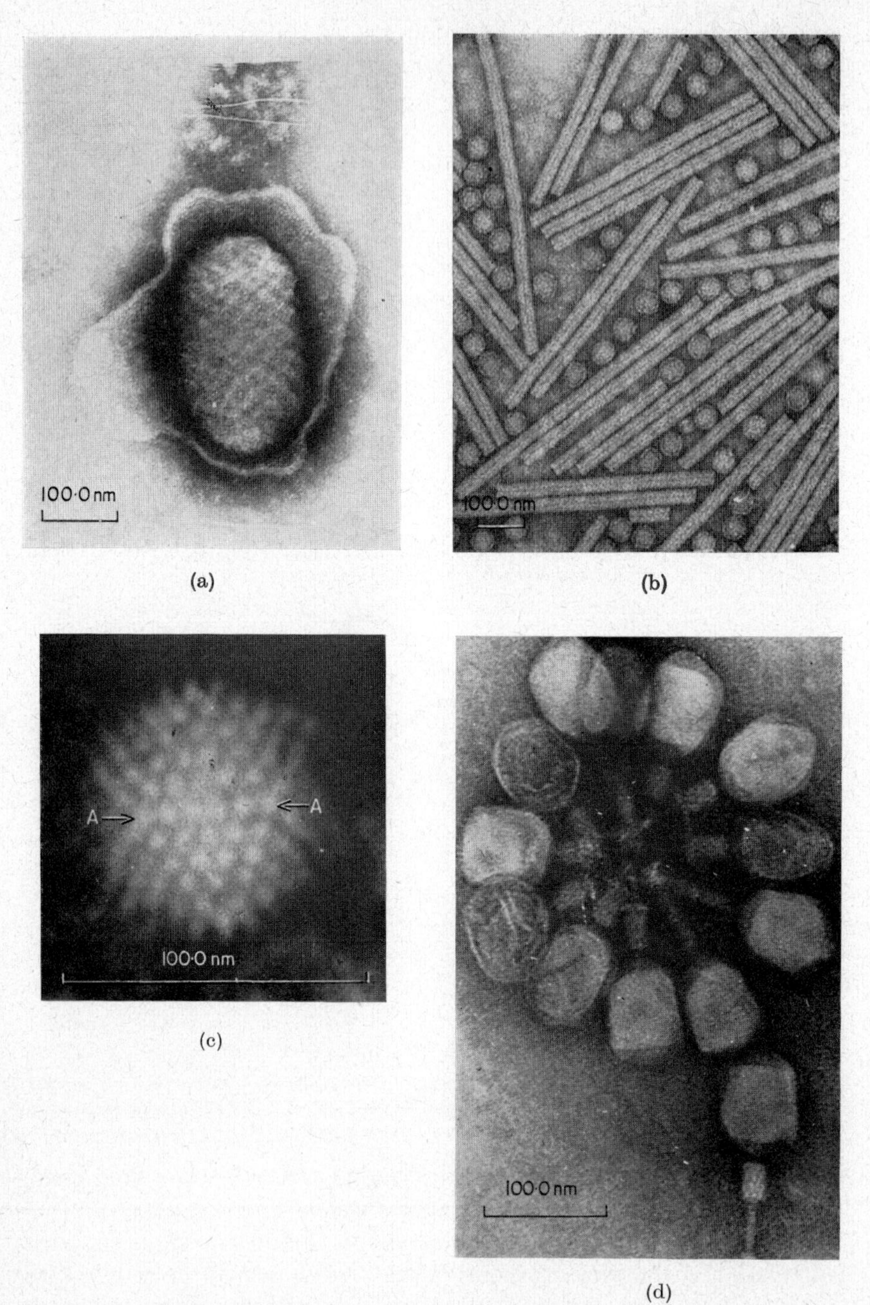

(a) (b) (c) (d)

FIG. 4. Some examples of viruses prepared in the presence of negative stains. (a) Orf (pustular dermatitis virus). (b) Mixed preparation of tobacco mosaic virus rods and brome mosaic virus. (c) Adenovirus type 5. (The arrows "A" indicate two axes of fivefold symmetry). (d) T-even bacterial viruses.

A third method, which has the additional advantages of allowing particle counting and deposition of pathogenic virus suspensions, makes use of droplet pattern spraying with the aid of special glass nebulizers as shown in Fig. 3. For a more detailed description of the spraying techniques the reader is referred to the publications of Williams (1953), Horne (1967), Horne and Nagington (1959) and Sharpe (1965).

The methods described up to this point for mounting negatively stained viruses apply to purified or partially purified particles and, subject to the correct spreading, staining and pH conditions, it is a matter of only a few minutes before the specimen grids are transferred to the electron microscope for examination. Several typical examples of the type of image expected from viruses prepared by the above methods are shown in Fig. 4. It can be seen that there is a high order of preservation and contrast present in viruses ranging from those possessing complex protein components to particles of the most basic form and size.

C. *Different Types of Negative Stains and their Properties*

Nothing has been said so far about the various negative staining solutions currently being applied to virus suspensions and other small biological particles. It is now becoming clear from the large amount of published literature that although negative staining is a simple method, some thought must be given to the selection of a given negative stain and the possible effects it may have on the virus material and its appearance in the final electron microscope image. Moreover, it does not necessarily follow that the most aesthetically pleasing image as seen from negatively stained specimens is the one which will provide the most useful information. For these reasons it is necessary to apply a considerable range of negative stain solutions at different concentrations and pH values in order to extract the maximum detail.

Much of the information seen in the electron micrographs will depend to a large extent on the penetration of the stain into the hydrated regions of the individual proteins or lipoproteins. This in turn will depend on the hydrophobic and hydrophilic properties of the molecules forming the structure. It is reasonable to consider that certain structural features will be enhanced by some negative stains as opposed to others on the basis of interaction between the protein molecular charges and negative stain material. A striking example of these effects was reported in relation to the band patterns of collagen prepared in the presence of a range of negative stains (Cox *et al.*, 1967). Elastin is another protein which has been found to be extremely difficult to prepare by any negative staining procedure because of the hydrophobic properties of the molecule (Gotte *et al.*, 1974).

To improve and assist with the spreading of both negative stains and particles it is possible to add a very small volume from a suspension containing low molecular weight protein or other material as a wetting agent either to the solution containing the stain and virus, or by placing a droplet on to the support film before the final suspension is added (see Reissig and Orrell, 1970; Gordon, 1972; Gregory and Pirie, 1972; Gregory and Pirie, 1973a, b). The disadvantage with this approach is twofold; first it is possible that there may be some interaction between the virus surface protein and the low molecular weight material or wetting agents used. Second, the presence of small particles of relatively low molecular weight could result in obscuring very small structural detail in viruses at very high resolution. This point is discussed later in the section of this article dealing with the negative staining-carbon method. However, a considerable improvement and reproducibility in spreading can be achieved by the careful use of wetting agents.

A further comment should be made here relating to the interaction of some negative stains with certain biological specimens. In a recent study by Hills *et al.* (1973), it was possible to show that binding of ammonium molybdate (AM) to components in the cell wall material of *Chlamydomonas* occurred and that the stain could not be removed following prolonged washing. Other evidence for the interaction between some negative stains and viruses comes from experiments on certain types of plant viruses which are almost completely disrupted by the application of the wrong negative stain (G. J. Hills, private communication). The same staining solution, on the other hand, can provide optimum preservation and contrast for another group of plant viruses (Fig. 4b). In the case of turnip yellow mosaic virus the stain KPT has the effect of releasing the viral nucleic acid and converting the preparation to a type of "top component" or empty particles (G. J. Hills and R. Markham, personal communication).

It should also be mentioned that the mechanism of disruption by a negative stain can be used to some advantage in the study of viral components. Conditions can be found which will enable a given virus to dissociate slowly on the specimen grids and the released viral components can then be resolved in more detail. These experiments will be discussed in more detail later (see Adenovirus in this article, p. 267).

A list of negative stains in current use is shown in Table I together with references to some typical applications of these solutions to biological specimens.

TABLE I

Some negative staining solutions in current use

Negative stain	Concentration %	pH range	References*
Ammonium molybdate	0·5–5·0	5·0–8·0	Nagington *et al.* (1964); Muscatello and Horne (1968); Munn (1968)
Potassium phosphotungstate	1·0–3·0	5·0–8·0	Brenner and Horne (1959)
Sodium phosphotungstate	1·0–3·0	5·0–8·0	Valentine and Horne (1962)
Silicon tungstate	1·0–3·0	5·0–8·0	Horne (1967)
Lithium tungstate	1·0–3·0	5·0–8·0	Bradley (1962); Rothfield *et al.* (1966); Horne (1967)
Tungstoborate	1·0–2·0	6·0–7·0	Rowe (personal communication) See Horne (1965)
Lanthanum acetate	1·0–2·0	5·0–7·0	Bangham and Horne (1962, unpublished)
Uranyl acetate	0·5–1·0	4·0–5·0	Huxley and Zubay (1960)
Uranyl formate	0·5–1·0	4·0–5·0	Leberman (1965)
Uranyl aluminium	0·5–1·0	4·0–5·0	Unwin (1972)
Uranyl acetate/EDTA	0·5–1·0	4·5–7·0	Bessis and Breton-Gorius (1960); Van Bruggen *et al.* (1960)
Uranyl oxalate	0·5	5·0–6·5	Mellema *et al.* (1967)

* The references listed are only intended as a guide for further reading and are therefore incomplete.

D. *The Application of Negative Stains to Fixed Virus Material*

There is a relatively large number of viruses which are particularly sensitive to changes in osmotic conditions or when placed in unsuitable buffer solutions. Many of the buffers used in the preparation and purification of viruses are unsuitable for electron microscopy when applied to whole mount specimens, because of the tendency for dried buffer to form small crystallites. The presence of small buffer crystallites will not

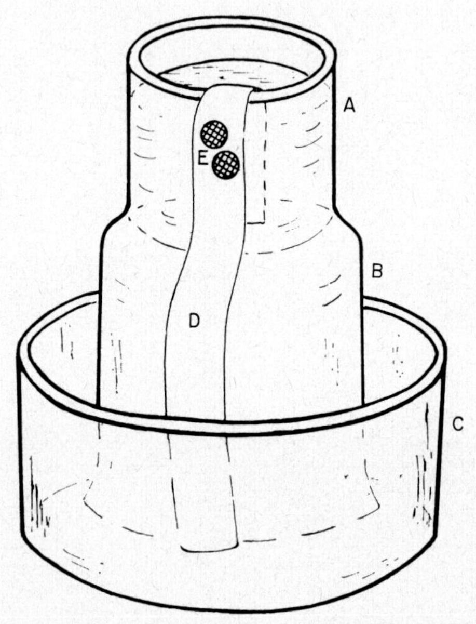

FIG. 5. Simple apparatus for removing sucrose or buffers present in liquid suspensions deposited onto electron microscope specimen supports. (A) Small beaker containing water or volatile buffer solution. (B) Inverted support beaker. (C) Collecting dish. (D) Strip of filter paper. (E) Specimen grids. (After Webb, 1973. *J. Microscopy.*)

only obscure viral detail, but will also interfere with the negative staining mechanism. This problem has also been encountered in material separated by sucrose density centrifugation and can present quite serious difficulties when attempting to study virus and other specimens banded into fractions in the ultracentrifuge (see Horne, 1967; Horne and Whittaker, 1962; Muscatello and Horne, 1968).

Under normal conditions the buffered virus suspensions can be dialysed against a large volume of 1–2% ammonium acetate or ammonium carbonate, which are volatile buffers and much more suitable for high resolution studies (see Williams, 1953). It is also possible to remove buffers or sucrose successfully by the method described by Webb (1973), illustrated in Fig. 5. However, the effect of the removal of buffer can result in the collapse of certain viruses. The lipoprotein enveloped particles for instance, or their nucleocapsids are particularly prone to changes in buffering conditions (i.e. Friend leukaemia virus, parainfluenza viruses, influenza and other viruses). Some of the effects of negative stains on unfixed membrane systems have been discussed in detail by Bangham and Horne (1964), Rothfield and Horne (1967), Glauert and Lucy (1969) and Muscatello and Guarriero-Bobyleva (1970).

A comparison between fixed and unfixed viruses has recently been made by Nermut (1972) who applied various negative stains to a range of viruses following treatment with osmium tetroxide and glutaraldehyde. One of the striking observations was the appearance of the Friend strain of leukaemia virus (MULV) before and after fixation. Some earlier reports of negatively stained MULV indicated that these enveloped particles possess a tail-like structure or extension of the envelope and this morphological feature was considered to be a structural characteristic (see Dalton *et al.*, 1962; Nowinski *et al.*, 1970). However, following treatment with 5% glutaraldehyde or direct staining with uranyl acetate the MULV particles were seen as approximately spherical forms devoid of any tail-like structures (Nermut, 1972).

There are two points which should be made here and are relevant to the studies by Nermut (1972). Firstly, the modification to lipids or lipoprotein envelopes by certain negative stains can result in considerable shape changes in many membrane-bound viruses, and some caution is necessary in using gross morphological features for taxonomic purposes. The second point to mention here is that, although uranyl acetate acts as an effective negative stain, it clearly has some cross-linking properties and tends partially to "fix" many viruses and their components. For this reason it is likely that good preservation and contrast can be achieved with uranyl acetate when applied to certain viruses, but cross-linking may produce its own type of artefact in other virus components, coupled with a type of pseudo-negative staining effect particularly at the low pH of uranyl salt solutions.

E. *Freeze-Drying of Negatively Stained Specimens*

Although fixation procedures can be applied to a number of "soft" viruses to retain their structural features, some collapse coupled with distortion frequently results. In order to reduce the collapse of these "soft" viruses during the final dehydration stages it is possible to apply the principles of freeze-drying to mixtures of virus and negative stain, as shown in Fig. 6. (Nanninga, 1968; Nermut and Frank, 1971; Nermut, 1972, 1973; Nermut *et al.*, 1972). Some of the techniques are relatively sophisticated and have to be carried out in a special apparatus; the reader is referred to the above publications for the full experimental details. It should be stressed that the freeze-drying techniques are especially valuable in the case of enveloped or sensitive viruses and their components, but the author has found little advantage with the method when applied to a large range of the more stable viruses. Experimental evidence for this will be provided later in the section of this article dealing with the negative staining-carbon film technique.

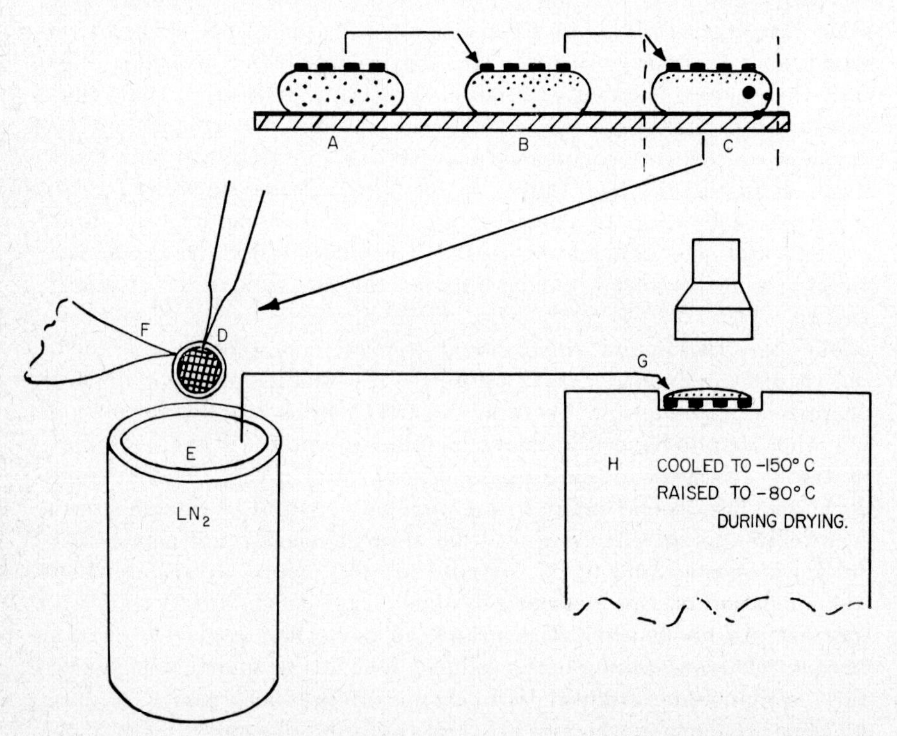

FIG. 6. Diagram showing the stages in freeze-drying using the techniques of Nermut (1972, 1973). Droplets are deposited as shown at (A), and filmed grids placed at the surface to allow adsorption of virus particles. Grid washed on droplet (B), with water or volatile buffer. Negative stain droplet added at stage (C). The stained grid (D) is drained with filter paper (F) and immersed into coolant (E) for 10 minutes. Final drying is carried out in a suitable high vacuum plant with a cooling block held initially at −150°C and raised to −80°C during the high vacuum drying.

III. Methods for the Analysis and Interpretation of Images from Negatively Stained Preparations

A. *Some General Comments on the Structural Features of Viruses Determined from Negatively Stained Preparations*

Before describing some of the recent methods used in the analysis of electron micrographs with particular reference to virus particles, some general comments may be of use here for the reader unfamiliar with the way in which viruses are put together and the various symmetry patterns associated with these small biological structures.

There is very good evidence to show that a number of biological structures are built up by the assembly of proteins in a regular or repetitive fashion, and viruses are a good example to illustrate this efficient system. The experimental data showing evidence for the high order of regular features in viruses has come from X-ray diffraction, biochemical analysis, hydrodynamic studies and electron microscopy. In the case of electron microscopy there were several preparative procedures being applied to the study of viruses, including thin sections of infected cells, shadowed isolated particles and some attempts to stain positively certain viruses with a view to locating their nucleic acid. Each of the above techniques presented the virologist with a different morphological view of viruses. As a result the terminology was somewhat confusing, owing to the various viral components being described according to the images derived from one or other of the technical approaches. In addition, very little reference was made to the X-ray diffraction work, which clearly indicated that a number of viruses were built from regularly arranged protein units and nucleic acid.

B. *Terminology Applied to Viruses and their Components*

For the reasons mentioned above some common ground had to be found which would allow virus structures to be described regardless of the experimental techniques used and whether they were seen inside infected cells, as isolated particles photographed at high resolution in the electron microscope, or as crystals analysed by X-ray diffraction. It has now been generally accepted that the terminology shown in Figs 7 and 8 should be used when describing the structure of viruses (see Lwoff *et al.*, 1962; Wildy, 1962; Caspar *et al.*, 1962). There are other more sophisticated viruses which still require further detailed analysis and where the precise arrangement and composition of the viral components are not fully understood. For this reason the above terminology is applied generally to those viruses possessing icosahedral symmetry or helical symmetry, and to the bacteriophages.

Whatever the advantages or disadvantages associated with the negative staining of viruses may be, the method has allowed electron microscopy to be brought much closer to the other fields of research associated with the study of virus structure, assembly and genetics. It has also contributed to the direct analysis of many virus architectural features which would have been difficult if not impossible to study by X-ray diffraction or other experimental procedures.

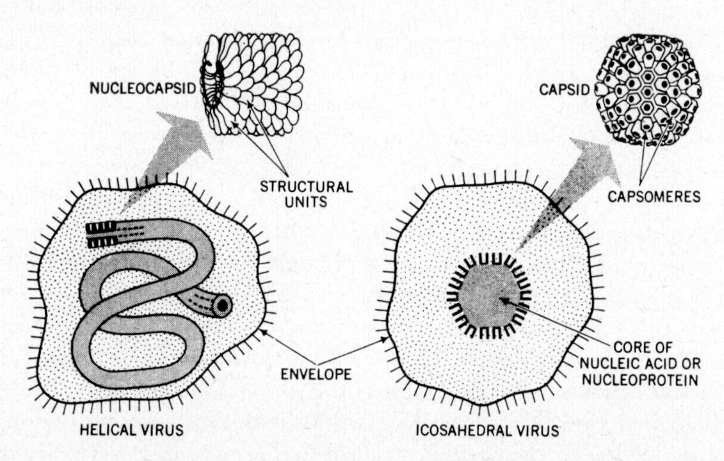

FIG. 7. Schematic diagram illustrating the terminology used to describe regular (icosahedral) viruses and those with helical symmetry (rod-like or filamentous nucleoprotein structures). (From R. W. Horne. "Structure of Viruses." Academic Press, 1974.)

It was shown by Horne and Wildy (1961) that the virus structure observed in the electron microscope, after being prepared by the negative staining method, could be divided into three main symmetry designs: icosahedral viruses, helical viruses and particles possessing a combination of symmetries. These early electron micrographs also suggested that the same symmetry patterns could be applied to all viruses whether they were derived from animal, plant or bacterial hosts. The possible designs and the theoretical considerations relating to virus shells generally have been extensively investigated by Caspar and Klug (1962, 1963), Caspar (1966), Klug *et al.* (1966), Klug (1967), Kushner (1969).

Four examples of viruses as seen in negatively stained preparations are illustrated in Fig. 4. It can be seen that the roughly hexagonal-shaped adenovirus has a protein shell or capsid arranged in accordance with icosahedral or 5.3.2 symmetry. It has been shown that by retaining the symmetry axes as indicated and changing the number of morphological units between the fivefold members, a series of virus capsids can be constructed of different size. It should be said at this point that the surface units seen in the electron micrographs correspond to morphological components named capsomeres (see Fig. 7), and that the capsomeres themselves are composed of smaller asymmetrical structure units (protein molecules).

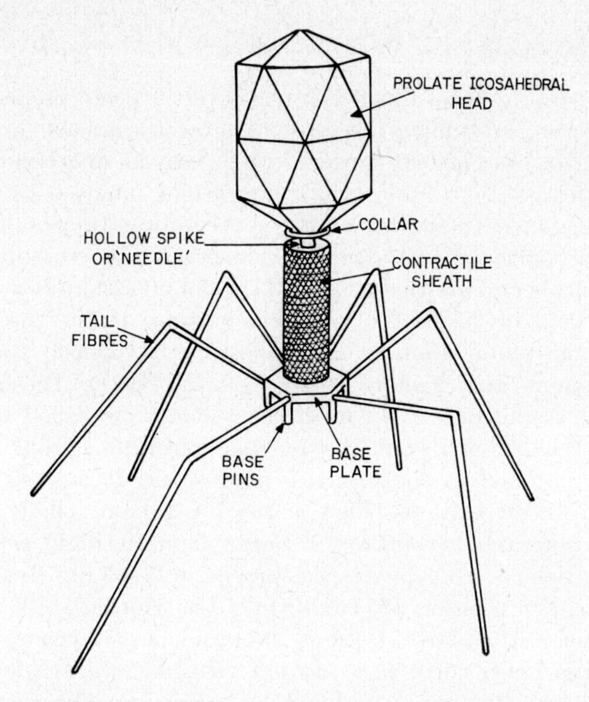

FIG. 8. Drawing of a T-even phase showing the terminology used to describe th various protein components associated with this group of viruses.

The rod-shaped and filamentous viruses are built from nucleic acid and protein units arranged in helical array. The length of the infective rod or filament will depend on the length of the nucleic acid strand carrying the genetical information (see Figs 7 and 32). Much of the rigidity (or flexibility) of the helix will be determined by specific binding sites between the individual structure units forming each turn of the helix around the particle axis and neighbouring turns along the axis.

In the case of several types of bacterial virus the protein components are assembled as helical arrays (tail structures), icosahedra or prolate icosahedra (head structures) (see Figs 4 and 8). The tailed bacterial viruses are among the most complex virus structures observed in the electron microscope and present special problems of assembly within infected bacteria as they are built from a variety of protein molecules. For a more detailed account of virus architecture, assembly and function the reader is referred to the review publications of Kellenberger (1973) and Wood *et al.* (1973).

C. *Resolution from Negatively Stained Virus Preparations*

The structural features imaged in negatively stained preparations of isolated viruses, and many other biological specimens, are mainly limited by the presence of noise in the electron microscope image, coupled with radiation damage and limitations imposed by the total specimen thickness. These restrictions on the information available from "amorphous" specimens studied in the electron microscope present serious problems and are the subject of special investigations which fall within the field of electron optics. For further details the reader is referred to the publications of Ruska (1965), Kobayashi and O'Hara (1966), Williams and Fisher (1970), Glaeser (1971), Riecke (1971), Cosslett (1971) and others. From measurements published by several workers who have analyzed the average resolution from biological specimens examined in the electron microscope, it is clear that the visible detail rarely extends below about 2·5–3·0 nm. This value is far above the potential resolution of most modern high performance instruments, which are capable of achieving 0·2–0·3 nm (see Williams and Fisher, 1970; Glaeser, 1971; Unwin, 1972; Horne, 1973).

In addition to the above, there is the problem of superimposition of the upper and lower surfaces of regular virus structures which results in the production of moiré patterns in the final image. These interference patterns tend to obscure the surface detail, unless the imaging or separation of one-sided particles can be achieved. In the case of negative staining procedures the majority of viruses present two-sided images, and no reliable method has been developed so far to ensure that only one surface is consistently resolved. However, the fact that most viruses have manifold symmetry makes it possible to enhance and extract repeating features from these particles with the aid of photographic or optical techniques. Moreover, images can be reconstructed in such a way that the upper and lower surface detail can be separated. These photographic and optical procedures are also useful when attempting to improve the signal-to-noise ratio present in the recorded electron micrographs.

D. *Photographic Averaging Applied to Electron Micrographs of Negatively Stained Specimens*

The principles of applying photographic averaging were explored by Sir Francis Galton (1878) for demographic studies and for investigating the visual appearance of people by superimposing a large number of photographic images in order to establish any average characteristics. In these particular investigations Galton used the distance between the eyes as the constant scale factor.

In applying these methods of photographic averaging to electron microscopy it should be stressed that by superimposition there is the possibility of improving the signal-to-noise ratio by reinforcing repeating common features in the image. The non-repeating or random details tend to cancel out and the "signal" is enhanced in comparison to the "noise". However, there will be no gain in resolution as this is not equivalent to an increase in the "signal" itself. It is possible to demonstrate the technique in a simple way by referring to the two-dimensional regular lattice shown in Fig. 9. Integration or averaging is achieved by moving one transparency over another, the latter being derived from the same lattice and used as a reference. For each interval of translation a photograph is taken and the final image recorded will have averaged N_1 intervals in one direction of the lattice and N intervals in the other (see Horne and Markham, 1972).

These basic techniques were extended and applied to electron micrographs of viruses by Markham *et al.* (1963; 1964), with considerable success. They were able to demonstrate hidden periodicities in virus particles possessing rotational symmetry and linear repeating features. The apparatus for rotational photographic integration is shown in Fig. 10, and that for linear integration in Fig. 12. Some indication of the enhancement obtainable from radially arranged features in tobacco mosaic stacked disc protein can be seen in Fig. 13 and from turnip yellow mosaic virus in Fig. 14. An alternative arrangement for reinforcing elements of rational symmetry with the aid of stroboscopic methods is illustrated in Fig. 11. In the case of the stroboscopic technique the centration of the illuminated prints obviously requires critical adjustment.

The superimposition of detail in linear objects was also utilized by Markham *et al.* (1964), in a study of tobacco mosaic virus protein, using the apparatus shown in Fig. 12. The electron micrographs of part of an integrated image are illustrated in Fig. 15.

From both the above examples it is clear that some considerable reinforcement of repeating radial and linear features present in an

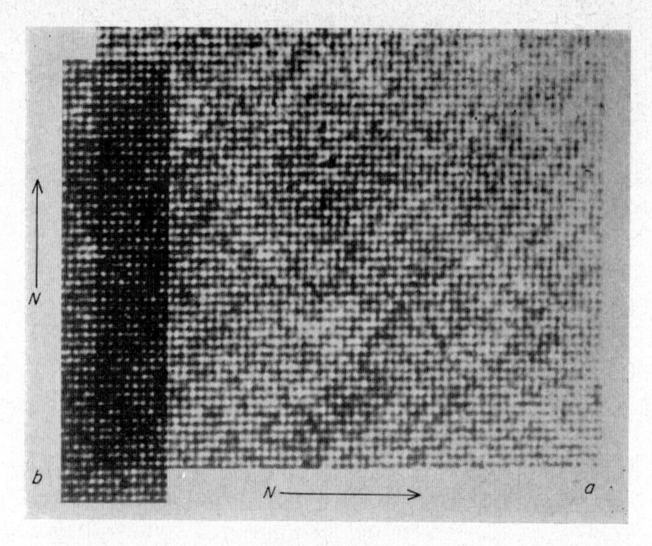

(a)

(b)

FIG. 9(a). Simple photographic procedure for integrating a two-dimensional atomic lattice of gold. The transparency shown at (a) is moved in the lattice directions by spacings determined by a reference strip (*b*) cut from (*a*). An exposure is made for each interval.

(b). The photograph was the result of averaging the lattice in (a) at $N = 9$ intervals in one direction and $N_1 = 9$ intervals in the other. (From Horne and Markham (1972), "Practical Methods in Electron Microscopy," North-Holland, Amsterdam.)

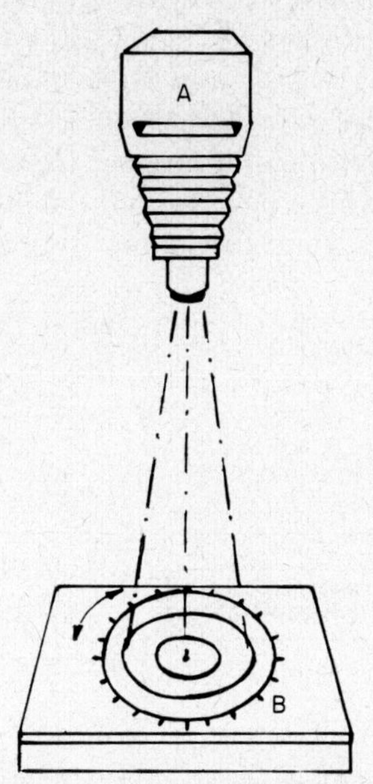

FIG. 10. Simple method for obtaining integrated prints from electron micrographs with images containing rotational symmetry. (A) Photographic enlarger, (B) calibrated table for rotating photographic paper at known angles. The image rotational axis must be coincident with the concentric rings indicated on the table. (Markham *et al.*, 1963.)

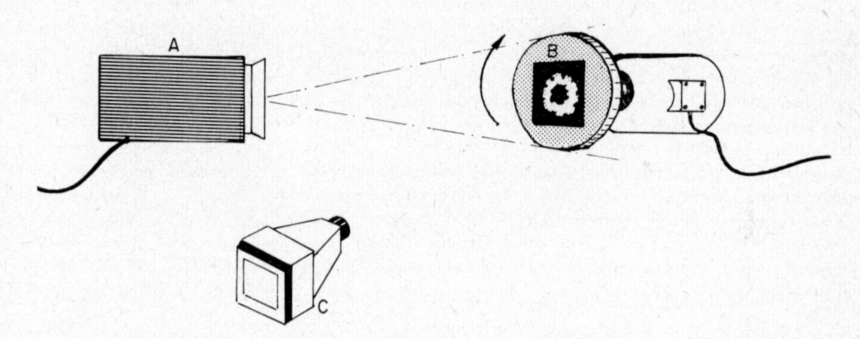

FIG. 11. Integration of objects with rotational symmetry with the aid of a Strobe flash illuminator, by adjusting the flash frequency (A) to synchronize with the rotating object (B). The integrated image can be obtained by direct observation and then recorded in the camera (C).

object, and subsequently recorded on the micrograph, can be achieved by the use of photographic procedures. The apparatus of Markham *et al.* (1964) is relatively complex, but can be constructed from components available in the average experimental laboratory; the only expensive item of equipment is the strobe lamp device. An alternative and simple method for integrating linear periodicities in electron micrographs was described by Warren and Hicks (1971). Their approach was to make use of the principle that when a transparent film or plate possessing a periodic lattice is illuminated from a point source of light, a shadow will be cast which has the same density distribution as the transparency. It follows that two point sources suitably placed will

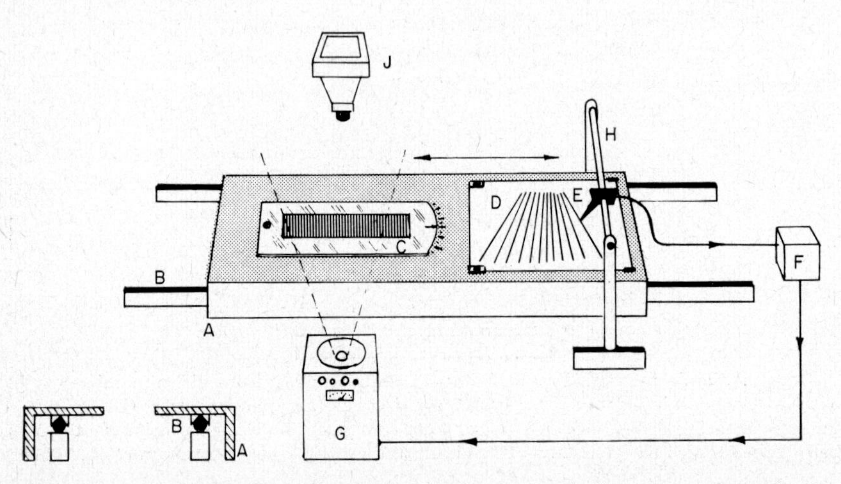

FIG. 12. Apparatus developed by Markham *et al.* (1964) for linear integration of electron micrographs. (A) sliding table on Teflon supports, (B) Electron micrograph transparency, (C) Calibrated fan-shaped lines on card, (D) Electronic pick-up head, (E) Electronic relay, (F) Strobe flash lamp, (G) Horizontal translation screw for pick-up head, (H) The electron micrograph transparency is fixed to the table as shown and fine adjustment along its axis is adjusted in relation to the direction of the sliding table. By sliding the table in the direction of the arrow, the conducting lines ruled on the card cause the pick-up head to operate the electron relay switch followed by the Strobe flash lamp. Each line on the card will produce an illuminated image of the transparency to be recorded on the camera (J) and if the correct intervals are selected any periodic detail will be integrated in the final camera image. The number of flash intervals will depend on the number of lines scanned by the pick-up head (see Fig. 15). (From Horne and Markham (1972). "Practical Methods in Electron Microscopy". North-Holland, Amsterdam.)

produce two overlapping shadows, and the lattice periods can be adjusted on a viewing screen to give precisely one lattice period. By providing a row of equally spaced point sources a relatively large number of superimpositions can be made by moving the transparency, the point sources or the viewing screen relative to each other. This arrangement is illustrated in Figs 16a and b, and a diagram of apparatus in Fig. 17.

In any photographic integration procedure there is bound to be a certain loss in contrast and some thought must be given to the type of film or plate emulsion used for these methods. Several practical hints were discussed by Warren and Hicks (1971) in their original publication. It is also clear that some indication of the basic spacings and distances

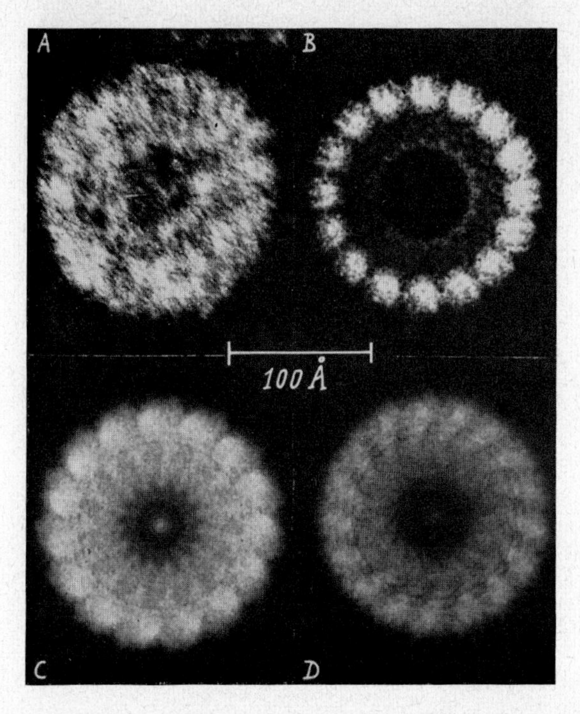

FIG. 13. "End on" images of tobacco mosaic virus stacked disc protein subjected to image rotation. (A) original electron micrograph. The image at (B) shows the pronounced reinforcement of the protein subunits when integrated at 16 rotational intervals. At values of rotation C = 15 and D = 17, the images are blurred. (From Markham *et al.*, 1963.)

relating to the repeating structures in the original image have to be determined to some degree of accuracy. This lack of information may result in the image being reinforced at the wrong frequency and possible artefacts being produced, together with a considerable loss of time due to trial and error. These points have been discussed by Markham *et al.* (1964) in relation to virus particles and other objects subjected to rotational and linear integration.

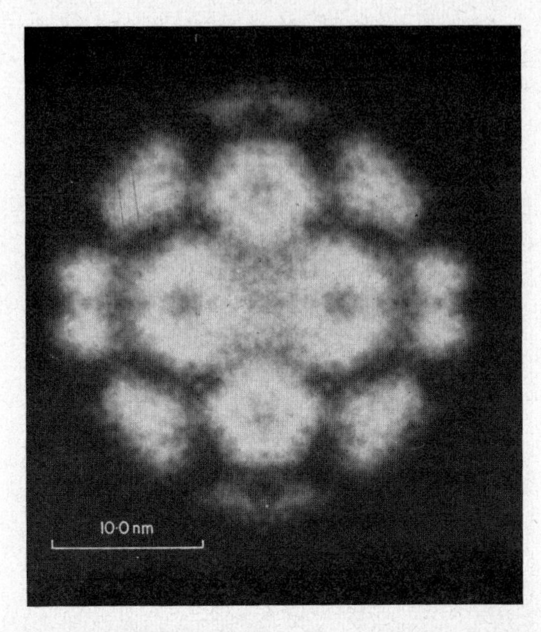

Fig. 14. Photograph of a single particle of turnip yellow mosaic virus which was "symmetrized" about its twofold rotational axis. Following a first exposure, the micrograph was rotated by 180° in the enlarger and a second exposure made. A third reinforcement was made by reversing the micrograph. It is clear that all the views of the particle shown are equivalent due to the virus image having two mirror planes at right angles to each other. (From Horne and Markham, (1972) "Practical Methods in Electron Microscopy", North-Holland, Amsterdam.)

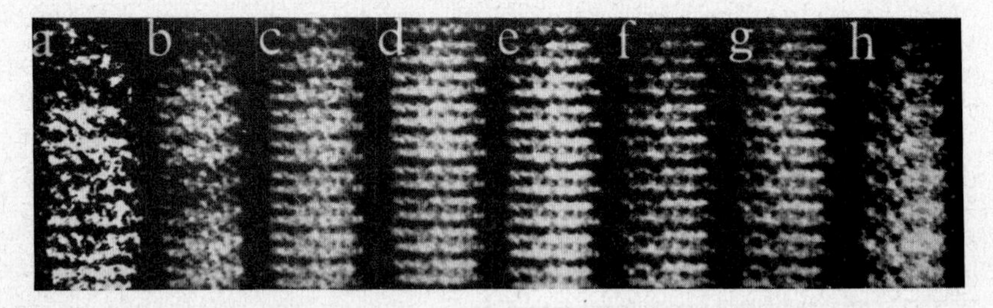

Fig. 15. The photograph shows a series of integrated images of tobacco mosaic stacked disc rods obtained from the apparatus illustrated in Fig. 12. At "e" the integrated distance was equivalent to 4·54 nm which was considered to be the best approximation to the stacked disc structure. (From Markham *et al.*, 1964.)

The most accurate and direct way to determine the basic repeating features and distances in any electron micrograph is to subject the negative to analysis by optical diffraction. This technique is described in more detail below and can be used in close conjunction with rotational or linear photographic methods for image enhancement.

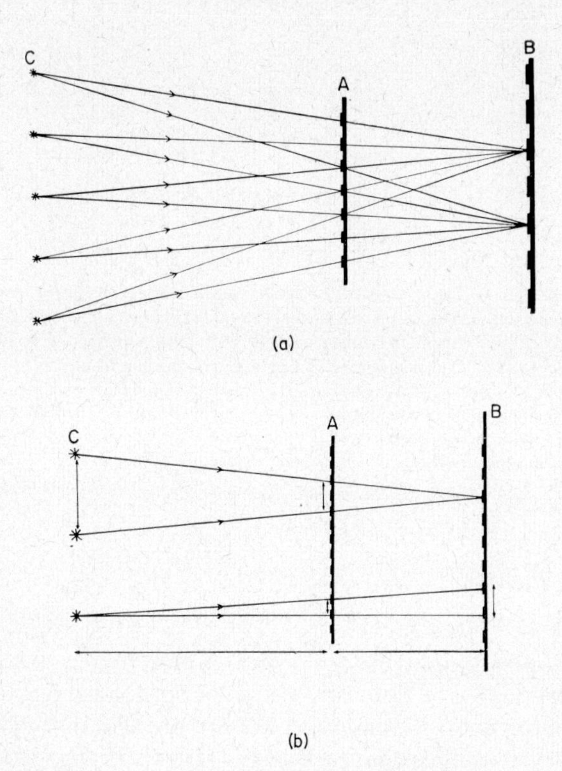

FIG. 16. The principles and geometry of the device described by Warren and Hicks (1971), for linear integration of electron micrographs. (a) The shadows from the periodic lattice in the transparency (A) are cast on to the screen (B), such that the subunits of the lattice are in register. The point light sources are shown at (C). (b) The diagram shows the light rays through the transparency. The upper rays illustrate conditions required for shadows to overlap in register and the lower rays show the magnification of the lattice. (From Warren and Hicks, 1971. *J. Ultrastruct. Res.*)

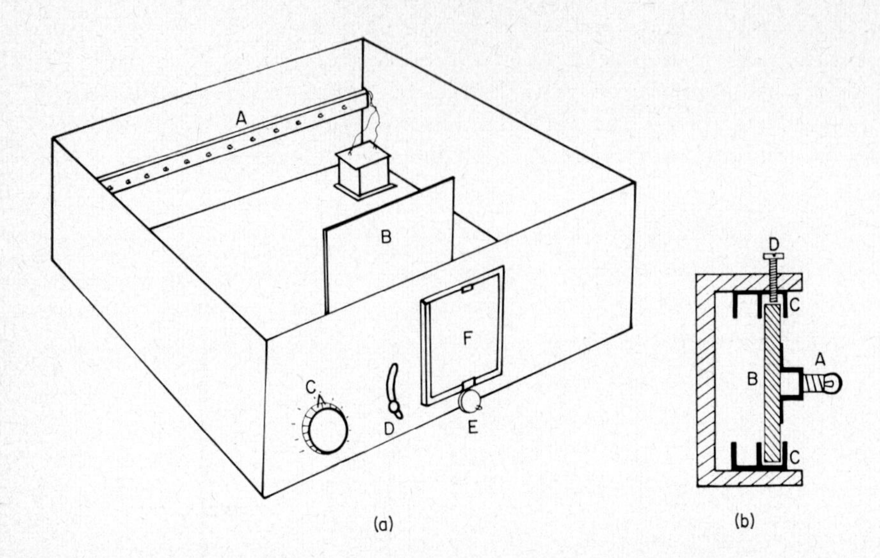

FIG. 17(a). The apparatus described by Warren and Hicks (1971) for linear integration of transparencies from electron micrographs. A row of equally spaced 12 V illuminating bulbs are shown at (A) supplied from a transformer. The transparency is held between two perspex sheets (B). The position and angle of the transparency is controlled by the lead screw knob (E) and rotation setting (D). Control (C) varies the input to the transformer and light intensity of the bulbs (A). A 5″ × 4″ Polaroid camera back or viewing screen can be placed in the position (F).

(b). Practical arrangement for locating the positions and spacings of the 12 V illuminating bulbs. (A) Bulb on its mount. (B) Supporting batten sliding in plastic runners (C) and locking screw (D).

E. *The Practical Application of Optical Diffraction and Image Reconstruction Techniques to Electron Micrographs*

The development and use of optical diffractometers has been the subject of a number of publications covering a considerable period of time. For a more detailed historical account of these methods the reader is referred to the publications of Buerger (1941), Bragg (1944), Wyckoff *et al.* (1957), Wilkins *et al.* (1950), Taylor and Lipson (1951), (1964), Horne and Markham (1972), and Hahn (1972). In recent years valuable use has been made of this technique of analysing electron micrographs in optical diffractometers of various types. These studies followed the work of Klug and Berger (1964), Klug and De Rosier (1966) and Bancroft *et al.* (1967) who applied optical diffraction and filtering techniques to virus structures.

In principle the formation of an optical diffraction pattern is essentially the same as an electron diffraction pattern or X-ray diffraction pattern. By comparison with the apparatus required for electron or X-ray diffraction studies, the optical diffraction camera is a simple device. The basic components needed to obtain an optical diffraction pattern are shown in Fig. 18 and Fig. 19.

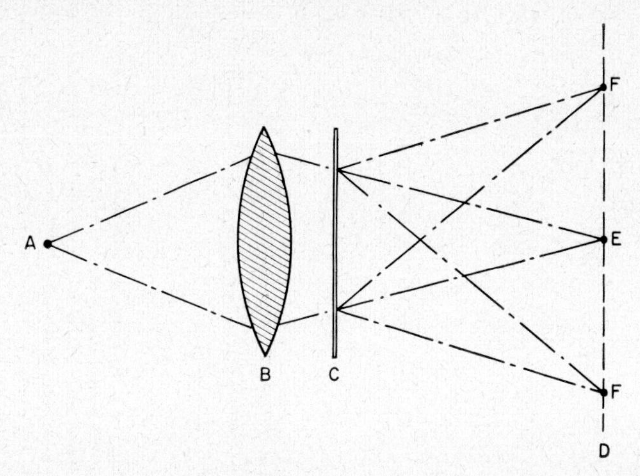

FIG. 18. Diagram illustrating the formation of an optical diffraction pattern using a monochromatic light source (A). Diffraction lens (B). Electron micrograph (C). The optical diffraction pattern formed in the diffraction plane (D), will be directly related to the features recorded on the micrograph (C). An amorphous image at (C) will produce an optical diffraction pattern consisting of a random spectrum of noise radially distributed about the central spot (E). In the case of a micrograph or grating containing periodic features discrete spots will be formed at (F) in addition to the "noise" spectra. The distances and orientations can be measured and analysed in terms of reciprocal lattice space and is essentially similar to the methods of electron diffraction or X-ray diffraction images. (See Figs 20, 22, 27 and 28.)

An electron micrograph containing only amorphous detail will produce an optical diffraction pattern containing diffuse spectra or noise radially distributed around the central direct beam. If the detail recorded in a micrograph is periodic or crystalline, the optical diffraction pattern will show a series of spots that can be analysed to give periodic information relating to the specimen. As mentioned below, the optical diffraction pattern will also provide information about the resolution and performance of the electron microscope, together with an indication of the degree of over- or under-focusing of the objective lens (see Fig. 20).

One of the main advantages of using the optical diffraction camera is that patterns can be obtained from relatively small areas, and in the case of some viruses from single particles, but it does suffer from other limitations which are discussed briefly below.

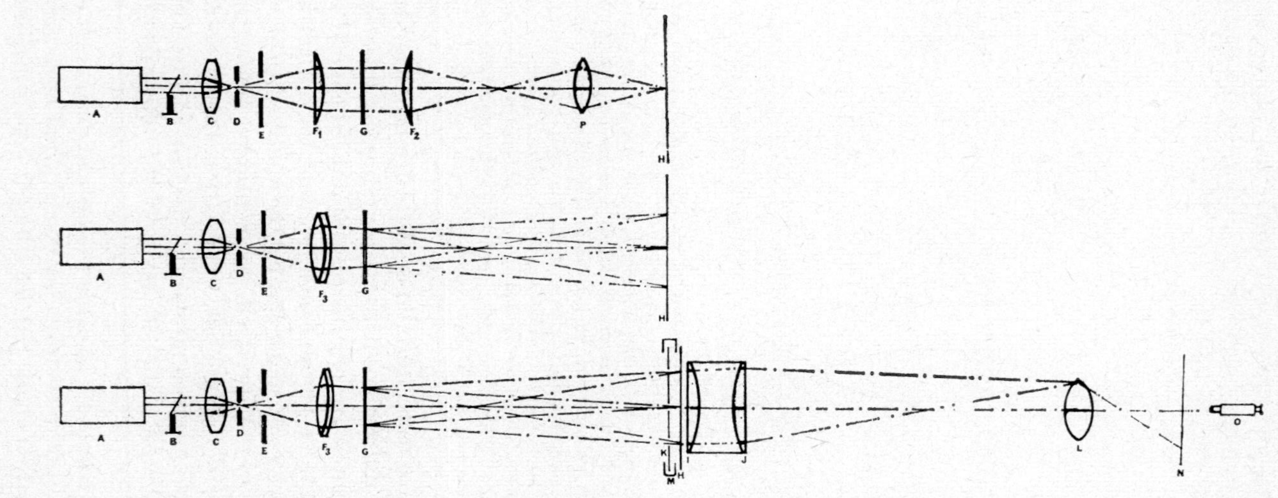

FIG. 19. The diagram illustrates the practical geometrical arrangement for forming an optical diffraction pattern and image reconstruction. A. Laser light source; B. Electronic relay for switching beam output "on or off"; C. Expanding lens; D. Beam aperture (100 μm diam. electron microscope objective lens contrast aperture); E. Adjustable diaphragm; F_1 and F_2. Diffraction lenses; F_3. Diffraction lens; G. Electron micrograph; H. Optical diffraction image plane; I. J. Image reconstruction lens system; K. Filter mask; L. Magnification lens (if required); M. Polaroid camera back or viewing screen; N. Reconstructed image plane; O. Focusing telescope. (From Horne and Markham, 1972. "Practical Methods in Electron Microscopy". North-Holland, Amsterdam.)

A typical practical arrangement for obtaining optical diffraction patterns and image reconstruction from electron micrographs is shown in Fig. 19. The construction and experimental details relating to this type of diffractometer were described by Horne and Markham (1972) together with the necessary alignment procedures. It can be seen in Fig. 19 that it is a relatively simple matter to obtain optical diffraction patterns from an electron micrograph possessing repeating features, by placing the photographic negative in a suitable position relative to the diffraction lens and illuminating it under the appropriate coherent conditions with a laser beam. The diffraction image is recorded on a 5″×4″ Polaroid camera back. The camera length of the optical system

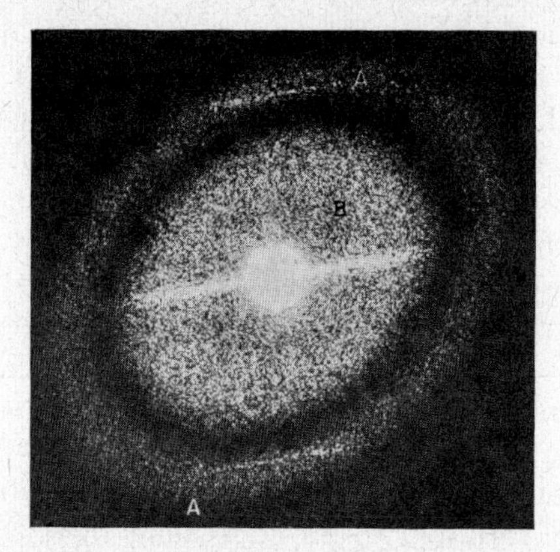

Fig. 20. Optical diffraction pattern obtained from roughly parallel arrays of tobacco mosaic virus rods. The layer lines at (A) are derived from the 2·3 nm pitch of the protein helix. Due to the high degree of under-focusing, information has been lost in the region due to the phase change (dark ring), but has appeared in the second noise spectra ring at (A). Note the ellipsoidal shape of the noise spectra indicating the presence of astigmatism in the electron microscope image.

has to be considered in terms of the spacings and details recorded at a given electron optical magnification on the original electron micrographs. In the case of very small repeating features appearing at low magnification, the diffraction spots could fall outside the Polaroid camera for a given optical bench camera length. Images recorded at high magnifications, on the other hand, will form diffraction spectra too close to the direct illuminating beam and will be obscured. For these reasons, and taking into account the lens apertures selected for a given diffractometer bench, some thought must be given to both the optimum magnifications used for particular electron microscope applications and the design of the diffractometer bench system (see Markham, 1968; Horne and Markham, 1972).

Much of the work involving the application of optical diffractometers to electron micrographs has come from the analysis of negatively stained images recorded at high resolution, where small repeating details are close to the electron microscope image noise level or where superimposition of the specimen structural detail has occurred. Many of the published optical diffraction patterns of viruses were obtained by selecting single suitably oriented particles or assemblies and recording the relatively simple patterns from such areas. Typical examples of the diffraction spectra produced from an electron micrograph are shown in Figs 20, 27 and 28. The diffuse area radially distributed around the central spot is the spectrum from the noise of amorphous detail present in the micrograph, and its shape together with the distance it extends from the central spot will depend on several electron optical and instrumental factors. If the electron microscope image was astigmatic or the image was recorded with stage drift, the noise spectra will appear ellipsoidal in shape as illustrated in Fig. 20. In addition, the degree of underfocusing by the objective lens is also indicated in the noise spectra and is shown by the presence of one or several sets of concentric diffuse rings (Fig. 20). Some indication of the resolution of the image can be determined from the noise spectra present in any electron micrograph, and in the case of Fig. 20 the first order of noise spectra extends to a distance equivalent to 3·0 nm (in object space). This is followed by a region showing no detail or spectra, due to underfocusing and consequent phase change where information has been lost, but a second diffuse ring can be seen extending to an equivalent distance of 2·0 nm. The area from which the optical diffraction spectrum shown in Fig. 20 was taken contains a rod-shaped virus (Tobacco mosaic virus) which is known to have a helical periodic structure of 2·3 nm along its major axis. This periodicity is reproduced in the optical diffraction pattern (Fig. 20) as discrete lines or spots. The optical diffraction data can be compared directly with X-ray diffraction measurements made from three-dimensional crystals formed from the same virus material (see Finch and Holmes, 1967).

From this simple example of optical diffraction it can be seen that the method will extract all the information contained in the electron micrographs including the positions of the diffraction spectra, the amplitude and phase of diffraction spectra, together with the angles of any discrete spots recorded in the diffraction plane. It should be stressed here that the amount of information present in the electron micrograph, with particular reference to negatively stained objects, will be related to the degree of underfocusing by the electron microscope objective lens (see Erickson and Klug, 1971). Moreover, it must be remembered that the optical diffraction spectrum is entirely dependent on the contrast and details recorded on the actual micrograph, including any defects in the film or emulsion. As mentioned earlier the contrast in the case of negatively stained images depends on surrounding and penetrating a relatively electron transparent area with an opaque material. The fine detail will also be revealed or obscured by the degree of penetration by the stain coupled with its preservation properties. Consequently, the analysis of the image by optical diffraction is limited by the characteristics of the film or plate, and the details recorded may only represent part of the original specimen.

F. *Calibration*

Some mention should be made here about the calibration of the micrographs and optical diffraction system for the purposes of making measurements from viruses and other specimens. One of the most convenient and reliable methods for the calibration of biological specimens over a wide magnification range is the use of catalase crystals mixed with virus samples which allows an internal standard to be used on the same grid or in the specimen area. There are several publications which give values for the catalase lattice spacings prepared under differing conditions of stains and drying (see Wrigley, 1968). The currently accepted mean spacings for negatively stained catalase are 6·21 nm (A spacing) and 8·7 nm (B spacing). These crystals have the advantage over other materials in that they can be examined by low-angle electron diffraction methods and accurately calibrated against non-biological standards (Ferrier and Murray, 1966; Wrigley, 1968). Once the electron optical magnifications have been determined the catalase crystals can then be used to calibrate the optical diffractometer.

G. *Image Reconstruction*

Considerable interest has been generated in the possibility of using the optical diffractometer as an optical or spatial frequency filtering system with a view to the separate display of the two sides of a thick object or for removing the noise spectrum from the diffraction spots in the optical transform (Klug and De Rosier, 1966; Bancroft *et al.*, 1967). The principle of introducing a mask into the optical system which will allow selected diffraction spots to be used in the process of image reconstruction from electron micrographs is at first sight a simple procedure (K in Fig. 19). One example of image reconstruction directly from the optical diffraction pattern is shown in Figs 21 and 22. The

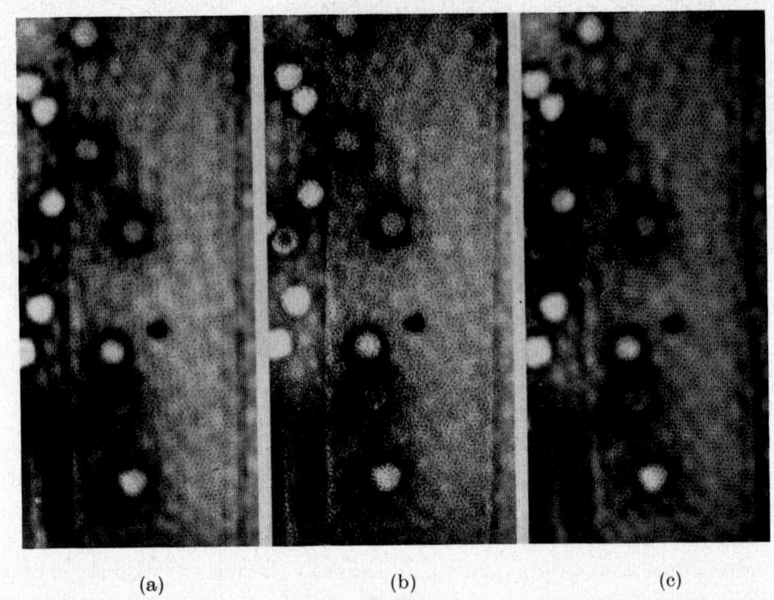

(a) (b) (c)

FIG. 21. Electron micrograph of a virus protein tube prepared in the presence of a negative stain. The print at (b) shows the presence of interference patterns resulting from the superimposition of the upper and lower parts of the tube structure. The print at (a) was obtained by selecting one set of hexagonal spots from the optical diffraction pattern in Fig. 22(a) and (b), thus reconstructing one side of the tube image. By selecting the second hexagonal array it was possible to reconstruct an image of the other side (c). The diffraction pattern and image reconstruction was obtained using the diffractometer illustrated in Fig. 19.

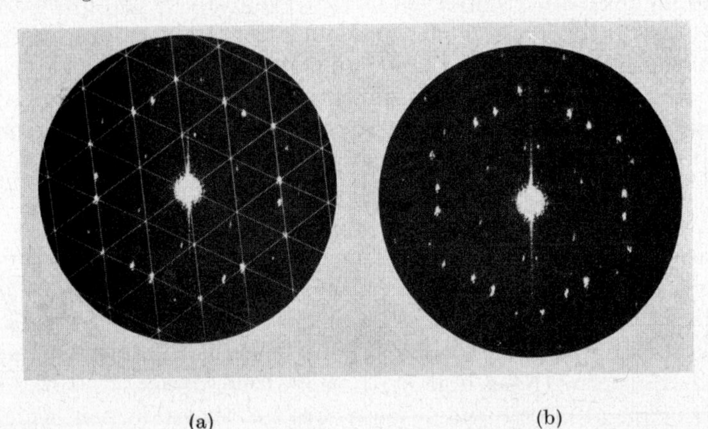

(a) (b)

FIG. 22(a). Optical diffraction pattern obtained from Fig. 21(b) showing two superimposed patterns rotated by 14°. (b). For lattice identification purposes and mask construction, lines have been drawn through one set of spots corresponding to spectra from one side of the image in Fig. 21(b). The reconstructed images in Fig. 21(a) and (c) were formed by allowing one set of the hexagonal spots to pass through the image reconstruction lens and then selecting the second set of spots. (From Horne and Markham. (1972) "Practical Methods in Electron Microscopy", North-Holland, Amsterdam.)

electron micrograph was recorded from a negatively stained virus protein which was assembled to form hollow tubes (Bancroft *et al.*, 1967). Examination of the micrographs revealed that the upper and lower parts of the tube-like structure had collapsed with the result that interference patterns were seen in the final image. By selecting a suitable area from the electron micrograph it was possible to obtain the optical diffraction pattern illustrated in Fig. 22, which shows the distribution of spots originating from the upper and lower parts of the object. The spots fall into two separate groups and are rotated with respect to each other by an angle of about 14°. The different sets of spots can be identified by drawing lines through the related diffraction spectra as shown in Fig. 22, and a suitable mask or filter made which will allow one set or other of the diffraction spots forming the hexagonal arrays to be used in the reconstruction optical system (Fig. 19). Two reconstructed images corresponding to the upper and lower parts of the viral protein tube are illustrated in Fig. 21 and show that the morphological structure is composed of hexagonal units with their centres penetrated by the electron dense stain.

Obviously it is possible to use the same filtering procedures for the separation of noise from the repeating features in a single-sided object where only one set of diffraction spectra will be presented. However, the question of making a suitable filtering mask raises a number of important problems as the optical diffraction pattern is the result of a complex interaction process where every individual modulation in the recorded image on the plate or film takes part. In other words, the repeating structural features appearing as differences in contrast on the emulsion are rarely completely independent of the noise spectra and when seen close to the limit of resolution from biological objects may be difficult to separate from the noise level present in the diffraction pattern. The situation illustrated in Figs 21 and 22 can be considered to be a relatively simple case and the construction of a mask to filter the diffraction spots merely consists of punching holes with sufficient aperture in an opaque card or similar material (see Horne and Markham, 1972). The filtered spots passing through the mask can then be used to reconstruct the final image as shown in the diagram (Fig. 19). It goes without saying that the mask must be accurately located in the diffraction plane from which the original pattern was formed.

To achieve sharper and well separated diffraction spots it is advantageous to record electron micrographs with regular structural features distributed over large areas. The techniques described later in this article relating to the negative staining-carbon method were developed with this aim. However, it is now clear that the diffraction spectra recorded from virus crystals prepared by these techniques are so complex that the construction of a filter mask by the simple mechanical procedures described above is almost impossible (see Figs 27, 28 and 29). Moreover, the introduction of a filter mask possessing a large number of regularly arranged holes in the optical system appears to introduce an artefactual interference effect in the final reconstructed image. This problem is obviously one of a fundamental nature and presents an interesting subject for further investigation using laser optics applied to optical diffractometers.

H. *Computer Processing of Images and Diffraction Patterns*

There are several approaches to analysing optical diffraction and electron diffraction patterns mathematically with the aid of computer methods (as Fig. 23), which are outside the general scope of this article. However, it should be mentioned here that these procedures using computer and data processing technology offer considerable scope with particular reference to reconstruction of three-dimensional images from a series of two-dimensional electron micrographs. Such experiments have been carried out mainly on images of negatively stained virus particles. The reader is referred to the publications of De Rosier and Klug (1968), De Rosier and Moore (1970), Erickson and Klug (1971), Hoppe (1971), Crowther (1971), and Klug (1971).

There are obvious advantages associated with computer processing over direct image reconstruction in the optical diffractometer, but the former requires considerable technological facilities to take full advantage of this type of image analysis and reconstruction. The recent advances in the application of photographic averaging, optical diffractometers, image averaging and reconstruction by computer methods has allowed a much more accurate analysis and quantitative interpretation to be made of the structural details resolved in negatively stained virus particles.

I. *Analogues for Negative Staining*

It was mentioned earlier that the majority of negatively stained virus particles are two-sided and that interference patterns between the upper and lower surfaces frequently appear in the final image. In order to assist with the interpretation of these images and the construction of suitable three-dimensional models, several studies have been carried out on the development of analogues for negatively stained virus particles. Almeida *et al.* (1965) applied X-rays to suitable rubber models

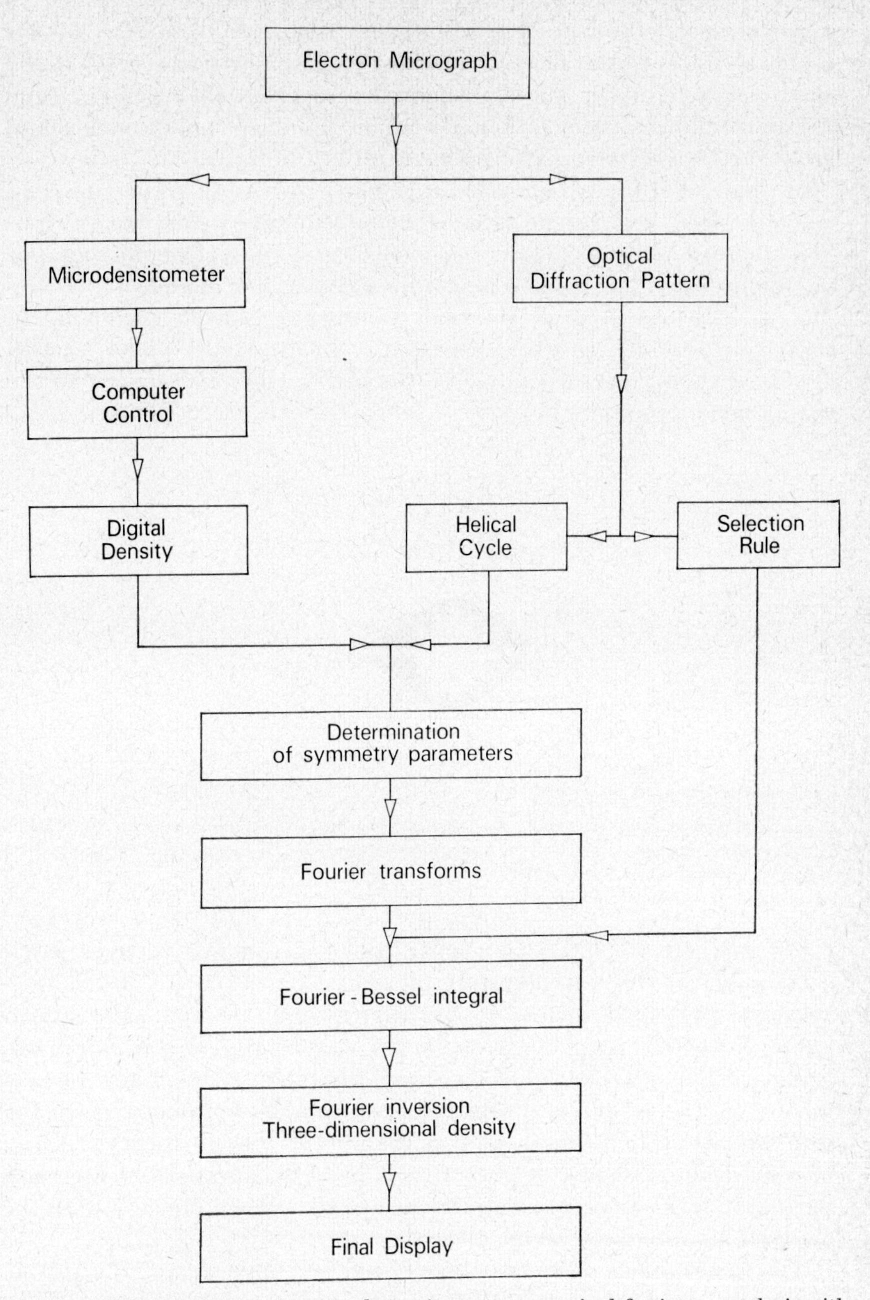

FIG. 23. Block diagram showing the various stages required for image analysis with the aid of computer and data processing techniques.

representing helical nucleocapsids and reported that they could duplicate some of the images corresponding to those observed in the electron micrographs. These authors also reported that this analogue technique allowed them to X-ray a range of helical rubber models and check the transmission patterns for a variety of helical arrays.

In the case of the icosahedral viruses an X-ray radiographic method was described by Caspar (1966). The technique made use of constructing three-dimensional virus models based on electron micrographs and embedding the models in a suitable medium opaque to X-rays (Fig. 24). It was pointed out that X-radiographs will correspond to negatively stained electron microscope images only under certain restricted conditions, depending on the distribution and opacity of the opaque media used.

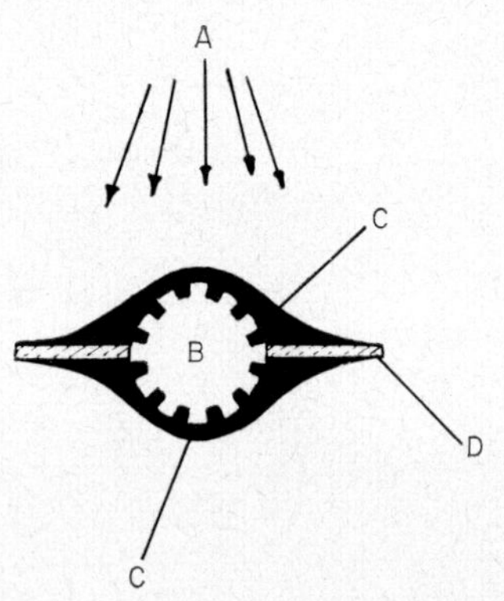

FIG. 24. Diagram illustrating the negative staining analogue described by Caspar (1966) for the interpretation of electron micrographs. (A) X-ray source. (B) Polythene model of virus. (C) Calcium sulphate/barium sulphate mixture as X-ray opaque "stain". The model and opaque media were supported on a hardboard sheet (D).

The method developed by Caspar (1966) was shown to be of particular value in the interpretation of virus particles appearing as one-sided or two-sided images in the electron microscope with special reference to the morphology of human warts virus. For X-ray radiographic analogues to be applied successfully there are two basic requirements: (a) the model used must be a reasonable representation of the virus and its protein capsid components and (b), the control of the X-ray opaque medium should be such that it provides an approximate comparison in terms of contrast to the negatively stained image seen in the electron microscope (Caspar, 1966).

The virus models were constructed from polyethylene and the opaque X-ray "stain" consisted of surrounding or embedding the model in calcium sulphate. It was pointed out that calcium sulphate when mixed with a small amount of barium sulphate, as a dry powder, gave X-ray radiographs possessing a granularity very close to that of electron micrographs. The use of dry powder mixtures, on the other hand, presented difficulties in ensuring good contrast and penetration between opaque material and virus model components. Experiments were carried out by applying the calcium sulphate and barium sulphate as a wet paste which could be manipulated to greater advantage. Optimum conditions for the best analogue results were obtained empirically by varying the voltage of the X-ray tube and the opacity of the analogue stain material.

IV. Preparation of Highly Concentrated Virus Material in the Presence of Negative Stains for Electron Microscopy and Subsequent Analysis by Optical Diffraction

In the previous sections of this article dealing with negative staining applied to virus particles the methods described were mainly concerned with obtaining information from individual viruses or small groups of viruses. Much of the published data from optical diffraction studies has subsequently come from the analysis of single virus capsids photographed in the electron microscope at suitable orientations. Some considerable advantage could be gained from analysing in the diffractometer electron micrographs in which the virus particles were in regular crystalline arrays and in the same orientation. Under these conditions the diffraction spectra obtained would be averaged over many virus particles within a given area.

There are several practical problems associated with the mounting of highly concentrated virus suspensions (1·0–30 mg/ml virus) for electron microscopy. First, direct deposition of droplets containing large quantities of virus results in the disruption of the specimen supporting films, unless they are relatively thick and free from thermal drift in the electron beam. Secondly, large aggregates of virus tend to dry down over a relatively small area on the grid and, thirdly, it is very difficult to obtain optimum negative staining conditions.

In a recent study by Horne and Pasquali-Ronchetti (1974), a negative stain-carbon film technique was developed for the preparation of highly concentrated virus as crystalline arrays in the presence of negative stains. In addition, the method has allowed a much more favourable ratio between the carbon film thickness and virus dimensions to be established, with some improvement in the resolution from biological particles possessing repeating structural features. The underlying principles of the original negative staining method of Brenner and Horne

(1959) were used, but the procedure for depositing the virus mixed with negative stain and carbon on mica was different and followed three main stages of preparation which included the use of a second negative stain.

It has been previously reported that freshly cleft mica presents a more favourable hydrophilic surface for the purpose of spreading liquid suspensions containing biological particles (Hall, 1966). These mica surfaces were found to be ideally suited for the negative stain-carbon

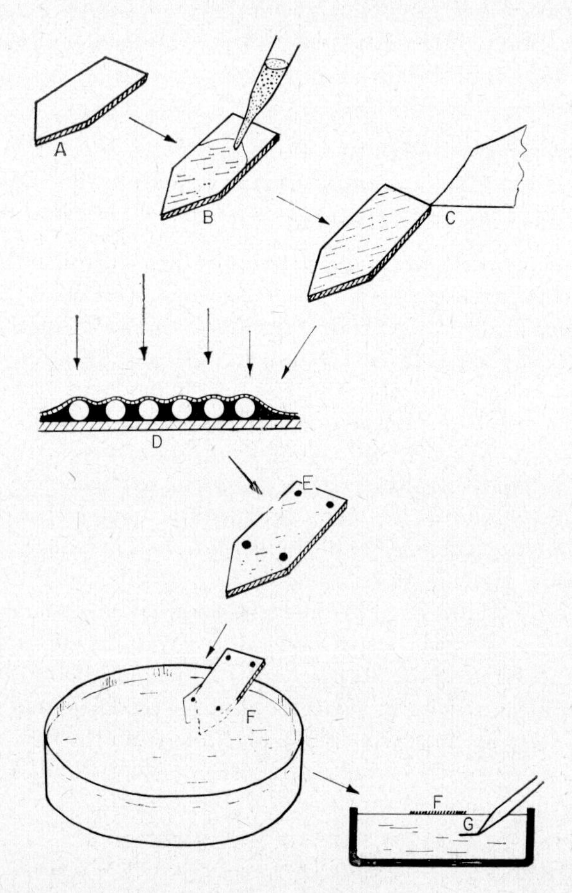

FIG. 25. Diagram showing the negative staining-carbon procedure for preparing specimens. (A) Freshly cleft mica sheet. (B) Mixture of highly concentrated virus suspension and first negative stain deposited by Pasteur pipette. (C) Excess fluid drained with pointed filter paper. Section through a specimen on mica after being coated with thin carbon layer (D). Carboned specimen marked with black spots (E), to indicate position of film and virus preparation when released on to surface of second negative stain (F). Specimens are picked up with holey-filmed grids (G). (From Horne. Negative staining techniques, in "Practical Methods in Electron Microscopy". North-Holland. In press.)

method described below and in practice is the only material which will allow the technique to work satisfactorily. The mica was cut into small pieces of the shape shown in Fig. 25 and about 2 cms long by 1 cm wide. One end of the mica sheet was pointed to facilitate the release of the specimen from the surface.

Suspensions of highly concentrated virus were mixed with an equal volume of 3% ammonium molybdate acting as the first stain, previously adjusted to the appropriate pH value (5·2–7·5) according to the type of virus being studied and crystalline packing required. Other negative stains were used in the experiments and included 2% sodium tungstate (NaPT), 2% phosphotungstate (KPT), uranyl acetate (Ua), uranyl formate (Uf). It was found that ammonium molybdate was the most suitable material as the first stain, as difficulties were encountered when releasing the specimen from the mica in the presence of the other stains.

The mixture of virus and negative stain solutions was deposited by a fine Pasteur pipette onto the pointed mica pieces which had been cleaved with a razor blade, as illustrated in Fig. 25. Much of the liquid was removed from the surface with the aid of a pointed filter paper, leaving a wetted monolayer of virus and stain to dry at room temperature. The dried specimens were transferred to a high vacuum evaporator fitted with a liquid nitrogen-trapped Hg pump and liquid nitrogen-trapped mechanical pumping lines. Carbon was evaporated from a standard arc source controlled by a variable transformer system. It was essential that only a very thin layer of carbon should be deposited as shown in Fig. 25D. If the carbon was visible following removal from the mica it was considered to be too thick for high resolution studies.

In order to identify and locate the specimen once it was separated from the mica, it was found convenient to place four black dots on the carbon surface (Fig. 25E), using a soft marking pen.

Removal of the carbon-coated specimen was achieved by very slowly immersing the mica into the second negative stain solution as shown in Fig. 25F. The second negative stain was made up from a 1% solution of uranyl acetate or a 1% solution of uranyl formate. The specimens were then picked up on to previously prepared grids covered with holey carbon support films and transferred to the electron microscope specimen holders.

The above technique has been applied to concentrated suspensions of several plant viruses including cowpea chlorotic mottle virus (CCMV), brome mosaic virus (BMV) (Fig. 26), turnip yellow mottle virus (TYMV), southern bean mosaic virus (SBMV), apple chlorotic leafspot virus (CLSV) (Fig. 32a), citrus tristeza virus (CTV), potato virus X (PVX) and tobacco mosaic virus (TMV) (Fig. 33). In addition, the

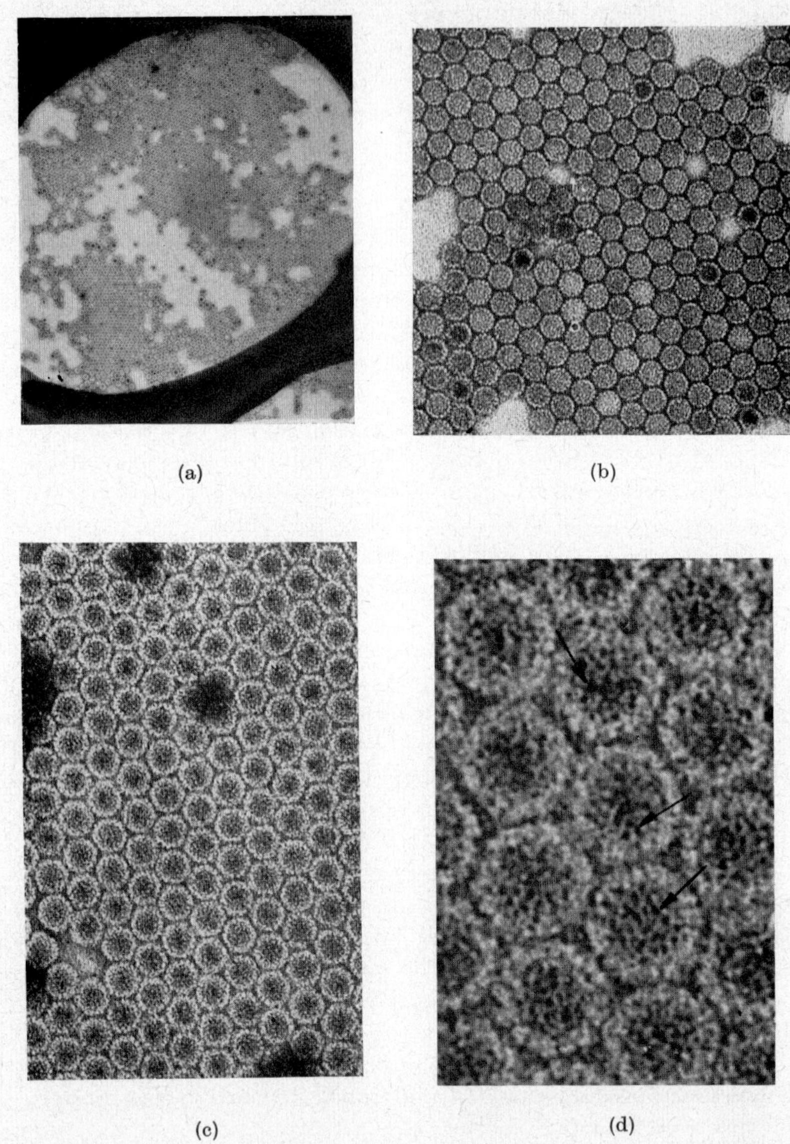

FIG. 26(a). Low magnification area showing an aperture in a holey support film with two-dimensional arrays of brome mosaic virus. Prepared by the method illustrated in Fig. 25(B). The general appearance of the virus material is essentially similar to conventionally negatively stained virus, but large areas containing hexagonally packed particles (see Fig. 27), can be prepared depending on the first stain pH and buffering condition of the virus suspension (see Fig. 28). (c) Hexagonally packed cowpea chlorotic mottle virus prepared by the method illustrated in Fig. 25. (d) Highly magnified area from C, showing small crystallites (arrows) of electron dense stain mainly associated with the virus particles. (From Horne and Pasquali-Ronchetti, 1974. *J. Ultrastruct. Res.*)

technique was used in a study of crystalline human adenovirus type 5, which is a relatively large icosahedral virus. (See Figs. 4c, 29, 30 and 31).

The electron micrograph illustrated in Fig. 26a shows an area at low magnification containing a two-dimensional crystalline array of BMV virus particles extending across a hole in the support film. It was found under suitable concentrations of virus, negative stain and pH, that the arrays extended across several microns or over entire grids. At higher magnifications the packing of the viruses within the lattice were seen in more detail (Fig. 26b). A number of features were noted, together with the information contained in the optical transforms recorded from the same areas (Fig. 27). The general appearance of the individual virus

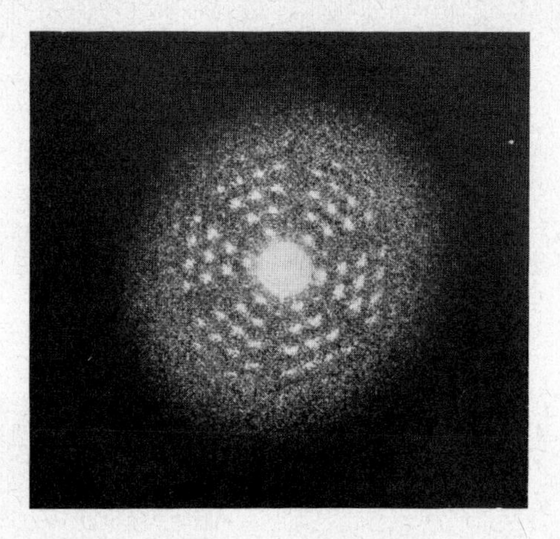

FIG. 27. Optical transform from the micrograph shown in Fig. 26(b). The precise packing of the brome mosaic virus particles in the hexagonal lattice is indicated by the diffraction spectra. Note: The first order diffraction spots are obscured within the bright central spot region.

particles was considered to be essentially similar to those prepared by conventional negative staining methods showing particles embedded in, or surrounded by an electron dense stain. However, the precise packing of the particles within the lattice is indicated in the optical diffraction patterns as shown in Figs 27, 28 and 29.

It was shown from a number of experiments that the technique was not a replica method resulting from deposition of the carbon over the dried mixture of virus and stain. Several specimens prepared by the above method were placed in strong solutions which dissolved the biological material present. When the remaining films were examined in the electron microscope there were no virus particles visible and only

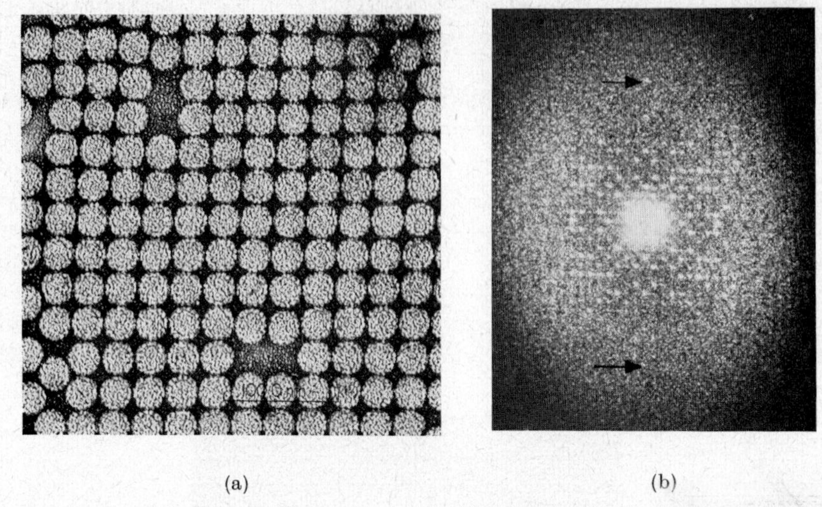

(a) (b)

FIG. 28(a). Underfocused image of cowpea chlorotic mottle virus in approximately square array prepared by changing pH of the first negative stain. (b). Optical transform of A, showing spectra extending into the noise level (arrows) at a distance equal to about 1·7 nm. The first order spectra are obscured by the central spot. (From Horne and Pasquali-Ronchetti, 1974. *J. Ultrastruct. Res.*)

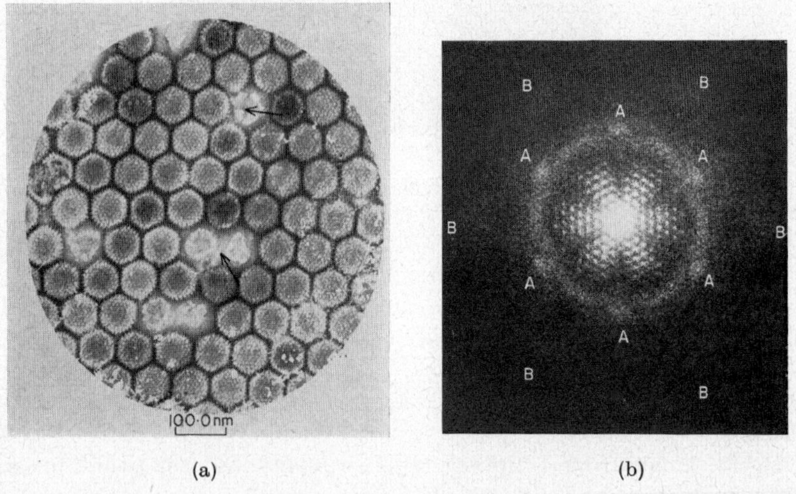

(a) (b)

FIG. 29(a). Electron micrograph of a two-dimensional lattice of human adenoviru type 5 prepared by the negative staining-carbon technique. In the areas indicated by the arrows some of the capsids have disrupted. (b). Optical diffraction pattern from (a). Note the two diffuse arrays of spectra indicated at (A) and (B). The diffuse areas at (B) extend to a distance of 2·8 nm. (From Horne and Pasquali-Ronchetti, 1974. *J. Ultrastruct. Res.*)

in a small number of areas was it possible to find carbon replica-type films. These replica regions only indicated the crystal area outline and gave poor resolution.

In another series of experiments carried out on the disruption of adenovirus, it was clearly shown that the images were derived from both the negative stain and virus material. When highly concentrated

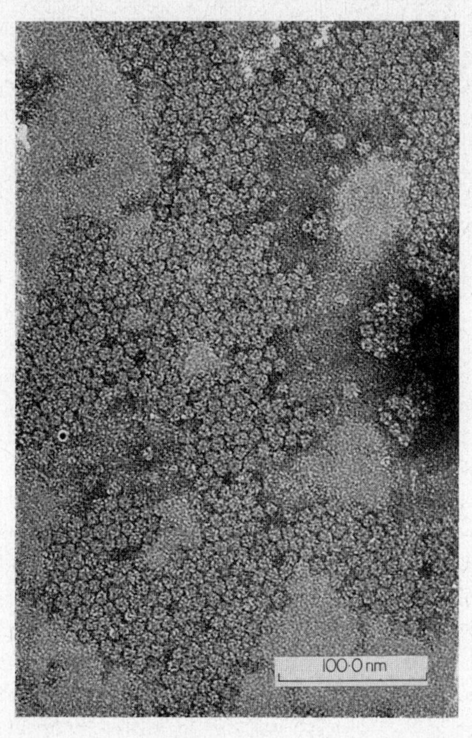

FIG. 30. Crystalline arrays of capsomeres from adenovirus following disruption after 36 hours at the surface of a solution of 3% ammonium molybdate used as the second negative stain in place of uranyl acetate. The hexamer capsomeres spontaneously reform as two-dimensional crystals on the film surface. (From Horne *et al.*, 1975.)

suspensions of adenovirus were prepared by the negative staining-carbon film method, and separated from the mica by the use of 3% ammonium molybdate as the second negative stain instead of uranyl acetate, it was noted that the virus crystalline arrays became disorganized at the surface of the second stain. Moreover, following a period of 12–48 hours in contact with the ammonium molybdate second stain, the individual virus particles slowly disrupted into their components. One striking observation was the recrystallization of the adenovirus hexon capsomeres on the same specimen films used in the original virus preparation (Fig. 30). From these results it may be possible to use the method for the purposes of slowly dissociating certain viruses directly on to the electron microscope specimen films.

Several additional studies were made to compare the preservation of the virus material and stain when air dried at room temperature with those carefully prepared under freeze-dried conditions. The results obtained from using the above mentioned viruses indicated that there was no advantage to be gained by removing water from the specimens with the aid of freeze-drying methods, as the optical diffraction patterns and measurements were found to be similar.

V. Resolution from Virus Material Prepared by the Negative Staining-Carbon Technique

Close examination of virus capsids and nucleocapsids at high magnification showed pronounced interference patterns resulting from the superimposition of the upper and lower structural protein components forming the virus particles (Fig. 31). These interference images are similar to those discussed by Klug and Finch (1965), Caspar (1966), but were interpreted by Horne and Pasquali-Ronchetti (1974) as being more complex. The complexity of the interference patterns could result from the extra detail being made visible in the specimen due to a considerable reduction in the total specimen thickness.

In addition to the interference patterns mentioned above, it was possible to resolve details of the negative stain distributed over the surface of the virus capsids. The electron dense material was clearly seen as small crystallites (Fig. 26d), and was present regardless of the negative stain used in the preparation. The existence of the small electron dense crystallites raises questions about the possible interaction between the negative stains and buffering material used in the virus samples and the structural correlation of the crystallites to viral proteins.

The negative staining-carbon technique also revealed the presence of low molecular weight (estimated to be about 20 000–40 000 m.w.) material forming a monolayer in the background of the viruses studied. In many areas these low molecular weight components were seen extending over or were associated with the virus particles (Fig. 28a). Measurements suggested that this material was probably derived from virus particles disrupted during the negative staining or purification process. Experiments on support films and negative stain alone failed to reveal any evidence of this small particular material.

The electron micrographs and their optical diffraction patterns (as in Fig. 28) clearly show spectra extending out to spacings below 2·0 nm which suggested that the repeating features in the specimens could be resolved down to this limit. In electron micrographs from any biological specimen there are bound to be non-repeating structural features present which will merge with the noise spectra in the optical diffraction patterns. It follows that the non-repeating information present may well be below 2·0 nm.

Fig. 31. Preparation of human adenovirus virus type 5. Note the pronounced interference effects resulting from superimposition of the upper and lower particle surfaces in the same image. The hollow elongated form of the capsomeres is clearly indicated at the periphery of the particles. (From Horne and Pasquali-Ronchetti, 1974. *J. Ultrastruct. Res.*)

There has been much discussion about the limitations in resolution from biological specimens studied in the electron microscope, with particular reference to radiation damage, specimen thickness, noise in the image and phase contrast effects (see Ruska, 1965; Kobayashi and O'Hara, 1966; Williams and Fisher, 1970; Glaeser, 1971; Riecke, 1971; Erickson and Klug, 1971; Horne, 1973 and others). The recent observations by Horne and Pasquali-Ronchetti (1974) have indicated that the resolution limits were also closely linked to problems associated with the specimen itself. Removal of the low molecular weight components from virus suspensions or other particles may well improve the present situation. Moreover, monolayers formed on the support films from impurities in the distilled water and solvents used in the various stages of specimen preparation were considered to be equally important. In the

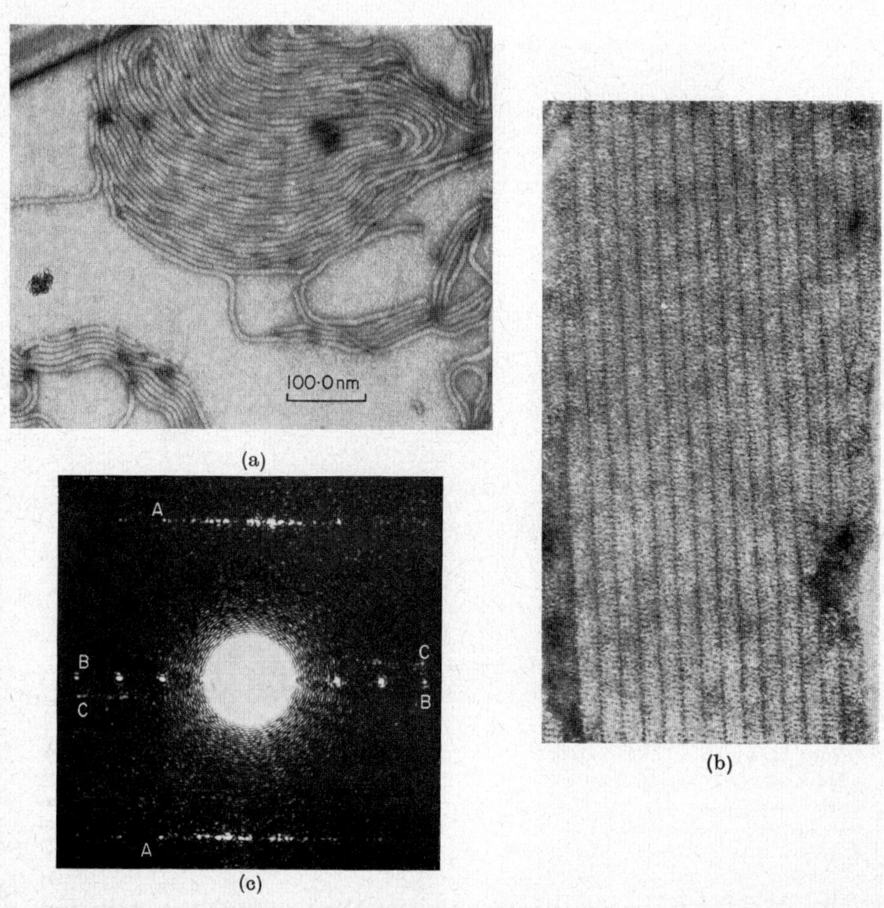

FIG. 32(a). Apple chlorotic leafspot virus filaments formed into approximately parallel arrays. The external shape of the aggregates could be varied by changing the pH of the first negative stain. Some of the aggregation patterns observed were essentially similar to those found in infected tissues seen in thin sections. (b). Part of a large area showing filaments of potato virus X in roughly parallel array. (c). Optical transform from (b). The layer-line at (A) is derived from the filament helix with a basic pitch of about 3·65 nm, and the interparticle distance of 12·6 nm is shown at (B). The unsymmetrical layer-line (C) results from a repeat of the basic helix at every 8 turns. (From Horne and Pasquali-Ronchetti, 1974. *J. Ultrastruct. Res.*)

case of negative stains there is the additional risk of the formation of small electron dense crystallites which, when added to the above problems, may seriously limit information obtainable from the specimen in the region of 1·0 to 2·0 nm.

FIG. 33. Relatively low magnification micrograph showing a small area from a preparation of a highly concentrated (25 mg/ml) suspension of tobacco mosaic virus after mixing with an equal volume of a 3% ammonium molybdate as the first negative stain and 0·5% uranyl acetate as the second stain.

From the current experimental work being undertaken on the application of negative staining techniques coupled with modifications to the method for special purposes of optical diffraction analysis, it should be possible to extend the limit of resolution from the specimen to below 1·2–1·5 nm, where a considerable amount of structural detail would be of value in relating electron microscope measurements to X-ray diffraction data and analysis from other techniques.

In conclusion it is worth recording that although there is a good deal of literature available concerning the theoretical aspects of electron scattering from amorphous objects such as electron microscope specimens, only a relatively small number of publications deal in practical terms with the limitation of information imposed by the specimen itself. It is in this experimental area that there is still considerable scope for improvement, independent of the basic instrumental problems.

References

Almeida, J. D. and Howatson, A. F. (1963). *J. Cell. Biol.* **16**, 616.

Almeida, J. D., Waterson, A. P. and Fletcher, E. W. L. (1965). *Nature, Lond.* **206**, 1125.

Bancroft, J. B., Hills, G. J. and Markham, R. (1967). *Virology* **31**, 354.

Bangham, A. D. and Horne, R. W. (1964). *J. molec. Biol.* **8**, 660.

Bessis, M. and Breton-Gorius, J. (1960). *C. r. hebd. Séanc. Acad. Sci., Paris* **250**, 1360.

Bradley, D. E. (1962). *J. Gen. Microbiol.* **29**, 503.

Bradley, D. E. and Kay, D. (1960). *J. gen. Microbiol.* **23**, 553.

Bragg, W. L. (1944). *Nature, Lond.* **154**, 69.

Brenner, S. and Horne, R. W. (1959). *Biochim. biophys. Acta* **34**, 103.

Brenner, S., Streisinger, G., Horne, R. W., Champe, S. P., Barnett, L., Benzer, S. and Rees, M. W. (1959). *J. molec. Biol.* **1**, 281.

Buerger, M. J. (1941). *Proc. natn. Acad. Sci. U.S.A.* **27**, 117.

Caspar, D. L. D. (1966). *J. molec. Biol.* **15**, 365.

Caspar, D. L. D. and Klug, A. (1962). *Cold Spring Harb. Symp. quant. Biol.* **27**, 1.

Caspar, D. L. D. and Klug, A. (1963). *In* "Viruses, Nucleic Acids and Cancer". p. 27. Williams & Wilkins, Baltimore.

Caspar, D. L. D., Dulbecco, R., Klug, A., Lwoff, A., Stoker, M. P. G., Tournier, P. and Wildy, P. (1962). *Cold Spring Harb. Symp. quant. Biol.* **27**, 49.

Cosslett, V. E. (1971). *Phil. Trans. R. Soc.* **B. 261**, 35.

Cox, R. W., Grant, R. A. and Horne, R. W. (1967). *Jl. R. microsc. Soc.* **87**, 123.

Crowther, R. A. (1971). *Phil. Trans. R. Soc.* **B. 261**, 221.

Dalton, A. J., Haguenau, F. and Maloney, J. B. (1962). *J. natn. Cancer Inst.* **29**, 1177.

De Rosier, D. J. and Klug, A. (1968). *Nature, Lond.* **217**, 130.

De Rosier, D. J. and Moore, P. B. (1970). *J. molec. Biol.* **52**, 355.

Erickson, H. P. and Klug, A. (1971). *Phil. Trans. R. Soc.* **B. 261**, 105.

Ferrier, R. P. and Murray, R. T. (1966). *Jl. R. microsc. Soc.* **85**, 323.

Finch, J. T. and Holmes, K. C. (1967). Structural Studies of Viruses. *In* "Methods in Virology" (K. Maramorosch and H. Koprowski, eds.), Vol. 3, p. 351. Academic Press, New York.

Galton, F. (1878). *Nature, Lond.* **18**, 97.

Glaeser, R. M. (1971). *J. Ultrastruct. Res.* **36**, 466.

Glauert, A. M. and Lucy, J. A. (1969). *J. Microscopy* **89**, 1.

Gordon, C. N. (1972). *J. Ultrastruct. Res.* **39**, 173.

Gotte, L., Giro, M. G., Volpin, D. and Horne, R. W. (1974). *J. Ultrastruct. Res.* **46**, 23.

Gregory, D. W. and Pirie, B. J. S. (1972). *Proc. 5th European Conf. Electron Microscopy.* Inst. Physics. Lond. p. 234.
Gregory, D. W. and Pirie, B. J. S. (1973a). *J. Microscopy* **99**, 251.
Gregory, D. W. and Pirie, B. J. S. (1973b). *J. Microscopy* **99**, 267.
Hahn, M. H. (1972). *Optik* **35**, 326.
Hall, C. E. (1955). *J. Biophys. biochem. Cytol.* **1**, 1.
Hall, C. E. (1966). "Introduction to Electron Microscopy". 2nd Ed. p. 321. McGraw-Hill Pub. Co., London.
Hills, G. J., Gurney-Smith, M. and Roberts, K. (1973). *J. Ultrastruct. Res.* **43**, 179.
Hitchborn, J. H. and Hills, G. J. (1965). *Virology* **27**, 528.
Hoppe, W. (1971). *Phil. Trans. R. Soc.* **B. 261**, 71.
Horne, R. W. (1965). Negative staining methods. *In* "Techniques for Electron Microscopy". (D. Kay, ed.) p. 328. Blackwells, Oxford.
Horne, R. W. (1967). Electron microscopy of isolated virus particles and their components. *In* "Methods in Virology". (K. Maramorosch and H. Koprowski, eds.) Vol. 3, p. 521. Academic Press, New York.
Horne, R. W. (1973). *J. Microscopy* **98**, 286.
Horne, R. W. (1974). "Structure of Viruses." Academic Press, New York.
Horne, R. W. and Markham, R. (1972). Application of optical diffraction and image reconstruction techniques to electron micrographs. *In* "Practical Methods in Electron Microscopy". (A. M. Glauert, ed.) Vol. 1, p. 327. Elsevier-North Holland, Amsterdam.
Horne, R. W. and Nagington, J. (1959). *J. molec. Biol.* **1**, 333.
Horne, R. W. and Pasquali-Ronchetti, I. (1974). *J. Ultrastruct. Res.* **47**, 361.
Horne, R. W. and Whittaker, V. P. (1962). *Z. Zellforsch. mikrosk. Anat.* **58,** 1.
Horne, R. W. and Wildy, P. (1961). *Virology* **15**, 348.
Horne, R. W. and Wildy, P. (1963). *Adv. Virology* **10**, 102.
Horne, R. W., Hobart, J. M. and Pasquali-Ronchetti, I. (1975). *J. Ultrastruct. Res.* (In press).
Huxley, H. E. (1956). *Proc. First European Conf. Electron Microscopy.* Stockholm. p. 260. Almqvist and Wiksell, Stockholm.
Huxley, H. E. and Zubay, G. (1960). *J. molec. Biol.* **2**, 10.
Kellenberger, E. (1973). "The Generation of Subcellular Structures." First John Innes Symp. (R. Markham, J. B. Bancroft, D. R. Davies, D. A. Hopwood and R. W. Horne, eds.) p. 59. North-Holland, Amsterdam.
Klug, A. (1967). *In* "Formation and Fate of Cell Organelles". (K. B. Warren, ed.). Academic Press, New York.
Klug, A. (1971). *Phil. Trans. R. Soc.* **B. 261**, 173.
Klug, A. and Berger, J. E. (1964). *J. molec. Biol.* **10**, 565.
Klug, A. and De Rosier, D. J. (1966). *Nature, Lond.* **212**, 29.
Klug, A. and Finch, J. T. (1965). *J. molec. Biol.* **11**, 403.
Klug, A., Finch, J. T., Leberman, R. and Longley, W. (1966). CIBA Foundation Symposium on Principles of Biomolecular Organisation, p. 158.
Kobayashi, K. and O'Hara, M. (1966). *Proc. 6th Int. Cong. Electron Microscopy.* Kyoto. **1**, 579.
Kushner, D. J. (1969). *Bact. Rev.* **33**, 302.
Leberman, R. (1965). *J. molec. Biol.* **13**, 606.
Lwoff, A., Horne, R. W. and Tournier, P. (1962). *Cold Spring Harb. Symp. quant. Biol.* **27**, 51.
Markham, R. (1968). The optical diffractometer. *In* "Methods in Virology". (K. Maramorosch and H. Koprowski, eds.). Vol. 4. Academic Press, New York.

Markham, R., Frey, S. and Hills, G. J. (1963). *Virology* **20**, 88.
Markham, R., Hitchborn, J. H., Hills, G. J. and Frey, S. (1964). *Virology* **22**, 342.
Mellema, J. E., Bruggen, E. F. J. Van and Gruber, M. (1967). *Biochim. biophys. Acta* **140**, 180.
Munn, E. A. (1968). *J. Ultrastruct. Res.* **25**, 362.
Muscatello, U. and Guarriero-Bobyleva, V. (1970). *J. Ultrastruct. Res.* **31**, 337.
Muscatello, U. and Horne, R. W. (1968). *J. Ultrastruct. Res.* **25**, 73.
Nagington, J., Newton, A. A. and Horne, R. W. (1964). *Virology* **23**, 461.
Nanninga, N. (1968). *Proc. natn. Acad. Sci. U.S.A.* **61**, 614.
Nermut, M. V. (1972). *J. Microscopy* **96**, 351.
Nermut, M. V. (1973). Freeze-drying and freeze-etching of viruses. *In* "Freeze-etching Techniques and Applications". (E. L. Benedetti and P. Favard, eds.), p. 135. Soc. Française de Microscopie Electronique, Paris.
Nermut, M. V. and Frank, H. (1971). *J. gen. Virol.* **10**, 37.
Nermut, M. V., Frank, H. and Schäfer, W. (1972). *Virology* **49**, 345.
Nowinski, R. C., Old, L. J., Sarkar, N. H. and Moore, D. H. (1970). *Virology* **42**, 1152.
Parsons, D. F. (1963). *J. Cell Biol.* **16**, 620.
Reissig, M. and Orrell, S. A. (1970). *J. Ultrastruct. Res.* **32**, 107.
Riecke, W. D. (1971). *Phil. Trans. R. Soc.* **B. 261**, 15.
Rothfield, L. and Horne, R. W. (1967). *J. Bact.* **93**, 1705.
Rothfield, L., Takeshita, M., Pearlman, M. and Horne, R. W. (1966). *Fedn Proc. Fedn Am. Socs exp. Biol.* **25**, 1495.
Ruska, R. (1965). *Jl. R. microsc. Soc.* **84**, 77.
Sharpe, D. G. (1965). *In* "Quantitative Electron Microscopy" (G. F. Bahr and E. H. Zeitler, eds.) p. 93. Williams and Wilkins, Baltimore.
Taylor, C. A. and Lipson, H. (1951). *Nature, Lond.* **167**, 809.
Taylor, C. A. and Lipson, H. (1964). "Optical Transforms". Bell, London.
Unwin, P. N. T. (1972). *Proc. Fifth European Regional Cong. Electron Microscopy.* p. 232. Inst. of Physics, London.
Valentine, R. C. and Horne, R. W. (1962). *In* "The Interpretation of Ultra-structure". (R. J. C. Harris, ed.) p. 263. Academic Press, New York.
Van Bruggen, E. F. J., Wiebenga, E. H. and Gruber, M. (1960). *Proc. 2nd European Regional Cong. Electron Microscopy, Delft* **2**, 712.
Warren, R. C. and Hicks, R. M. (1971). *J. Ultrastruct. Res.* **36**, 861.
Webb, M. J. W. (1973). *J. Microscopy* **98**, 109.
Wildy, P. (1962). Microbial classification. *Symp. Soc. Gen. Biol.* **12**, 145.
Wilkins, M. F. H., Stokes, A. R., Seeds, W. E. and Oster, G. (1950). *Nature, Lond.* **166**, 127.
Williams, R. C. (1953). *Expl. Cell Res.* **4**, 188.
Williams, R. C. and Fisher, H. W. (1970). *J. molec. Biol.* **52**, 121.
Wood, W. B., Dickson, R. C., Bishop, R. J. and Revel, H. R. (1973). "The Generation of Subcellular Structures". First John Innes Symp. (R. Markham, J. B. Bancroft, D. R. Davies, D. A. Hopwood and R. W. Horne, eds.) p. 25. North-Holland, Amsterdam.
Wrigley, N. G. (1968). *J. Ultrastruct. Res.* **24**, 454.
Wyckoff, H. W., Bear, R. S., Morgan, R. S. and Carlstrom, D. (1957). *J. opt. Soc. Am.* **47**, 1061.

Scanning Electron Probe Microanalysis

KURT F. J. HEINRICH

Institute for Materials Research, National Bureau of Standards
Washington D.C. 20234, U.S.A.

I. Introduction

Electron probe microanalysis is based upon the interaction of an accelerated and focused electron beam with a microscopic region at the surface of a solid specimen. Among the signals generated in this process, the most important to the analyst is the emission of characteristic X-rays. The wavelengths of the X-ray lines identify the emitting elements (qualitative analysis), and their intensities can, with appropriate manipulation, be correlated with the concentrations of the elements (quantitative analysis). The technique thus has the characteristics of elemental chemical analysis (Castaing, 1951; Birks, 1971), and can be described in terms of the analytical parameters: accuracy, sensitivity, and specificity. At concentrations of elements above the trace level, relative errors in quantitation are typically 1–3%; limits of detection are, for elements of atomic number above ten, in the order of 0·01% or lower. Interferences seldom occur and can always be overcome. Thus, electron probe microanalysis compares reasonably well with other analytical techniques.

Specific to electron probe microanalysis is its capability to perform local analysis of a small portion ($\sim 10^{-11}$ g) of a solid specimen, without physical or chemical separation of this portion from its surroundings. The spatial selectivity of excitation of electron probe microanalysis distinguishes it from most classic microanalytical procedures. It establishes an affinity with those microscopic techniques which reveal distribution of composition, such as histological specific staining, etching, and decorating techniques. A closer relation exists with other

FIG. 1. Basalt. Scanned area: 80×100 μm, 20 kV. Specimen current: 3×10^{-8}A. The minerals present are plagioclase (pl), olivine (ol), and pyroxene (py). *Upper left*: target current image; *upper right*: AlKα (100 000 photons); *lower left*: MgKα (500 000 photons); *lower right*: CaKα (100 000 photons). The specimen current image provides topographic information which correlates with the X-ray scans.

probe techniques such as laser probe emission spectrography (Rasberry *et al.*, 1967) and ion probe mass spectrometry (Castaing *et al.*, 1960; Long, 1965; Andersen and Hinthorne, 1972). None of these techniques, however, are as accurate in a quantitative sense as electron probe microanalysis.

The spatially localized excitation of X-rays determines the character and the applications of electron probe analysis, which is appropriate when each local distribution of elements on a microscopic scale must be investigated. The use of a focused electron beam establishes a particularly close analogy with scanning electron microscopy (Oatley *et al.*, 1965; Oatley, 1972). This link was affirmed in practice when Cosslett and Duncumb incorporated beam scanning in the electron probe microanalyzer (Cosslett and Duncumb, 1956). Nowadays all instruments have provisions for scanning, and images obtained with backscattered, secondary, transmitted or absorbed electrons are an indispensable adjunct to electron probe X-ray analysis (Heinrich, 1967) (Fig. 1).

Various alternative procedures are available to the electron probe analyst. Depending on the problem to be investigated, he may either perform quantitative measurements with a stationary beam at a few selected points, scan along a line at the surface, or he may qualitatively investigate an entire microscopic area in one scanning operation. In either case, several elements may be observed or determined simultaneously by means of individual X-ray spectrometers.

A complete characterization of a scanned area would consist of the qualitative and quantitative analysis of every point. However, statistical requirements for precise measurement of X-ray intensities and the slow speeds of available detectors limit the information that can be gathered in a given time period. A fully quantitative investigation of every point of a typical area of scan (in the order of 10^{-7} m^2) could require a full year's work. Even if such an operation were economically feasible, it would produce much more information than required in any conceivable practical case. Furthermore, the usefulness of such a complete procedure would be limited by our inability to personally observe, and transmit to others, the enormous bulk of information obtained. Even in scanning experiments of a more moderate scope, it becomes necessary to select and synthesize a representation of relevant results from the raw data. These can assume the form of statistics, or of images showing the topographic distribution of composition ranges (isoconcentration contours).

If the specimen composition is completely unknown, the quantitative measurements must be preceded by a qualitative characterization. The complete inventory of elements present at all points of the scanned area is by no means a trivial undertaking, but it will not be discussed here in detail. We will merely remark that the combining of energy- and wavelength dispersive spectrometers, and the use of computer programs for spectrum analysis, hold promise of much progress in these areas.

The duration of a scanning procedure must thus be drastically reduced by compromises in the achievable accuracy, or resolution, or both. In most cases, the specimen is not a complete unknown to the analyst, and the desired information can be obtained with a relatively small number of measurements or scanning operations. These practical laboratory operations can be considered simplifications or abbreviations of the complete theoretical procedure. The potentialities of this complete procedure are, in turn, determined by the achievable limits of spatial resolution, analytical accuracy, and size of the scanned area. We will now discuss these limiting parameters in more detail.

II. Spatial Resolution and Width of Scan

The scanning pattern in which the electron beam moves on the specimen surface is usually a set of continuous parallel lines. The beam of a cathode-ray tube (CRT) sweeps in synchronism on the CRT screen, while the amplified X-ray signal is used to modulate the brightness of the CRT beam. The image formed on the screen is recorded by a camera. In the procedure commonly followed for X-ray images, the detector of a curved-crystal spectrometer produces signal pulses which, after amplification, appear as dots of light on an otherwise dark CRT screen. We will refer to this mode of operation as the standard procedure.

The magnification of the scanning image is equal to the ratio of the excursion of the CRT to that of the primary electron beam. The range of magnifications which are useful in X-ray area scanning is determined by the resolution of the CRT, the size of the electron probe, and the maximum permissible excursion.

Although display systems of high resolution can be used with the electron probe microanalyzer, in most instruments the CRT cannot resolve more than 400–500 lines. The width of the screen is usually 8 cm; hence, each line will occupy a width of 0·02 cm. A distinguishable picture element could not be smaller than this distance; if the size of the smallest picture element becomes as large as 0·1 cm, we would call the picture blurred.

The signals obtained from backscattered, transmitted, or secondary electrons, or from specimen current, have a spatial resolution close to the width of the electron beam, which in the electron probe microanalyzer is typically between 0·2 μm and 0·5 μm. X-rays emerge from a larger region, the dimensions of which depend mainly on the initial energy of the electrons which form the beam. According to Castaing (1960), the effective range of primary X-ray generation can be obtained by the equation

$$z_R = 3.3 \times 10^{-6} \frac{A}{Z} (V_o^{1.7} - V_q^{1.7}) \text{ g/cm}^2. \tag{1}$$

In this equation, A and Z are the atomic weight and number of the target, and V_o and V_q denote the operating voltage and the critical excitation voltage for the emitting shell of the target atom (both in kilovolts). The width of the excited volume is of the same order of magnitude as z_R; for typical materials and operating conditions, it is 1–3 μm. If the operating voltage were dropped from 20 kV to 10 kV, the cross-section of the excited region would drop to about one third. Unfortunately, the intensities of emitted X-ray lines also diminish very fast. This limitation frequently precludes working at low voltages.

With a lateral signal dispersion of 2 μm, a picture element smaller than 0·1 cm, and a CRT of 8 cm width, the area scanned on the specimen must be more than 160 μm wide if the picture is to be clear. Therefore, a reasonably sharp image cannot be obtained unless the magnification is less than 500. Secondary—or fluorescent—characteristic radiation is generated at even larger distances from the point of impact of the electron beam, and where such radiation is present at significant levels, it will further reduce the sharpness of the X-ray image.

There are also limits to the maximum size of the scanned area. If crystal spectrometers are used, the width of the scan is limited by the losses of signal intensity which occur at large beam excursions due to defocusing of the spectrometer (Fig. 2). The characteristics of the spectrometers and the required accuracy of the intensity measurement determine how large a beam excursion can be tolerated. As indicated by Malissa (1966), the intensity losses are severest with crystals having good focusing properties. Even with crystals of poor resolution—such as most commercially available lithium fluoride analyzers—the loss becomes substantial for scanning excursions larger than 200 μm. Hence, for such crystals, the useful magnification would be limited to a value equal to, or above, 400.

The two limitations we have discussed above leave the analyst with a surprisingly narrow range of useful magnifications. In fact, the criteria of very accurate measurement intensity and sharpness of image contours are mutually exclusive as long as fully focusing crystal spectrometers are used. Some analysts are seemingly unaware of such limitations. They do not try to obtain sharp X-ray scanning images; frequently they do not collect a sufficiently large number of pulses to obtain a definition of contours even within the above mentioned limits, and usually they correlate the distribution shown on the X-ray scan with the topography of the specimen by means of scanning micrographs based on electron signals. An unsharp micrograph may satisfactorily convey the desired analytical information; nevertheless, the analyst should find it rewarding to make the most efficient use of the tools at his service.

The limitations as to the size of the scanned area are particularly frustrating since they do not apply equally to the scanning electron micrographs frequently observed in the initial survey of the specimen. They severely limit the quantitative evaluation of the X-ray scanning image. It is difficult to devise reliable procedures of correction for the intensity variations due to spectrometer defocusing, because the effect can vary somewhat with the conditions of spectrometer alignment.

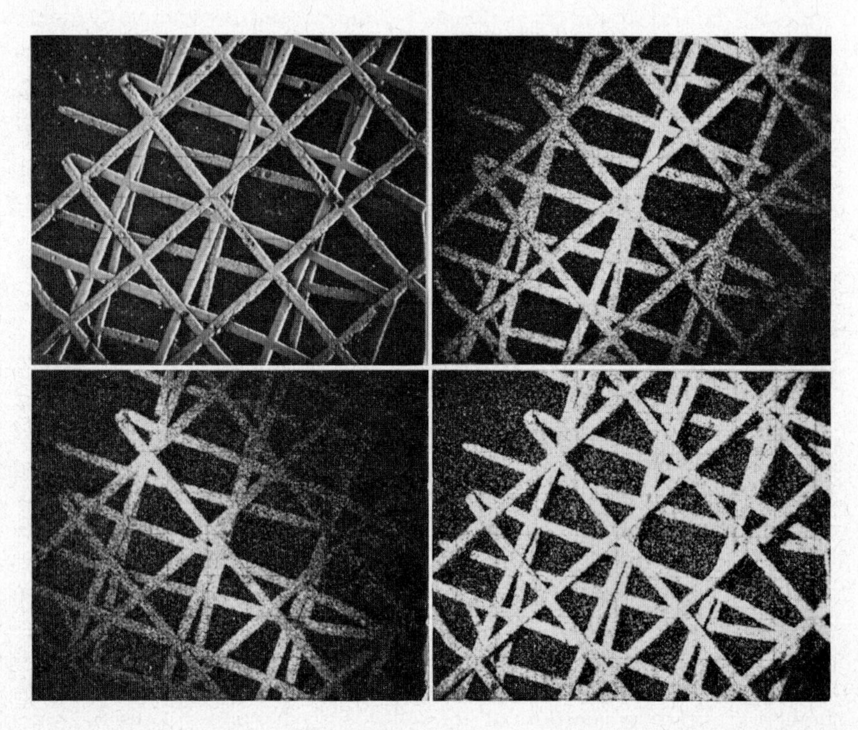

FIG. 2. Gold grid, folded, on a brass disc. Area: 990×770 μm. *Upper left*: inverted target current images; *upper right*: AuMα, observed with an ammonium dihydrogen phosphate crystal of 10 cm radius; *lower left*: AuLα, lithium fluoride crystal of the same radius; *lower right*: AuMα on silicon detector.

It is possible to avoid the spectrometer defocusing by the use of semi-focusing spectrometers (Duncumb, 1957). Such devices, however, have a poor line-to-background ratio for some regions of the X-ray spectrum which limits their use for the detection and analysis of some elements when present at low concentrations.

A more attractive solution is the combination of fully focusing spectrometers and of an energy-dispersive solid-state detector system

(Russ, 1970) which is insensitive to the beam excursions used in scanning procedures (Fig. 3). Such a system is most valuable for rapid qualitative characterization of specimen areas and can also be employed for quantitative purposes (Beaman and Solosky, 1972). It is now widely used in both electron probe microanalyzers as a complement of crystal spectrometers and in scanning electron microscopes as an exclusive means of X-ray detection (Kimoto, 1972). As in the semifocusing system, the high background levels limit the usefulness of the energy-dispersive detector for the formation of scanning images. Recent improvement in the energy resolution of available detectors has considerably enhanced the obtainable line-to-background ratios, and thus the use of solid-state detectors for scanning is more attractive, particularly if an alternative crystal spectrometer can be used for trace elements. Interferences of X-ray lines are also possible with such a system and must be recognized and avoided.

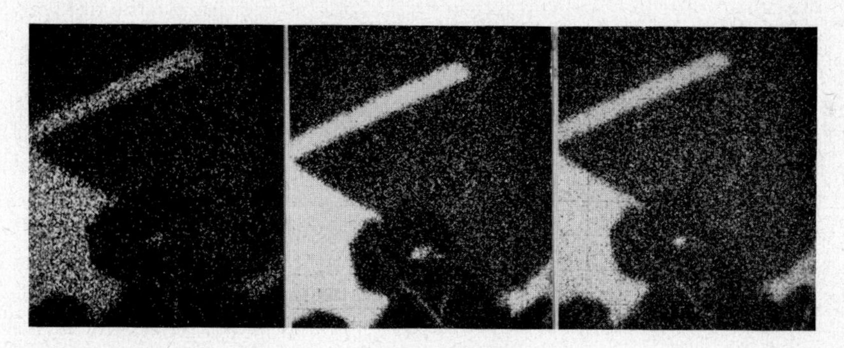

FIG. 3. AlKα X-ray scans over the regions of the mineral shown in Fig. 1 (70×80 μm). From left to right: 1. ammonium dihydrogen phosphate crystal spectrometer (10 cm radius), 13 000 photons; 2. Same spectrometer, 100 000 photons; and 3. Lithium-drifted silicon detector, 100 000 photons. The figure to the left lacks in contour definition, due to the insufficient number of accumulated photons. The figure to the right shows the effects of the high background of the silicon detector in the olivine (bottom), which contains no aluminium.

It should be noted that a displacement of the specimen normal to its surface also produces defocusing in crystal spectrometers which are mounted with the focusing circle perpendicular to the specimen surface. This effect can become a significant source of error in instruments in which the specimen elevation is not monitored by means of an optical microscope.

When electron beam scanning causes large beam deflections with respect to the optical axis of the objective lens of the electron probe microanalyzer, the resolution of the beam is adversely affected by the field aberrations (Duncumb, 1957). Usually, the resulting slight deterioration of resolution at the edge of scanning images can be tolerated.

The defocusing of crystal spectrometers can be avoided completely if the raster is obtained by moving the specimen instead of the electron beam. This technique was first published by Rouberol *et al.* (1962). Scans of large areas can be performed, and well-focusing spectrometers can be used so that low concentration levels are observable. However, mechanical scans proceed at rather slow speeds, and usually consist of a single-raster sweep. For this reason it is desirable that provisions exist for rapid beam scans with electron signal display, even if mechanical specimen movement for X-ray scans is available.

The spatial resolution of X-ray microanalysis, as well as the intensity of the signal relative to the background, depend on the choice of X-ray lines and operating voltages. In general, lines of lower photon energy (long wavelengths) have higher line-to-background ratios and are thus appropriate for low concentration levels. Such lines can also be obtained at low operating voltages with the corresponding gain in spatial resolution. This advantage of long wavelengths was demonstrated by Andersen (1966) who compared the spatial resolution obtainable with the FeKα line with that obtained by means of the FeLα_1 line. One must remember, however, that the use of very long wavelengths may lead to large X-ray absorption losses which render the quantitative evaluation of the intensities more difficult (Yakowitz and Heinrich, 1968a).

The diffusion of electrons and the production of secondary X-rays can be drastically reduced by analyzing a specimen which is substantially thinner than the range in depth of the electron beam. This approach is very useful in the analysis of biological specimens (Hall, 1968). Due to the low density of soft tissue the range of electrons (on a linear scale) is large, and as the elements of interest are usually present at low concentrations, the X-ray intensities emitted at low voltages are usually insufficient. Yet, in the study of biological material spatial resolution is often important. Soft tissues can be cut into thin slices, and biologists routinely prepare and study such cross-sections. Electron images of satisfactory contrast can be obtained from tissue sections of thickness below 10^{-7} m (1000 Å) when scattered electrons are used as the signal (Kimoto, 1973). Since, in such thin specimens, an increase in electron energy does not cause an observable increase in lateral electron diffusion, it is advantageous to use high operating potentials to obtain a high X-ray signal output (Strojnik, 1973). In other types of specimens, particularly in minerals and ceramics, this technique apparently has not been used, because it is difficult to prepare unsupported thin specimen slices.

III. Statistical Limitations to the Measurement of X-ray Intensities

The precision of the measurement of X-ray intensities is limited by the statistics of unrelated events which are expressed by the Poisson distribution (counting statistics) (Liebhafsky *et al.*, 1955; Espejo, 1972). In the absence of errors other than those due to counting statistics, the standard deviation of a measurement is equal to the square root of the mean number of events (i.e. photons) counted, $\bar{N}$. An unbiased estimate of the Poissonian standard deviation can be obtained by substituting the square root of the number of counts in a single counting period, N:

$$\sigma_N = \sqrt{\bar{N}} \simeq \sqrt{N}. \tag{2}$$

It would be optimistic to assume that sources of imprecision other than counting statistics are never present. Poissonian statistics provide, however, an estimate for the smallest number of pulses required for a desired precision; this number is a useful limiting criterion in selecting the experimental strategy. If, for instance, a relative standard deviation larger than 0·1% is inadmissible, then, from Eqn. (2), it follows that N must be equal to or larger than 10^6. At low counting rates, stringent statistical requirements lead to excessive counting times; if the counting periods are made too large, imprecision due to the instability of the instrument dominates. When only a few quantitative measurements are needed to characterize a specimen, the problem is usually manageable. But counting fluctuations becomes a serious limitation when a large array of locations must be investigated, as is the case in the area-scanning mode of analysis.

The duration of a measurement can be shortened by increasing the count rate, usually by raising the operating current or the operating voltage. Loss of spatial resolution or damage and contamination of the specimen may set a limit to these recourses; otherwise the use of high count rates is ultimately limited by the effects of detector dead-time (counter paralysis). In energy-dispersive systems, overly high count rates also produce shifts in the position and deterioration of the resolution of X-ray lines.

The discussion of counting statistics of the area scan is simplified if we consider each scanning line of the frame as an array of discrete points or image elements. If these points are closely spaced, the difference between a continuous and a discontinuous scan is not discernible. Hence, the conclusions derived from the counting statistics of a scan in discrete steps are equally applicable to the usual scans in continuous lines.

It is plausible that the distance between discrete points along a scanning line should be equal to that between lines. Therefore, the equivalent of a square frame containing 200 lines is a matrix of 4×10^4 points. If we assume that the intensities observed at these points are not significantly affected by errors other than those due to counting statistics, we can determine how many X-ray photons must be collected to satisfy a predetermined criterion of precision. We will not adopt separate requirements for points at different concentration levels; rather, we will assume that we are satisfied if the pre-established precision criteria are met at the points of the highest signal intensity.

Let us assume that this intensity is 2×10^4 cts/sec, and that we wish to measure the count rate with a relative standard deviation not larger than 0·1%. Equation (2) requires the accumulation of 10^6 counts at each point, and the time to be spent on the entire frame would be 2×10^6 sec (more than 555 hours). Clearly, a compromise is needed. If we are satisfied with a relative standard deviation not larger than 1%, the number of photons for the total array is reduced to 4×10^8, and the time requirement to little more than $5\frac{1}{2}$ hours. Although this scanning time considerably exceeds that usually invested in a single scan, operations of such duration are feasible.

Strong statistical intensity variations adversely affect the apparent resolution of scanning images. Within a uniform area, the eye of the observer tends to disregard statistical fluctuations; this visual integration is, however, deficient or absent in regions of strong concentration changes, and particularly in the definition of phase limits and edges (see Fig. 3). In practice, this resolution limitation is frequently at least as significant as that established by the diffusion of the electrons in the specimen.

IV. Effects of Instability in Quantitative Area Scanning

In operations of long duration, the Poissonian counting error is probably not the only source of inaccuracy. The instability of the instrument imposes further limitations, and the effects of fluctuations or drift of the beam intensity must be taken into account.

The stability of the electron beam is dependent on the characteristics of the electron optics and the conditions of alignment and operation. In carefully controlled experiments, runs of 15 hours' duration were performed with a drift of less than 1% in the obtained X-ray intensity measurements (Yakowitz *et al.*, 1971). Such stability is not easily obtained. The main sources of instability are three: variations in the geometrical alignment due to warping of the filament or thermal expansion of instrument components, instability of power supplies, particularly for the condenser lens and for the operating potential, and the thinning of the filament due to evaporation and erosion of its heated tip region. Alignment variation after initial optimization always diminishes the beam intensity, while the thinning of the filament may cause its temperature, and hence its emission, to increase. The effects of instability of power supplies are unpredictable, although often cyclic.

In the conventional procedures of point analysis, the effects of instability are observed and corrected by measuring the intensity of X-rays from a known standard at regular intervals. If the drift between measurements is assumed to be linear, a simple drift correction can be applied; otherwise, the measurements of both specimen and standard within each period are averaged. Such a procedure is difficult or impossible in the context of area scanning. At best, a drift correction proportional to the time from the start of operation can be applied; there is no good reason, however, to assume that large drifts can be effectively compensated by a linear drift correction.

In certain types of characterization, such as homogeneity studies, it may be assumed that the mean composition of the specimen is known, and the mean intensity can therefore be used as a standard measure. However, if the scan is performed, as most are, in a regular pattern in time, it is impossible to separate a continuous drift from a real variation of concentration as a function of the slower scan direction. In such cases it is advisable to use a pseudo-random pattern in the sequence of points to be analyzed, rather than progressing row by row. Alternatively, the pattern can be scanned repeatedly, in a sequence of superposed frames, so that any drift is distributed over all points of the array.

V. Deadtime

After the detection of a photon, the X-ray detector and associated electronic components require a brief period to return to their initial state. Therefore, after each detection there is a short interval during which the detector system is inactive (Schiff, 1936). The subject has been discussed in detail by Beaman *et al.* (1969).

The fraction of time for which the detector is inactive increases with count rate. Hence, the efficiency of the system also decreases as the count rate is increased. If we call τ the deadtime, N the true count rate, which would be observed in the absence of deadtime, and N' the observed count rate, the following relation holds for the useful range of count rates (up to 3×10^4 cts/sec in typical proportional detector systems) (Ruark and Brammer, 1937; Heinrich *et al.*, 1965):

$$N = N'/(1-N'\tau) \quad \text{or} \quad \tau = 1/N'-1/N. \qquad (3)$$

The value of τ can be experimentally determined (Heinrich *et al.*, 1965); the uncertainty in τ produces an uncertainty in the value of N which increases with the count rate. Furthermore, at very high count rates the behaviour of the detector is not accurately described by Eqn. (3), or predictable in any other reliable way. For this reason there is a need, which can be filled with presently obtainable commercial equipment, for detector systems with deadtimes shorter than the usual values of 2–3 μsec.

VI. Background

In properly working detector systems, the background level, i.e. that part of the signal which is not due to the X-ray line of interest, is

Fig. 4. Same object as Fig. 2, depicted by means of continuous X-rays (1·4–5·4 Å). The gold wires emit more continuous radiation than the brass. The shaded areas are due to the blocking of emergent X-rays by the gold mesh.

mainly determined by the intensity of continuous radiation which varies with the atomic number of the emitting target (Ware and Reed, 1973). With crystal spectrometers, which have good energy resolution, the background effects are important only in the investigation of low concentration levels. When energy-dispersive systems are used, the background is higher and cannot be neglected even at higher levels of concentration (see Fig. 4).

In the conventional quantitative procedure, the background can be estimated by measuring the signal intensity at a wavelength close to that of the line. This measurement would be impractical in scanning procedures. If the photon counts at each point of the scan can be subjected to corrective manipulation, the background can be approximately calculated by means of Kramers' law (1923). However, in the more conventional image-forming procedures, such as the standard area scan, there is no provision for separating the background level from the net signal produced by the X-ray line of interest. For this reason, the choice of lines and spectrometers conducive to a high line-to-background ratio is more important for such scanning techniques than for quantitative measurements in which corrections can be applied. With such techniques and with spectrometers of poor resolution, variations in background due to topography or atomic number differences can be misinterpreted as evidence for the presence of elements which are not contained at all in the respective specimen areas (Fig. 4).

VII. Corrections for the Effects of Absorption, Fluorescence, and Atomic Number

The need in quantitative evaluation of X-ray emission measurements to correct for the effects of absorption, fluorescence, and primary emission as a function of atomic number is well documented in the literature (Heinrich, 1968; Heinrich, 1972; Henoc *et al.*, 1973). The mathematical procedures will not be discussed here. Considering their complexity, including iterative approximation, we must conclude that they cannot be fully applied to an area scan which is equivalent to the measurement of thousands of points. The empirical technique of correction developed by Ziebold and Ogilvie (1963) is much simpler, and could be adapted if the number of points to be investigated is low and the specimen has few components. In most scans, the application of such corrections is not possible; the operator should choose the excitation conditions in such a manner as to minimize the absorption losses which can cause particularly strong deviation from linearity of the analytical calibration (Yakowitz and Heinrich, 1968a). It should also be recognized that a quantitative evaluation of X-ray emission by the usual means is impractical when the specimen is thin or if its surface deviates considerably from flatness. In extreme cases, shadows may appear in areas in which the path from the specimen to the detector is blocked by other features of the specimen (see Fig. 4).

The importance of the errors discussed above, including those due to spectrometer defocusing and fluctuations, depends on the purpose of the scan and the particular technique used. We will now investigate the characteristics and limitations of some of these techniques in more detail, with references to the effects of the errors which we have previously treated in a general way.

VIII. The Standard X-ray Area Scan

In the standard X-ray area scan, the speed of the area scan, which consists of a line raster, may vary over a wide range, including high speeds at which an almost continuous visual image is formed. The pulses produced by the detector are amplified and stretched, to about 10^{-4} sec, and then applied to modulate the brightness of the CRT which in the absence of a signal is dark. The length of the pulses which form the image on the CRT must be adjusted so as to permit the formation of a dot of light which is not stretched into a short line at the highest intended scan speed. The image is formed in a camera which receives all the dots of light as they appear on the screen. It is practically an orthographic projection of the specimen surface; the signal intensity does not affect the location of a point of the specimen on the image plane.

The standard X-ray area scan has several advantages which account for its popularity. It only requires equipment incorporated in practically all instruments, and the signal transforms are held at a minimum so that the performance and interpretation of the scan are, apparently, simple. The only variables that must be defined for a specimen region are the element and line to be observed, the magnification, the number of lines per frame, the total number of photons to be collected, and the level of brightness on the photographic image produced by a single pulse. As mentioned, the scanning speed may be varied between wide limits; at high speeds, many frames can be superposed in one film exposure. The duration of the scan is defined by the mean count rate of the signal and the number of photons to be accumulated.

The possible use of high scanning speeds and superposition of several frames on the photographic image render the method extremely flexible when compared with other image formation schemes. The localization of an element in certain spots, or the rough distribution patterns of major components, can be observed in a few seconds so that the visual observation of the pattern can be used in the selection of areas to be studied. Conversely, if the statistical variations observed in the pattern due to counting statistics prove to be excessive, the scan can be repeated under essentially identical conditions, while the number of frames collected for the image formation is increased. The simplicity and flexibility of the technique account for its use in the overwhelming majority of applications.

The method has, however, two important and inherent limitations. The first limitation is the impossibility of manipulating in any way the experimental count rates, which is due to the fact that every single pulse acts independently upon the photographic emulsion. The second limitation arises from the difficulty or impossibility of a quantitative measurement or evaluation even of the uncorrected pulse intensities.

Since every pulse contributes individually in the image formation, we cannot correct, in the standard procedure, for the background contribution or for the effects of atomic number, fluorescence, absorption, deadtime, or line interference. Nor is it possible to subtract a pre-set intensity level from the signal in order to represent with increased contrast a limited range of concentration variations. All such signal manipulations require that an electrical equivalent of the instantaneous count rate be obtained, either by digital counting on each picture element, or, in an analog mode, by the use of a ratemeter.

The most serious shortcoming of the standard scan is the impossibility of discriminating against the background level. As mentioned before, it is the background which limits the observation and identification of line signals from trace components. Although a rigorous treatment of the background level by almost any variation of area scanning is almost impossible, one would frequently like to eliminate the background noise from a scanning image even if this would imply an increase in the limit of detection. Such a cosmetic procedure is possible only with more complicated techniques of signal treatment.

The background level would be less distracting if the signal intensities observed on the image could be evaluated on a quantitative scale. However, the representation of signal intensity in terms of grey levels provides a very narrow range of observable signal levels, is notoriously non-linear, and, unless unusual measures are taken, it is difficult to maintain stable conditions which would allow a quantitative interpretation.

The response of the grey level of the standard scan image to the signal intensity depends decisively on the combination of the brightness setting of the CRT, the aperture of the camera, and the film sensitivity. The simplest situation arises when these parameters are adjusted so that a signal pulse produces a tiny white spot on the film. In this case, the number of white dots in any region of the image is proportional to the uncorrected mean signal intensity in this region. As the visual perception of grey tones is closer to a logarithmic scale, the smallest concentration difference observable from the grey tone thus increases with increasing concentration. This is perhaps not an undesirable feature. However, the smallest size for a white dot is approximately the same as the size of an image element in a sharp picture. Therefore, at each point the image reveals a binary bit of information: the element is either "present" or "absent". If the background level is negligible, a sufficient number of detected photons will result in a map on which the region containing the element is white, against an essentially black background. Within the range of detectability, no gradations can be observed. If one intends to modify this condition by reducing the number of photons, the statistical distribution of the photons results in a very grainy image with poor edge definition (Fig. 5, right).

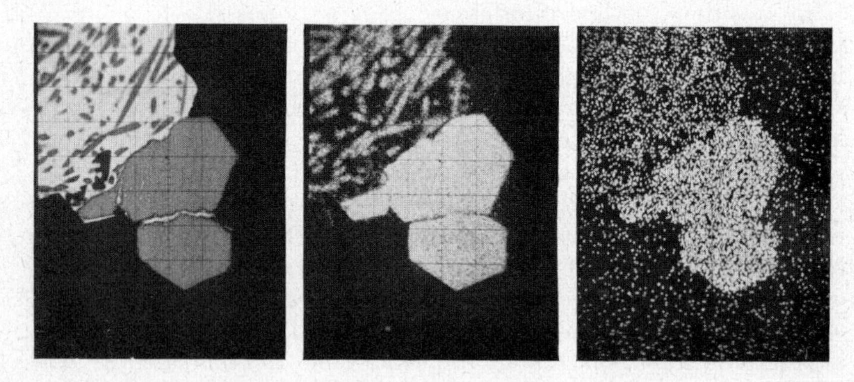

FIG. 5. Quartz inclusions in galena. Area of scan: 100×135 μm. *Left*: target current; *centre*: SiKα (250 000 counts); *right*: OKα (11500 counts). The upper left corner consists of diallylphthalate with glass fibre fillings.

When the brightness of the single-pulse image is reduced, one can obtain several grey levels by superposition of individual pulses at the same location. This results in better spatial definition of the patterns, and less graininess; it also requires a larger number of pulses per exposure to reach the same average grey level, and thus the effects of statistical fluctuations are diminished. This type of scanning image is much to be preferred, and to obtain it the operator should accumulate at least 5×10^4 pulses per image, unless the element of interest is concentrated in a small fraction of the scanned surface. Under such conditions, however, a low level of film exposure produces a very weak response or none at all. The latent exposure at the lower intensity level tends to bias out low signal levels. This effect is used, sometimes unintentionally, to mask the background signal; the proper manipulation of exposure intensity can, therefore, result in a remarkable cosmetic improvement. The same phenomenon can, however, wipe out low signal levels from low element concentrations just as readily.

Another variation of the standard technique is obtained when the pulses are reproduced on a CRT, the beam of which is slightly defocused. This procedure avoids the graininess which is esthetically objectionable, but such a cosmetic improvement is paid with a deterioration of spatial resolution (McMillan, 1967). The seriousness of this deterioration depends on the magnification and the electron energy, as well as on the resolution uncertainties introduced by counting statistics.

FIG. 6. Grey scales obtained from areas of uniform pulse density at three levels of CRT brightness. The scale of pulse densities is the same for each series; adjacent fields differ in pulse density (count rate) by a factor of two.

It should be noted that the response to count rate in terms of grey levels widely varies with the brightness setting of the CRT, which in practice is seldom calibrated or even reset to a standard level (Fig. 6). Further uncertainties in the interpretation of grey levels are possible due to the imperfections of the photographic process (Fig. 7). In summary, the standard method is deservedly popular due to its simplicity, and provides an excellent means of qualitatively surveying the distribution of elements. If more quantitative evaluations become necessary, the limitations of the recognition of intensity levels, of corrections for background and non-linearity of the analytical calibration curve, of defocusing of spectrometers, and of counting statistics, must all be overcome. Clearly, there is not a single answer to all these limitations, and the scanning technique with quantitative evaluation is a difficult and complicated task.

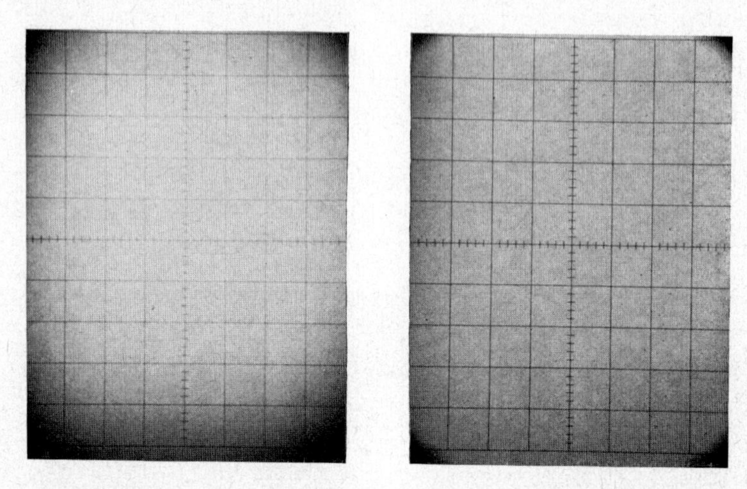

FIG. 7. Variation of grey levels within photographs of CRT screens of uniform brightness. A commercially available camera for CRT screens was used. *Left*: f = 1·9. *Right*: f = 16.

IX. COLOUR COMPOSITES OF X-RAY SCANS

As shown in Fig. 1, it is useful to correlate the information of the X-ray scan which is specific to one element to that of the target current image, which offers a more general compositional and topographic information (Heinrich, 1967). Where several elements are present, their distributions can also be correlated. This is best achieved by means of composites in which, as first suggested by Duncumb and Cosslett (1957), each element is shown in one primary colour. The areas of distribution of the elements usually overlap. Therefore, as long as mixed colours are to be interpreted in terms of the primaries, one is constrained to three primary colours: red, blue, and green. The binary colour mixtures which may appear are cyan (green+blue), magenta (red+blue), and yellow (green+red). The effect of the colours on the observer (attention value) varies with the colour, falling from red to blue in the primaries, and in the mixtures from yellow to cyan. The analyst can emphasize or de-emphasize certain features of the scanned area through the choice of colours.

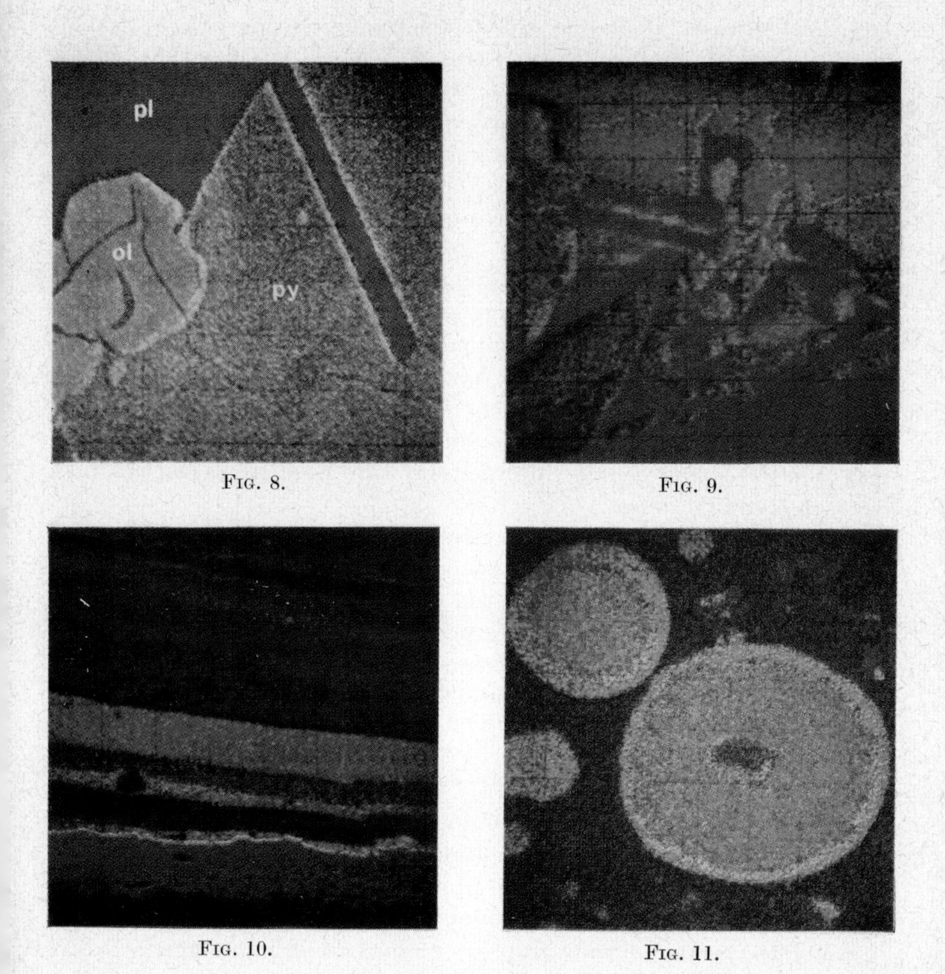

Fig. 8.

Fig. 9.

Fig. 10.

Fig. 11.

Fig. 8. Colour composite of the images shown in Fig. 1. The colours chosen for the X-ray scans are: Al: red, Ca: blue, Mg: green. Hence, we obtain for olivine: green (Mg), pyroxene: cyan (Ca+Mg), plagioclase: magenta (Al+Ca).

Fig. 9. Image of a more complex region of the same basalt (86×86 μm). Red: iron, green: calcium, blue: aluminium.

Fig. 10. Cross-section through a flake of painting from an old house. The oldest layers, at the bottom, are characterized by the presence of lead (red) which presents a health problem to children if the paint peels. Layers above the lead contain zinc (black), barium compounds (green), and titanium (blue). Other elements not shown here include sulphur (with barium) and calcium. Area: 200×200 μm.

Fig. 11. Phases in an alloy containing silver (red), mercury, copper (green), and tin (blue). The most important phases are yellow (Ag+Cu), cyan (Cu+Sn), and dark red (Ag). The distribution of mercury (outside the spherical structures) is not shown. Area: 70×70 μm.

Several devices, including colour television screens (Ficca, 1968), can be used to produce X-ray scans in colour. We have described elsewhere (Yakowitz and Heinrich, 1969) a simple and versatile technique based on copying, on fast development colour film, the black and white originals through colour filters.

As with the standard X-ray scan, the main limitation of the colour composites is the lack of sharpness of boundaries and of reference to topographic details. Both these limitations can be removed if, in addition to the three images in colours for the distribution of three elements, a target current or secondary electron scan is added in grey (Figs. 8–11). The relation of the intensities of X-ray scans and electron image can be varied within wide ranges, depending on the purpose of the representation. One may either wish to barely mark topographic details on an image which contains mainly X-ray information, or one may faintly colour X-ray information into what is essentially an electron scanning image.

Once the proper length of exposure for each colour has been established, an operator with little training can rapidly learn to prepare such colour composites; the electron probe microanalyzer itself is not involved in the process after preparation of the conventional black and white images. For details on exposures, films and filters, the reader may consult the reference (Yakowitz and Heinrich, 1969).

X. Area Scanning with Ratemeter Signals

A simple and well-known way to manipulate the X-ray signal is through a ratemeter. The conventional ratemeter produces a signal proportional to the rate of arrival of detector pulses. The *time constant* of a ratemeter system is an important parameter; it is equal to the time in which the system output drops, in absence of an input signal, to 1/e times its initial level (e is the base of the natural logarithms). If the time constant is too short, the rapid fluctuations in count rate due to statistics of the pulse arrival prominently appear in the output. The jitter is reduced by choosing a larger value for the time constant; in this case, however, the signal change after a sudden change in the pulse rate is undesirably slow.

The after effects of large changes in the X-ray signal can be shortened if the continuous ratemeter is replaced by a periodic integrator (Heinrich, 1963). In this case, the output is proportional to the number of pulses which have arrived in discrete periods; it is the analog representation of a continuously repeating scaling operation. After the output of one period is recorded, the effects of the X-ray signal produced during this period are deleted. Therefore, the signal cannot smear over more than one period.

The length of the integrating period determines the speed with which the output signal follows a change in the X-ray signal; if the periods are shortened, the statistical fluctuations increase. The integrating period therefore has a role analogous to that of the time constant in the continuous ratemeter.

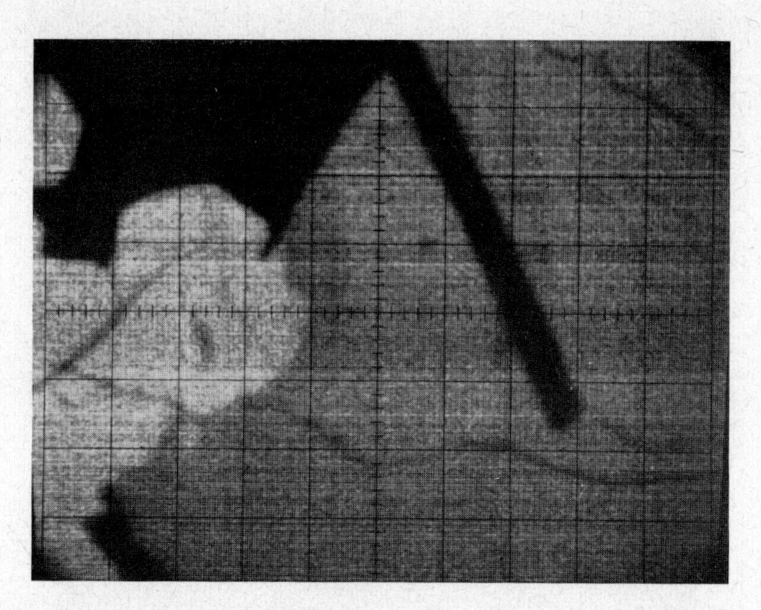

FIG. 12. The MgKα ratemeter signal was used to obtain this scanning image of the specimen area shown in Fig. 1. A bias was applied to increase the effect of differences in concentration. This bias, and the necessity to scan in a single frame for 90 minutes, causes some noise to appear. The spatial resolution of this image is far superior to the corresponding image in Fig. 1 (lower left).

If the ratemeter signal is applied to modulate the brightness of the CRT screen on which the X-ray image is formed, then the slow speed of the ratemeter may produce a loss of spatial resolution in the direction of the scanning lines. The adjustment of the integrating period or of the time constant is critical since it trades statistical intensity scatter for loss of spatial resolution. For this reason, it is necessary to scan slowly when ratemeter signals are used, and to form the image with a single frame scan rather than superimposing several frames as is usually done in the standard procedure (Fig. 12).

As the ratemeter produces a continuous signal, its use would appear particularly attractive for low signal levels at which the discrete nature of the X-ray emission becomes evident in the standard procedure. The apparent improvement is, however, mostly cosmetic unless a larger number of photons is collected. The real advantage of the ratemeter is the possibility of further manipulation of the signal.

XI. Linear Scanning

Since the conventional area scanning technique provides a clear though qualitative idea of the topographic distribution of an element, it is a good compromise to combine it with a single linear scan with ratemeter output. This line scan can be graphically superposed on either an X-ray scan or a target current image (Fig. 13). It can be

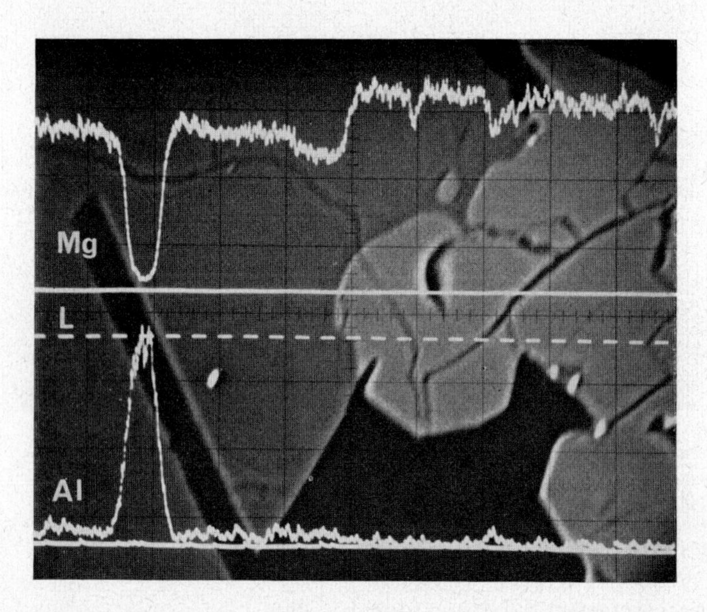

Fig. 13. Basalt. Scanned area: 80×100 μm, 20 kV, specimen current 3×10^{-8}A. Inverted target current image: superposed line scan at location L, and ratemeter signals for aluminium and magnesium with their respective zero levels.

calibrated by indicating the signal intensity corresponding to the pure element or to a standard of known concentration; the background, and its variation with the specimen composition, can also be shown. The scan can be performed in a continuous fashion, or consist of a series of point analyses across a line. The large reduction in the number of image points achieved by reducing data collection to a single line permits measuring the signal intensity along the scanned line with high precision. It is also possible to combine line scans of several elements on one image. In an image combined with a line scan, the background, or the zero level, as well as the location of the X-ray scan with respect to the two-dimensional area image should be indicated. The scale of the signal intensity can be expanded so that the zero level falls outside the image (differential amplification). Although the corrections developed for quantitative analysis are not applied to line scans at present, it is quite possible that they will be incorporated in the technique with the help of small computers. In such cases, the line scan can, in principle, be just as accurate as the analysis of a single location by a static beam. As an alternative, the line scan can be calibrated on a non-linear scale based on correction calculations. This procedure is strictly valid within binary systems only.

XII. Analogue Manipulation of the Ratemeter Signal

The most obvious improvement to the ratemeter signal applied to area scans is subtracting a constant bias in order to suppress the effects of background or to enhance minor variations in the concentration level (Heinrich, 1967). Such an "expanded contrast" technique was used by Melford (1962), and by Birks and Batt (1963).

A further development is the "concentration mapping" technique (Heinrich, 1967 and Heinrich, 1964). In this procedure, the brightness of the oscilloscope changes abruptly at preset signal levels, so that topographic contour maps are obtained. The signal variation outside the pre-selected levels is excluded from the presentation (Fig. 14).

For several reasons, the techniques involving analogue manipulation of the ratemeter signal are not frequently used. In the first place, the effects of counting statistics seriously limit the precise representation of the concentration levels. The long duration of the single-scan procedures demands a careful presetting of the signal levels and of the time constant. The results can only be observed after the scan is finished, and if the parameters were chosen inappropriately, or if drift or other errors have affected the results, the entire operation must be repeated.

The effects of defocusing also became obvious in such procedures; they demand rigorous restrictions as to the size of the scanned area. An objection, which was raised by Campbell and Brown (1964), is that analogue techniques do not lend themselves to the complex corrections required to obtain quantitative results of high accuracy. A fully quantitative treatment is possible only if the analogue procedures are replaced by digital manipulations; hence, a quantitative scanning technique must be developed by application of static-beam procedures to all image points, rather than by refinement of the conventional raster-scan approach.

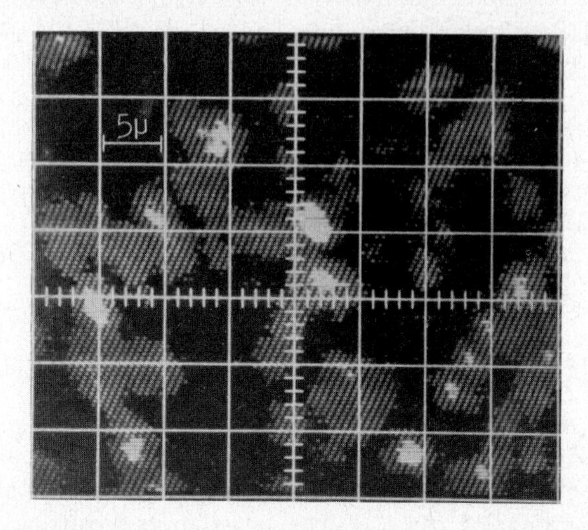

FIG. 14. A FeKα concentration map of Zircaloy-2. Area: 45×40 μm. The white area shows regions in which iron is above 1% (weight), and the concentration range of the dotted region is 0·1–1%.

XIII. THE MATRIX-SCANNING TECHNIQUE

In 1963 Birks and Batt (1963) described the use of a multichannel analyzer for data storage in two-dimensional quantitative X-ray area scanning. We have used a similar technique, in which the multichannel analyzer provides the driving voltage for a scan in discrete steps, forming a matrix of 40×40 points (Heinrich, 1967). We have also performed tests of homogeneity using a digital matrix generator capable of producing matrices of arbitrary size and number of points up to 100×100 (Heinrich *et al.*, 1971). The data obtained in these scans were treated by means of a computer program to yield both statistical and topographic information. The homogeneity of several NBS Standard Reference Materials, 481 and 482, for use in electron probe microanalysis and related techniques, was tested by means of this device and program. The same technique was also applied to the study of inclusions in alloys (Yakowitz and Heinrich, 1968b).

The testing for homogeneity of single-phase materials presents a particularly favourable case. The composition of the material is roughly the same over the entire field of scanning, and may be assumed to be known. The problem is reduced to studying variations of composition with respect to the known mean. This renders unnecessary the application of corrections for deadtime, background, atomic number, fluorescence, and absorption. In the course of these measurements, we observed the already discussed limitations concerning statistics, defocusing, and sometimes drift. It was, however, possible to perform scans lasting several hours without excessive difficulties. There is now no doubt that quantitative evaluation of area scans is a practical possibility.

XIV. Prospects for Quantitative Area Scanning

Most remaining difficulties in performing quantitative scans will probably be removed or alleviated in the near future. Several electron probe microanalyzers are already connected to computers which can read out and manipulate generated signals, and control diverse aspects of the analytical procedure such as stage movement, focusing of the spectrometers, and sequence of measurements (e.g. Kuntz *et al.*, 1971).

The cost and the physical volume of data storage devices for computer systems are decreasing rapidly; temporary storage on magnetic tapes and drums is becoming commonplace. There are also good prospects that the speed of X-ray detection systems can be increased nearer to the limits predicted by theory. Thus, the choice of the best technique of measurement and information presentation for a given analytical problem is becoming more important, although it has not as yet received the same attention as given to the theoretical aspects of quantitation. Further developments in the area of data handling are certain to come. Earlier analogue procedures will surely be replaced by digital techniques. For the final step, however, the best way to present the results of scanning electron probe microanalysis is usually the image. Far from eliminating the current usage of images produced by area scannings, more advanced techniques will render them more versatile and improve their quality.

References

Andersen, C. A. (1966). *In* "The Electron Microprobe" (T. D. McKinley, K. F. J. Heinrich and D. B. Wittry, eds.), p. 58. John Wiley & Sons, Inc. New York.
Andersen, C. A. and Hinthorne, J. R. (1972). *Science* **175**, 853.
Beaman, D. R. and Solosky, L. F. (1972). *Analyt. Chem.* **44**, 1598.
Beaman, D. R., Lewis, R. and Isasi, J. A. (1969). *In* "Proc. 5th Internatl. Congress on X-ray Optics and Microanalysis" (G. Möllenstedt and K. H. Gaukler, eds.), p. 84. Springer-Verlag, Berlin.

Birks, L. S. (1971). "Electron Probe Microanalysis", 2nd ed. Wiley-Interscience, New York.

Birks, L. S. and Batt, A. P. (1963). *Analyt. Chem.* **35**, 778.

Campbell, W. and Brown, J. (1964). *Analyt. Chem.* **36**, 323R.

Castaing, R. (1951). Doctoral Thesis, University of Paris.

Castaing, R. (1960). *In* "Advances in Electronics and Electron Physics" (L. Marton, ed.), Vol. 13, p. 353. Academic Press, New York.

Castaing, R., Jouffray, B. and Slodzian, G. (1960). *C. r. hebd. Séanc. Acad. Sci., Paris* **251**, 1010.

Cosslett, V. E. and Duncumb, P. (1956). *Nature* **177**, 1172.

Duncumb, P. (1957). Doctoral Thesis, University of Cambridge, (U.K.).

Duncumb, P. and Cosslett, V. E. (1957). *In* "X-ray Microscopy and Microradiography" (V. E. Cosslett, A. Engström, and H. H. Pattee, eds.), p. 347. Academic Press, New York.

Espejo, H. (1972). *Metallography* **5**, 449.

Ficca, J. (1968). *In* "Proc. Third Natl. Conf. on Electron Probe Analysis, EPASA", paper 15, Chicago, Ill.

Hall, T. A. (1968). *In* "Quantitative Electron Probe Microanalysis" (K. F. J. Heinrich, ed.), NBS Spec. Publ. 298, p. 269. Natl. Bur. Stds., Washington, D.C. 20234.

Heinrich, K. F. J. (1963). *In* "Adv. X-ray Anal." Vol. 7, p. 382. Plenum Press, New York.

Heinrich, K. F. J. (1967). "Scanning Electron Probe Microanalysis", NBS Tech. Note 278. Natl. Bur. Stds., Washington, D.C. 20234.

Heinrich, K. F. J. (1968). "Quantitative Electron Probe Microanalysis", NBS Spec. Publ. 298. Natl. Bur. Stds., Washington, D.C. 20234.

Heinrich, K. F. J. (1972). *Analyt. Chem.* **44**, 350.

Heinrich, K. F. J., Vieth, D. L. and Yakowitz, H. (1965). *In* "Adv. in X-ray Analysis", Vol. 9, pp. 208–20. Plenum Press, New York.

Heinrich, K. F. J., Myklebust, R. L. and Rasberry, S. D. (1971). "Preparation and Evaluation of SRM's 481 and 482 Gold–Silver and Gold–Copper Alloys for Microanalysis", Appendix 4, NBS Spec. Publ. 260–28. Natl. Bur. Stds., Washington, D.C. 20234.

Henoc, J., Heinrich, K. F. J. and Myklebust, R. L. (1973). "A Rigorous Correction Procedure for Quantitative Electron Probe Microanalysis (COR 2)", NBS Tech. Note 769. Natl. Bur. Stds., Washington, D.C. 20234.

Kimoto, S. (1972). *JEOL News* **10**, 2.

Kimoto, S. (1973). *In* "Proc. 6th IITRI Scanning Electron Microscopy Symposium", p. 2. Chicago, Ill.

Kramers, H. A. (1923). *Phil. Mag.* **46**, 836.

Kuntz, F., Eichen, E., Mathews, H. and Francis, J. (1971). *In* "Adv. X-ray Anal.", Vol. 15, p. 148. Plenum Press, New York.

Liebhafsky, H. A., Pfeiffer, H. G. and Zemany, P. D. (1955). *Analyt. Chem.* **26**, 1257.

Long, J. V. P. (1965). *Br. J. appl. Phys.* **16**, 1277.

McMillan, W. R. (1967). *In* "Proc. 2nd Natl. Conf. on Electron Probe Analysis", paper 58, EPASA, Boston, Mass.

Malissa, H. (1966) *In* "Elektronenstrahl-Mikroanalyse", p. 98. Springer-Verlag, Vienna.

Melford, D. A. (1962). *J. Inst. Metals* **90**, 217.

Oatley, C. W. (1972). "The Scanning Electron Microscope, Part I. The Instrument". University Press, Cambridge.

Oatley, C. W., Nixon, W. C. and Pease, R. F. W. (1965). *In* "Advances in Electronics and Electron Physics", (L. Marton, ed.), Vol. 21, pp. 181–247. Academic Press, New York.

Rasberry, S. D., Scribner, B. F. and Margoshes, M. (1967). *Appl. Optics* **6**, 81.

Rouberol, I. M., Tong, M., Weinryb, E. and Philibert, J. (1962). *Mem. Sci. Rev. Met.* **59**, 305.

Ruark, A. and Brammer, F. E. (1937). *Phys. Rev.* **52**, 322.

Russ, J. C. (1970). ASTM Spec. Tech. Publ. 485, American Society for Testing and Materials, Philadelphia, Pa.

Schiff, L. J. (1936). *Phys. Rev.* **50**, 88.

Strojnik, A. (1973). *In* "Proc. 6th IITRI Scanning Electron Microscopy Symposium", p. 18. Chicago, Ill.

Ware, N. G. and Reed, S. J. B. (1973). *J. Phys. E.* **6**, 286.

Yakowitz, H., Fiori, C. E. and Michaelis, R. E. (1971). National Bureau of Standards Special Publication, 260–22.

Yakowitz, H. and Heinrich, K. F. J. (1968a). *Mikrochimica Acta* **1968**, 182.

Yakowitz, H. and Heinrich, K. F. J. (1968b). *Metallography* **1**, No. 1, 55.

Yakowitz, H. and Heinrich, K. F. J. (1969). *NBS J. Res.* **73A**, No. 2, 113.

Ziebold, T. O. and Ogilvie, R. E. (1963). *Analyt. Chem.* **35**, 621.

Author Index

(*Numbers in italics refer to pages in the References at the end of each chapter*)

N

O

P

Subject Index

B

C

D

F

G

H

I

O

Q

R

S

T

W

X

Y

Z

Cumulative Index of Authors

Cumulative Index of Titles